Management and Industrial Engineering

Series editor

J. Paulo Davim, Aveiro, Portugal

More information about this series at http://www.springer.com/series/11690

Mangey Ram · J. Paulo Davim
Editors

Advances in Reliability and System Engineering

Springer

Editors
Mangey Ram
Department of Mathematics
Graphic Era University
Clement Town, Dehradun
India

J. Paulo Davim
Department of Mechanical Engineering
University of Aveiro
Aveiro
Portugal

ISSN 2365-0532 ISSN 2365-0540 (electronic)
Management and Industrial Engineering
ISBN 978-3-319-48874-5 ISBN 978-3-319-48875-2 (eBook)
DOI 10.1007/978-3-319-48875-2

Library of Congress Control Number: 2016955924

Printed on acid-free paper

This Springer imprint is published by Springer Nature
The registered company is Springer International Publishing AG
The registered company address is: Gewerbestrasse 11, 6330 Cham, Switzerland

Preface

Reliability and safety analysis are the important application in the mathematical engineering. Nothing last forever is so is the life of engineering systems. The main reason of the failure of engineering system ranges from minor inconvenience to signify to economic loss and deaths. Designers, manufactures and end users attempt by employing efforts to minimize the occurrence of the failure. This field creates a wide range of problems due to their practical importance and give rise to the new development to create methods and can contain an interesting mathematical procedure. Reliability deals with the failure concept, whereas safety deals with the consequence of failure. Reliability engineering explores the failures to improve the performance of engineering systems. It plays a vital role in different sectors such as chemical, nuclear facilities and aerospace and can impose potential hazards. It also gains much more importance among practicing engineers and manufacturers. The subject of reliability engineering also appearing in several institutions and universities. It should also be used in many fields and as a particular interest to all electrical, electronic, mechanical, chemical, and industrial and so on. Reliability and safety are the core issues to be addressed during the design, operation, and maintenance of engineering systems. Through this book entitled "*Advance in Reliability and System Engineering*" the engineers have to gain a great knowledge and help them in the reliability courses. The book is meant for those who to take reliability and safety as a subject of study. The material is intended for an audience at the level of postgraduate or senior undergraduate students. That's why reliability and safety is now as well recognized and rapidly developing branch of engineering.

Dehradun, India — Mangey Ram
Aveiro, Portugal — J. Paulo Davim

Acknowledgements

The Editors acknowledges Springer for this opportunity and professional support. Also, we would like to thank all the chapter authors for their availability for this work.

Contents

About the Editors

Mangey Ram received the B.Sc. degree in science from Chaudhary Charan Singh University, Meerut, India, in 2000, the M.Sc. degree in mathematics from Hemwati Nandan Bahuguna Garhwal University, Srinagar (Garhwal), India, in 2004, and the Ph.D. degree major in mathematics and minor in computer science from G. B. Pant University of Agriculture and Technology, Pantnagar, India, in 2008. He has been a Faculty Member for around eight years and has taught several core courses in pure and applied mathematics at undergraduate, postgraduate, and doctorate levels. He is currently *Professor* with the Department of Mathematics, Graphic Era University, Dehradun, India. Before joining the Graphic Era University, he was a Deputy Manager (Probationary Officer) with Syndicate Bank for a short period. He is Editor-in-Chief of *International Journal of Mathematical, Engineering and Management Sciences*; Executive Associate Editor of the *Journal of Reliability and Statistical Studies* and the Guest Editor & Member of the editorial board of many journals. He is a regular Reviewer for international journals, including IEEE, Elsevier, Springer, Emerald, John Wiley, Taylor & Francis and many other publishers. He has published 77 research papers in IEEE, Springer, Emerald, World Scientific and many other national and international journals of repute and also presented his works at national and international conferences. His fields of research are reliability theory and applied mathematics. Dr. Ram is a member of the IEEE, Operational Research Society of India, Society for Reliability Engineering, Quality and Operations Management in India, International Association of Engineers in Hong Kong, and Emerald Literati Network in the U.K. He has been a member of the organizing committee of a number of international and national conferences, seminars, and workshops. He has been conferred with "*Young Scientist Award*" by the Uttarakhand State Council for Science and Technology, Dehradun, in 2009. He has been awarded the "*Best Faculty Award*" in 2011 and recently "*Research Excellence Award*" in 2015 for his significant contribution in academics and research at Graphic Era University.

J. Paulo Davim received his Ph.D. in Mechanical Engineering in 1997, M.Sc. degree in Mechanical Engineering (materials and manufacturing processes) in 1991, Licentiate degree (5 years) in Mechanical Engineering in 1986, from the University of Porto (FEUP), the Aggregate title from the University of Coimbra in 2005 and a D.Sc. from London Metropolitan University in 2013. He is Eur Ing and Senior Chartered Engineer by the Portuguese Institution of Engineers with a MBA and Specialist title in Engineering and Industrial Management. Currently, he is Professor at the Department of Mechanical Engineering of the University of Aveiro. He has more than 30 years of teaching and research experience in Manufacturing, Materials and Mechanical Engineering with special emphasis in Machining & Tribology. Recently, he has also interest in Management/Industrial Engineering and Higher Education for Sustainability/Engineering Education. He has received several scientific awards. He has worked as evaluator of projects for international research agencies as well as examiner of Ph.D. thesis for many universities. He is the Editor in Chief of several international journals, Guest Editor of journals, books Editor, book Series Editor and Scientific Advisory for many international journals and conferences. Presently, he is an Editorial Board member of 30 international journals and acts as reviewer for more than 80 prestigious Web of Science journals. In addition, he has also published as editor (and co-editor) more than 80 books and as author (and co-author) more than 10 books, 60 book chapters and 350 articles in journals and conferences (more than 200 articles in journals indexed in Web of Science/h-index 35+ and SCOPUS/h-index 44+).

Reliability Analysis for Physically Separated Redundant System

HyungJu Kim, Stein Haugen and Ingrid Bouwer Utne

Abstract Redundancy has been a key concept to achieve high system reliability in technical systems, and the concept is central in modern technology regulations. One important aspect of successful redundancy is independence of the redundant components. Among several dependent failures and their causes, this chapter focuses on common cause failures that are caused by external environment (e.g. fire, explosion, flooding, etc.). These failures can be prevented by implementing physical separation of redundant components. The main objective of this chapter is to (1) present basic concept of redundancy and physical separation of redundant components, (2) explore requirements for physical separation in three industries (nuclear, maritime and aviation), and (3) study how to model and analyse physically separated redundant systems, quantitatively.

Keywords Physical separation · Redundancy · Quantitative reliability analysis · Location fault tree · Markov

H. Kim (✉)
Department of Production and Quality Engineering, NTNU,
Norwegian University of Science and Technology, Valgrinda 1.306B, S.P. Andersens veg 5,
7491 Trondheim, Norway
e-mail: hyung-ju.kim@ntnu.no

S. Haugen
Department of Marine Technology, NTNU, Norwegian University of Science
and Technology, Marinteknisk senter D2.098, Otto Nielsens veg 10,
7491 Trondheim, Norway
e-mail: stein.haugen@ntnu.no

I.B. Utne
Department of Marine Technology, NTNU, Norwegian University of Science
and Technology, Marinteknisk senter E2.222, Otto Nielsens veg 10,
7491 Trondheim, Norway
e-mail: ingrid.b.utne@ntnu.no

M. Ram and J.P. Davim (eds.), *Advances in Reliability and System Engineering*,
Management and Industrial Engineering, DOI 10.1007/978-3-319-48875-2_1

Abbreviations

CCF	Common cause failure
DD	Dangerous detected
DP	Dynamic positioning
DU	Dangerous undetected
EE	External event
EF	External failure
FTA	Fault tree analysis
LFT	Location fault tree
LRBD	Location reliability block diagram
MCS	Minimal cut set
MGE	Main generator engine
MOCUS	Method for obtaining cut sets
MSB	Main switchboard
MTTR	Mean time to repair
RBD	Reliability block diagram
RP	Redundant propulsion
SE	Survival after external event
SIF	Safety instrumented function
SRtP	Safe return to port
ZSA	Zonal safety analysis

1 Redundancy and Physical Separation

1.1 Redundancy in Technical System

There are two ways of achieving better technical reliability of systems: (1) use items with very high reliability, and (2) introduce one or more reserve items [1]. The latter is referred to as *redundancy*. More specifically, the term redundancy means that a system has two or more components so that if one component fails, the other component(s) enable the system to function continuously, and this design principle is also called as *fault tolerance* [2]. IEC 60050-191 [3] defines redundancy as follows:

> In an item, the existence of more than one means for performing a required function

It is self-evident that redundancy enhances the reliability of many technological systems [4]. Redundancy therefore has been a key concept to ensure high system reliability in engineering for over 50 years, and the concept is central in modern technology regulations [5].

Depending on its implementation, redundancy can be classified into two main categories: active redundancy and standby redundancy [2]. In active redundancy,

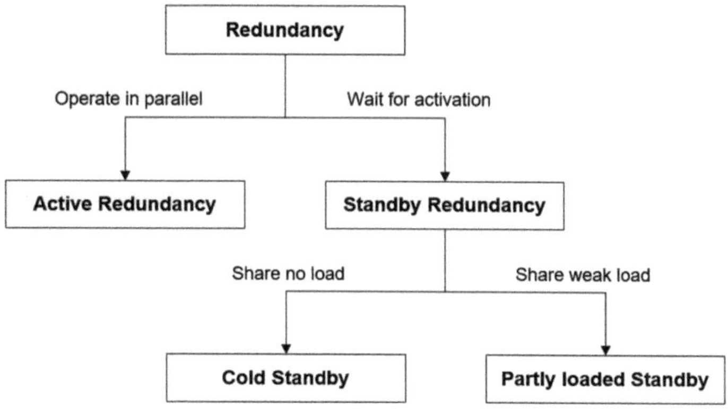

Fig. 1 Classification of redundancy

reserve components operate in parallel and share the load. Whereas in standby redundancy, reserve components are in standby and are activated when the ordinary component fails [1–3]. Standby redundancy can be further classified according to the load sharing. If the reserve components share no load in the waiting period, the redundancy is called cold standby. If the reserve components share a weak load in the waiting period, the redundancy is said to be partly loaded [1, 2]. The classification of redundancy is illustrated in Fig. 1.

1.2 Dependent Failure

One important aspect of successful redundancy is *independence*. If the ordinary and reserve components are dependent, a single failure may disable both of the components, and consequently, the entire system can be inoperable. Dependent failures can be classified in three main groups [1, 2]:

1. *Common cause failure* (*CCF*): two or more component fault states exist simultaneously, or within a short time interval
2. *Cascading failure*: a failure of one component results in multiple failure through domino effect
3. *Negative dependency*: a single failure reduces the likelihood of failures of other components

This chapter focuses on CCFs that can incapacitate redundant systems instantly (or within a short time interval). Negative dependency, which is not harmful (or maybe beneficial) to redundancy, and cascading failure, which may be modelled explicitly [6], are not within the scope of this chapter.

In accordance with NUREG/CR-6268 [7], CCFs can be caused by the following:

1. *Design/Construction/Manufacture Inadequacy* that represents causes related to actions and decisions taken during design, manufacture, or installation of components before and after the system is operational (e.g. design error, manufacturing error, installation/construction error, and design modification error).
2. *Operations/Human Error* that represents causes related to errors of omission and commission of operation staffs (e.g. accidental action, inadequate/incorrect procedure, failure to follow procedure, inadequate training, and inadequate maintenance).
3. *External Environment* that represents causes related to a harsh external environment that is not within component design specifications (e.g. fire/smoke, humidity/moisture, high/low temperature, acts of nature, contamination/dust/dirt, etc.).
4. *Internal to Component* that is associated with malfunctioning of something internal to the component (e.g. normal wear, internal environment, early failure, erosion/corrosion, etc.).
5. *State of Other Components* that represents causes related to failure of a supporting component or system (e.g. supporting system, interconnection, etc.)
6. *Unknown and Other* that represent causes of component state that cannot be identified or cannot be attributed to any of the previous cause categories.

The main focus of this chapter is CCFs caused by *External Environment*, which can be prevented (or the likelihood can be reduced) through *physical separation* of redundant components. Physical separation is further explored in following Sect. 1.3. Classification of dependent failures and causes of CCFs are illustrated in Fig. 2, together with the main scope of this chapter.

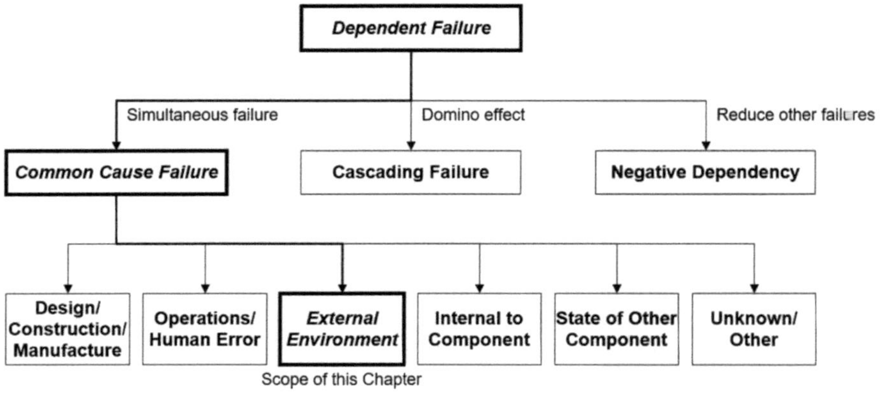

Fig. 2 Dependent failures and scope of this chapter

1.3 Physical Separation of Redundant Components

Various external environmental effects may lead to CCFs, and different systems focus on different external environmental effects. In the maritime industry, for instance, fire and flooding are the most significant external events that can result in simultaneous failure of redundant components [8], while in nuclear facilities, sabotage is a major concern, as well as fire and explosion [9]. Different external events in several industrial sectors are further explored in the following Sect. 2.

One effective and widely used measure to prevent CCFs in several industries caused by external events is to physically separate redundant components. Say that we have a cooling water supply system as shown in Fig. 3. The system is fully duplicated as system A and B, which is in a cold standby redundancy; cooling water system A carries 100 % of the load in normal operation, and system B is activated when system A fails.

By virtue of the reserve system B, a single failure of any component will not disable the cooling water supply, whereas an external event can make the entire system inoperable. For instance, if a fire occurs in the pump room, then the two redundant water pumps can be damaged simultaneously by fire. This is an example of a CCF that is caused by an external event. In order to prevent this CCF, the two redundant systems can be located far away from each other as shown in Fig. 4a or a fire resistant wall can be installed between the two redundant systems as shown in Fig. 4b. This is called *physical separation* or *physical segregation*, and the fire resistant wall is a *physical barrier*.

Figure 4a is a suitable solution for onshore process facilities, nuclear power plants, and other industries when there is no or little limitation to space. On the other hand, there is (nearly always) not enough space to separate redundant components by distance on offshore platforms, in marine vessels, aircrafts, etc. Figure 4b is therefore a more appropriate solution in these industries.

The concept of segregation can also be applied to other external events, like explosion, sabotage attack, flooding, etc. The following section explores

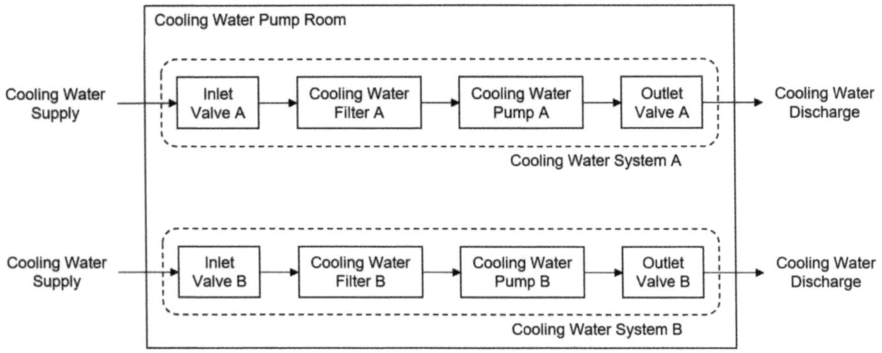

Fig. 3 Cooling water supply system

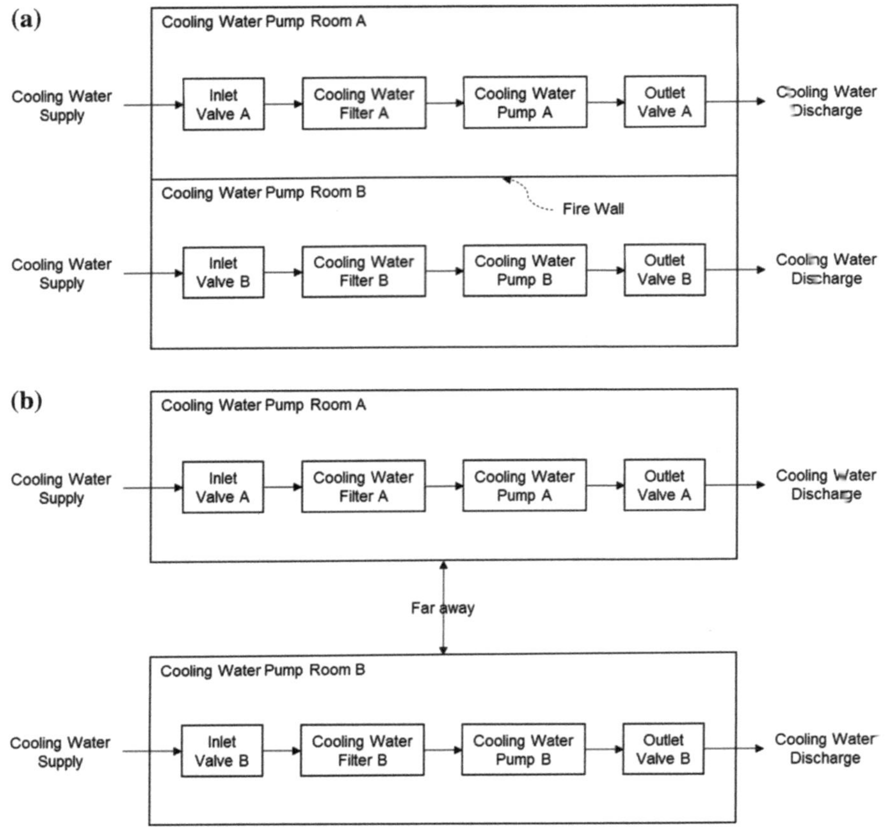

Fig. 4 Physically separated cooling water systems

requirements for segregation of redundant components in the nuclear, maritime and aviation industries.

2 Physical Separation Requirements in Several Industries

2.1 Nuclear Industry

Separation of redundant elements in the nuclear industry is called *equipment segregation* [10] or *physical separation* [11]. Equipment segregation is defined as [10]

> the separation of redundant and/or diverse components by distance or by barriers to prevent all or most of the components being damaged, particularly in the event of common hazards

Common hazards are further divided into internal hazards and external hazards. Examples of each hazard are provided by IAEA SSR-2/1 [11]

1. Internal hazards: fire, explosion, flooding, missile generation, collapse of structures, etc.
2. External hazards: natural external events, such as meteorological, hydrological, geological and seismic events, and human induced external events arising from nearby industries and transport routes.

This classification is slightly different from that of NUREG/CR-6268. Fire is classified as external environment in NUREG/CR-6268, whereas fire is internal hazard in ONR Guide. Fire and explosion are events that can be both internal and external. A diesel engine, for instance, can be the origin of a fire itself (internal hazard), but there can also be fires from other sources that can affect the engine (external hazard). Since the latter is our main concern in this chapter, fire and explosion are hereafter limited to external hazards, unless specifically noted.

Another important external hazard in a nuclear power facility is terrorist attacks against equipment or systems that normally maintain the facility in a safe state. The perception of potential terrorist attacks to nuclear facilities changed significantly after the attacks of 11 September 2001, and the International Atomic Energy Agency (IAEA) initiated an effort to develop guidelines on the security of nuclear facilities [12].

These common hazards can be prevented or mitigated by physical separation. For instance, if a safety system is duplicated, and the redundant system is installed in a separate building, we can prevent or at least reduce the likelihood of simultaneous failure of ordinary and reserve systems. ONR Guide [10] therefore states that

> 5.8 In a redundant system and despite diverse provisions, the threat of common cause failures particularly from hazards such as fire may be reduced by system segregation. This is the separation of components by distance or physical barriers … such barriers may also serve as barriers to other hazards

IAEA SSR-2/1 [11] further requires independence of safety system as

> Interference between safety systems or between redundant elements of a system shall be prevented by means such as physical separation, electrical isolation, functional independence …

More specifically for fire, IAEA NS-G-1.7 [13] specifies that

> 3.2 … the plant buildings should be subdivided into fire compartments and fire cells. The purpose is … to segregate redundant safety systems from each other. The aim of segregation is to … prevent common cause failures …
>
> 2.17 … a fire affecting one division of a safety system would not prevent the execution of the safety function by another division. This should be achieved by locating each redundant division of a safety system in its own fire compartment or at least in its own fire cell…
>
> 5.38 … fire pumps should be redundant and separated in the fire context in order to ensure adequate functionality in the event of equipment failure…

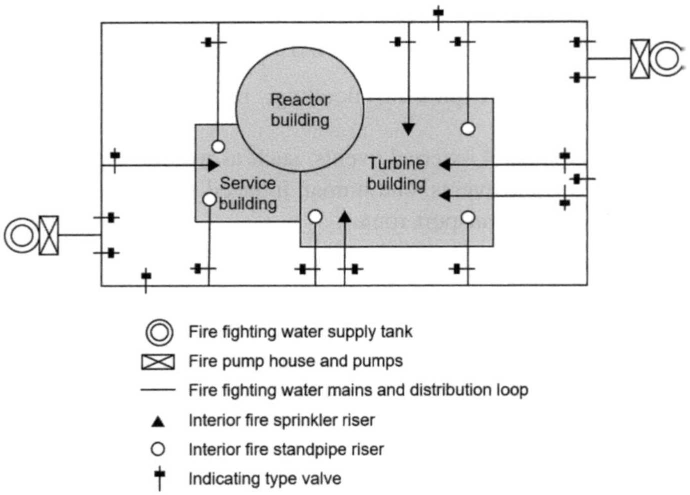

Fig. 5 Firefighting water main loop [13]

One possible layout of the water supply system for the fire extinguishing system is given in Fig. 5. Two independent fire pump houses and water tanks are physically separated by distance, so that the fire water can be continuously supplied in the event of a fire, explosion, sabotage, and other external events. As mentioned in Sect. 1.3, nuclear facilities have less limitation to space than offshore facilities and marine vessels, and therefore, the redundant systems or components can normally be located far away from each other, as shown in this layout.

2.2 *Maritime Industry*

Physical separation in marine vessels is specified by several regulations and guidelines of the International Maritime Organization (IMO) and Classification Societies: (1) equipment class of IMO and notations of Classification Societies for dynamic positioning (DP) capable vessels, (2) redundant propulsion (RP) notations of Classification Societies for merchant vessels, and (3) the IMO safe return to port (SRtP) regulation for passenger ships.

Dynamically positioned (DP) vessels are units or vessels that automatically maintain its position by means of thruster force. The DP-system is the complete installation comprising power system, thruster system and DP-control system. IMO equipment class defines three classes for a DP-system, depending on its redundancy and physical separation [14]

1. Equipment class 1: No redundancy. Loss of position may occur in the event of single fault.

2. Equipment class 2: Redundancy of all active components (generators, thrusters, switchboards, remote controlled valves, etc.). Loss of position shall not occur in the event of a single fault in any active component.
3. Equipment class 3: Redundancy of all components and physical separation of the components. In this class, single fault includes (1) items listed above for class 2, (2) any static component such as cables, pipes, manual valves, etc., (3) all components in any one watertight compartment, from fire or flooding, and (4) all components in any one fire sub-division, from fire or flooding.

Examples of the three classes are given in Fig. 6. System A consists of one thruster, one main switchboard (MSB), one main generator engine (MGE), and one set of utilities for the MGE. A single fault can therefore lead to loss of position; system A is equipment class 1. System B is fully duplicated so that a single fault of any active component will not result in loss of position; system B is equipment class 2. Even the duplicated components in system B may simultaneously fail by a single fire or flooding, if they are located in the same compartment. For instance, a

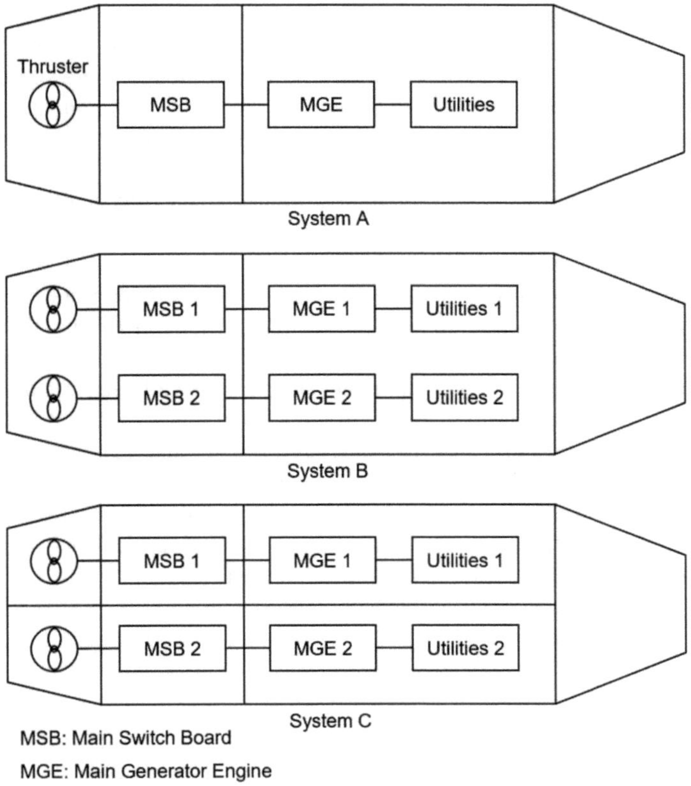

Fig. 6 Three simplified layouts of thruster system

flooding in the main switchboard room in system B can disable both MSB 1 and 2. Equipment class 3 therefore specifies that redundant components should be physically separated in different fire sub-divisions and/or different watertight compartments, like system C.

Other notations of Classification Societies for DP systems have similar categories with the three classes in IMO equipment class. For instance, Det Norske Veritas (DNV) classifies DP systems into four categories: DYNPOS-AUTS and DYNPOS-AUT that are equivalent to equipment class 1, DYNPOS-AUTR to equipment class 2, and DYNPOS-AUTRO to equipment class 3 [15]. Lloyd's Register (LR), American Bureau of Shipping (ABS), and other Classification Societies have their own categories and requirements for DP systems, which are in line with the equipment classes of IMO.

While DP notations refer only to dynamic positioning capable vessels, redundant propulsion (RP) notations can be applied to all kinds of merchant vessels that have a propulsion system. RP notations classify the grade of propulsion system and/or steering system into several categories depending on its redundancy and physical separation, which has a similar philosophy as IMO equipment classes for DP systems. The RP notation of DNV [16], for instance, has two grades: RP that is similar to IMO equipment class 2 and RPS to IMO equipment class 3. LR [17] categories propulsion systems and steering systems into six grades: PMR, PMR*, SMR, SMR*, PSMR and PSMR*. The three grades without "*" are similar to IMO equipment class 2 that specifies redundancy of all active components. The other three grades with "*" correspond to IMO equipment class 3 that requires redundancy of all components and physical separation of the components. Other Classification Societies have similar requirements for physical separation of propulsion and steering systems.

The Safe Return to Port (SRtP) regulation is stated in Chapter II-1, II-21, II-22 and II-23 of International Conventions for the Safety of Life at Sea (SOLAS) [18], and its explanatory notes are provided in MSC/Circ. 1369 [19]. The philosophy of this regulation is similar to IMO equipment class 3. Thirteen essential systems should have redundancy of all components and be physically separated against fire and flooding damage. The two main differences from DP and RP notations are (1) SRtP regulation refers only to passenger ships,[1] and (2) the scope of this regulation includes fire-fighting system, bilge/ballast system, etc. that are not frequently used, as well as propulsion and steering systems that are continuously operated during navigation. The latter is considered critical when modelling physical separation. This is further explored in Sect. 3.2.

[1]More specifically, passenger ships constructed on or after 1 July 2010 having length of 120 m or more or having three or more main vertical zones.

2.3 Aviation Industry

Physical separation of redundant components in an airplane[2] is stated in 14 CFR 25 [20]

> 25.795 Except where impracticable, redundant airplane systems necessary for continued safe flight and landing must be physically separated …

Especially for electrical wiring interconnection system (EWIS), it is required that

> 25.1707 For systems for which redundancy is required, … EWIS components associated with those systems must by designed and installed with adequate physical separation

AC 25.795-7 [21] suggests two approaches that will satisfy the requirements for physical separation: (1) system separation by distance, and (2) system protection. Separation distance varies depending on the airplane fuselage diameters, and a simple equation is given by this Advisory Circular. If this separation of redundant components is impracticable in a specific area, then one of the redundant components should be properly protected in accordance with several criteria. In most cases, system separation by distance is the preferred measure, because barriers themselves can damage redundant components and can cause maintenance errors [22]. System protection should therefore be applied only if system separation is impracticable [21].

Some examples of common cause failures that can be prevented or mitigated by physical separation are damage from localised fires, contamination by fluids, excessive voltage, physical or environmental interactions among parts, rapid release of energy from concentrated sources, non-catastrophic structural failures, etc. [23].

Separation of redundant components can be analysed through Zonal Safety Analysis (ZSA) that is described in ARP 4761 [24] and ARP 4754 [25]. ZSA is a systematic analysis of geographical locations of the components and interconnections of a system, which evaluates the potential system-to-system interaction and assesses the severity of potential hazards inherent in the system installation [26]. ARP 4754 [25] states that

> A Zonal Safety Analysis should examine each physical zone of the aircraft to ensure that equipment installation and potential physical interference with adjacent systems do not violate the independence requirements of the systems

---

[2]More specifically, airplanes for more than 60 persons or takeoff gross weight of over 100,000 lb (45,359 kg).

2.4 Summary

Physical separation requirements of three different industries have been explored in this section. There are some differences in background and practical application of physical separation, but they are not very different, principally. For instance, the physical separation can be realised by separation by distance or by physical barrier. While separation by distance is preferred by the nuclear power plant and airplane, redundant systems in maritime vessel should be separated only by physical barrier such as fire sub-division or watertight compartment.

The main purpose of these requirements is that even a duplicated system may fail simultaneously by a single external event (e.g. occurrence of fire near the system) unless the redundant system is not physically separated. The following section investigates how to model and analyse physically separated redundant systems quantitatively.

3 Modelling of Physically Separated Redundant Systems

3.1 Location Fault Tree

Fault tree analysis (FTA) is the most commonly used method for causal analysis of hazardous events [6]. A fault tree is a logic diagram with a top-down approach, which shows the interrelationships between a potential critical event (top event) and its causes. The causes at the bottom level are called *basic events*. Examples of basic events are component failures, environmental conditions, human errors, and normal events that the system may experience during its life span [6]. FTA can be qualitative, quantitative, or both, depending on the objective of the analysis [6]. The main objectives of FTA are [2]

1. *To identify all possible combinations of basic events that may result in a critical event in the system.*
2. *To find the probability that the critical event will occur during a specified time interval or at a specified time t, or the frequency of the critical event.*
3. *To identify aspects (e.g. components, barriers, structure) of the system that need to be improved to reduce the probability of the critical event.*

Location Fault Tree (LFT) is a specific type of fault tree used to examine and quantify the effect of external events, such as fire, explosion, flooding, etc. [27–29]. The main difference between the original fault tree and LFT is the types of basic events in the tree. Basic events in ordinary fault trees normally focus on equipment failures and/or human errors, whereas basic events in LFTs are limited to damages on the location where equipment is installed.

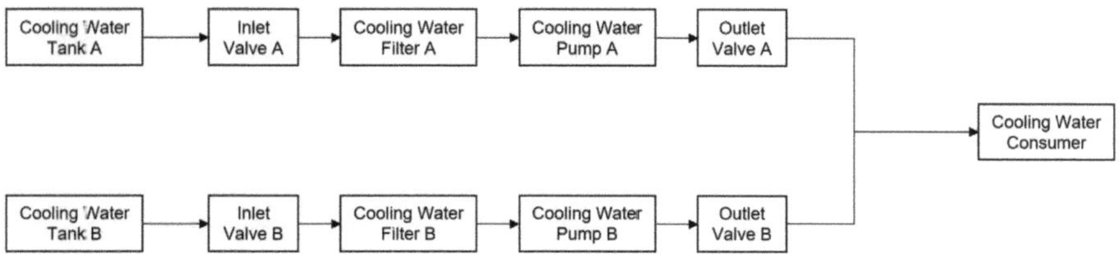

Fig. 7 RBD of cooling water supply system

Let us recall the cooling water supply system in Sect. 1.3 as an example. A simplified reliability block diagram (RBD) of the cooling water supply system is given in Fig. 7.

The cooling water system consists of water tanks, valves, filters, pumps and connecting pipelines. The system is fully duplicated and performs a cold standby redundancy; cooling water system A carries out 100 % capacity in normal operation, and system B is activated when system A fails. A single equipment failure, therefore, will not disable the system function.

An ordinary fault tree of this system can be built as shown in Fig. 8. The system can be divided into two sub-systems: cooling water system A and B. The top event is cooling water system failure, and basic events are equipment failures. The top event occurs when both system A and B fail. Cooling water system A fails if one of its components fails, and this is same for the system B. It is assumed that the frequency of pipe failure is extremely low so that pipe failures are negligible.

The minimal cut sets (MCSs) can be obtained by using the method for obtaining cut sets (MOCUS) that is described in Rausand and Høyland [1]. MCS in this fault tree are

{IVA, IVB}, {IVA, FB}, {IVA, PB}, {IVA, OVB},
{FA, IVB}, {FA, FB}, {FA, PB}, {FA, OVB},
{PA, IVB}, {PA, FB}, {PA, PB}, {PA, OVB},
{OVA, IVB}, {OVA, FB}, {OVA, PB}, {OVA, OVB}.

These MCSs show that the reserve system B prevents a single failure of any equipment from leading to the occurrence of the top event. However, there is still the possibility of a simultaneous failure of system A and B due to an external event. For instance, if system A and B are located in close proximity without any protection against fire, and a fire occurs near the systems, then redundant components may fail simultaneously, and consequently, the entire system can be disabled. This cannot be examined with the above RBD or fault tree. To analyse the effect of external events, the location of each component needs to be identified first. The RBD of the cooling water system including location information (Location-RBD, LRBD) can be drawn, for instance, as shown in Fig. 9.

The boundary of each location can be either a physical barrier or sufficient distance that can prevent escalation of the damage due to an external event. For

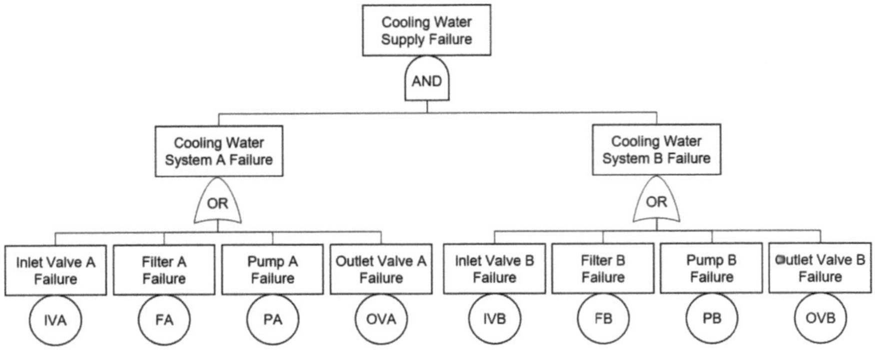

Fig. 8 Fault tree of the cooling water supply system

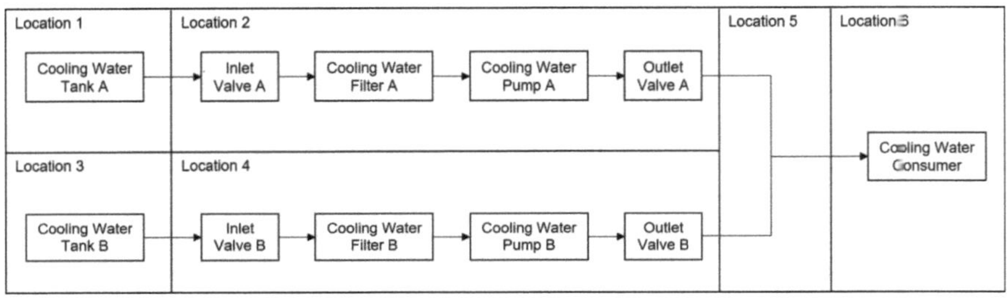

Fig. 9 Location-RBD of cooling water system

instance, for the occurrence of a fire, boundary of each location can be a firewall, or enough distance that a fire in one location cannot affect the other.

In this LRBD, cooling water tanks A and B are located in *Location 1* and *3*, respectively, and the two cooling water pumps, including necessary utilities, are located in *Locations 2* and *4*. Pipes from each pump are merged in *Location 5* and are routed to *Location 6*. The ordinary system A and reserve system B are therefore physically separated except in *Location 5*. A single damage in *Location 5* can disable the pipeline from both system A and B, and consequently, result in the entire cooling water supply system failure. This can also be examined by LFT.

A location fault tree of this system can be built as shown in Fig. 10. It is assumed that when an external event occurs in a location, every component in that location is always disabled simultaneously (complete damage), and the damage of the external event can never escalate to adjacent locations across the boundary of the location (perfect boundary).

MCSs of this LFT are

{L5},
{L1, L3}, {L1, L4},
{L2, L3}, {L2, L4}.

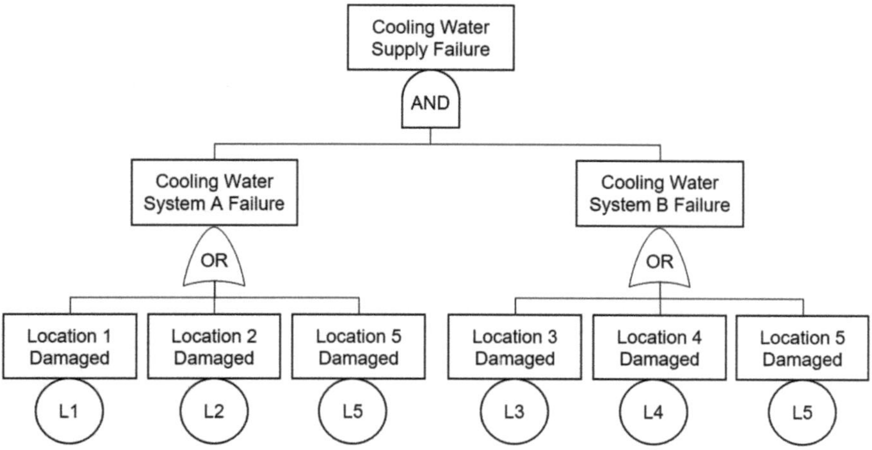

Fig. 10 LFT of cooling water supply system

From the MCSs above, we can identify that a single external event in *Location 5* can result in the entire system failing. The pipes in *Location 5* should therefore be rearranged in different locations, or sufficient protection measure(s) against external events should be provided. This cannot be identified through the ordinary fault tree in Fig. 8 that covers equipment failure only. While the system is fully duplicated so that a single equipment failure cannot disable the entire system, the system can fail due to an external event, depending on the layout of the system.

One important point in LFT is that the pipe and cable route should be included in the model. Normally, the frequency of pipe and cable failures are much lower than the failure of active components, like pumps, compressors, engines, remotely controlled valves, etc. Therefore, pipes and cables are, in some cases, ignored and excluded from the ordinary fault tree, as previously explored in Fig. 8. However, despite of the low failure frequency, an external event may damage pipes and cables unless pipes and cables are properly protected against the external event. Even though it is complex and time consuming to examine every relevant pipe and cable of a system, this is essential to identify external CCF, to ensure high reliability of the system, and to fulfil any requirements, as described in the previous Sect. 2.

Finally, we can build the whole fault tree that includes both the ordinary fault tree and the location fault tree, as shown in Fig. 11. The ordinary fault tree covers equipment failures, while the location fault tree examines the effect of external events. It is assumed that equipment failures and external events are independent. MCSs in this combined fault tree are

{L5},
{IVA, IVB}, {IVA, FB}, {IVA, PB}, {IVA, OVB},
{FA, IVB}, {FA, FB}, {FA, PB}, {FA, OVB},
{PA, IVB}, {PA, FB}, {PA, PB}, {PA, OVB},

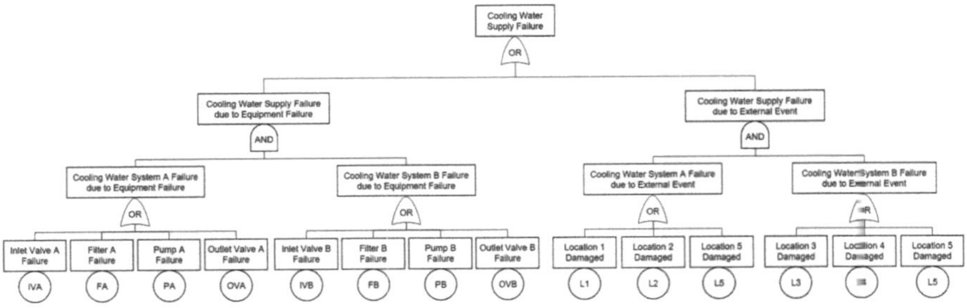

Fig. 11 Combined fault tree for cooling water supply system

{OVA, IVB}, {OVA, FB}, {OVA, PB}, {OVA, OVB},
{L1, L3}, {L1, L4},
{L2, L3}, {L2, L4}.

The top event frequency can be calculated using the *upper bound approximation* formula, which is well described by, e.g. Rausand [2]. Hence, quantification of the top event frequency is not further described in this chapter.

While FTA is a common method and used in a wide range of application areas, FTA is not well suited to examine dynamic systems with complex maintenance [6]. The requirements regarding binary states and Boolean logic is also a limitation that makes its application too rigid for some applications [6]. These weaknesses of fault trees apply to LFT, similarly.

3.2 Markov Approach

The main objective of the Markov approach is to examine how the state of a system can change with time. Markov methods are suitable for analysing small but complex systems with dynamic effects. Markov methods can therefore compensate for some of the weaknesses of FTA [6].

Another advantage of the Markov approach is that this method is suitable for reliability analysis including demands. A demand is defined as [2]

> An event or a condition that requires a safety instrumented function (SIF) to be activated (i) to prevent an undesired event from occurring or (ii) to mitigate the consequences of an undesired event.

A demand for some safety systems is closely related to the cause of system failure. For instance, an occurrence of a fire is a demand on the fire extinguishing system. The fire extinguishing system can mitigate the consequence of the fire. However, the fire extinguishing system can also be damaged and disabled by the fire, if the fire occurs in a location where the fire extinguishing system is installed. This location can be fire pump room, control station, or any other location where necessary pipes and cables pass through. In this case, the fire is a demand on the fire extinguishing system and, at the same time, a cause of fire extinguishing system failure. Demands should therefore be included as a cause of system failure when we model these systems, and various studies have shown that a Markov approach is suitable for reliability modelling with the demand rate [30–35]. This section therefore explores a Markov approach for modelling external events and physical separation. Firstly, this section introduces a Markov model for equipment failure that is introduced by Kim et al. [36]. A Markov model for external events with physical separation introduced by Kim et al. [37] follows, and lastly, a combined Markov model for both equipment failure and external event is presented.

Say that we have a safety system that needs to operate only on demand, such as a fire water supply system that simply consists of two fire water pumps, as shown in Fig. 12. Similar to the cooling water system in Sect. 3.1, the fire water supply system is duplicated so that a single equipment failure does not lead to the failure of the entire system. The main difference between the cooling water supply system and fire water supply system is that the former should operate continuously, while the latter need to operate only on demand.

The duplicated fire water system operates in a cold standby redundancy so that a single equipment failure cannot disable the entire system. It is assumed that (1) the two fire pumps are identical so that they have the same probability of equipment failure, (2) all transition rates are constant in time, (3) the system is restored to an *as good as new* state after the demand duration, and (4) the past state has no influence on the future state. It is also assumed that the system is exposed to dangerous undetected (DU) failures, only. A dangerous detected (DD) failure is assumed to be negligible because DD failure can be revealed and repaired immediately.

The fire water system may have six system states from State 0 to State 5, given in Table 1.

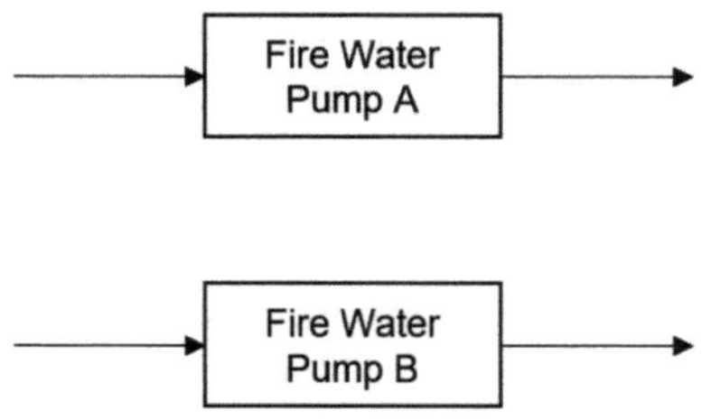

Fig. 12 RBD of simple fire water supply system

Table 1 Six system states

System state	Component state	Demand
5	2 Available	Non-demand
4	1 Available, 1 Functioning	On-demand
3	1 Available, 1 Failure	Non-demand
2	1 Functioning, 1 Failure	On-demand
1	2 Failures	Non-demand
0	2 Failures	On-demand

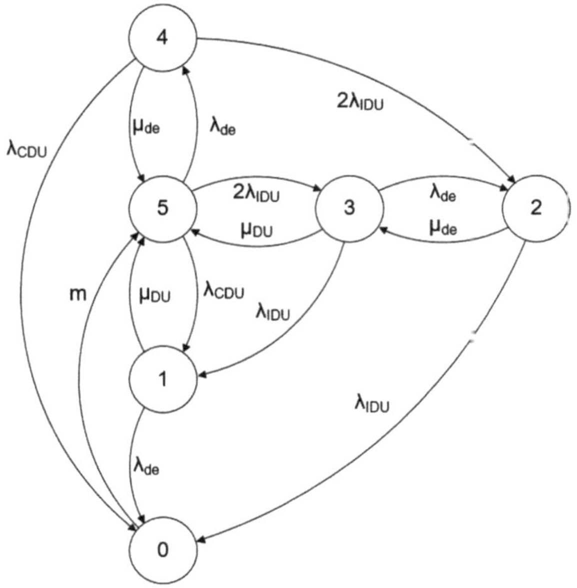

Fig. 13 Markov transition diagram for equipment failures

Available means that a component is operable and able to respond to a demand (fire). *Functioning* is a state where a component is responding to a demand at that moment, and *Failure* indicates that a component is inoperable and not able to respond to a demand.

State 5 is the initial state where (1) there is no fire, and (2) both pumps are operable and able to respond to a fire. State 0 is the hazardous state where (1) a fire occurs, and (2) both pumps are inoperable and not able to respond to the fire. If a fire occurs in the initial state, fire pump A operates and responds to the fire; the system state moves from State 5 to State 4. After the demand duration, the System A stops to operate and the system state moves back to State 5. If one pump fails in State 5, system state moves to State 3. The state goes back to State 5 after repair duration unless a fire occurs during the repair. If a fire occurs during the repair period, the system state moves from State 3 to State 2. The system state can be moved to State 1 either from State 5 by a common cause failure or from State 3 by an individual failure. Hazardous state, State 0, can be reached when (1) a demand (fire) occurs in State 1, (2) two pumps fail simultaneously by a common cause in State 4, and (3) an individual failure occurs in State 2. These transitions of system states are illustrated in Fig. 13.

λ_{IDU} and λ_{CDU} represents the individual equipment failure rate and common cause equipment failure rate, respectively. They can be obtained by

$$\lambda_{\mathrm{IDU}} = (1-\beta) \cdot \lambda_{\mathrm{DU}} \cdots \tag{1}$$

Table 2 Transition rates

Transition rate	Description
λ_{DU}	DU-failure rate
λ_{IDU}	Individual DU-failure rate
λ_{CDU}	DU-CCF rate
λ_{de}	Demand rate
μ_{de}	Demand duration rate
m	Renewal rate

$$\lambda_{CDU} = \beta \cdot \lambda_{DU} \cdots \quad (2)$$

where λ_{DU} is DU-failure rate, and β is a beta factor for DU-failures.

λ_{de} is demand rate, and μ_{de} is demand duration rate that is the reciprocal of mean demand duration. μ_{DU} is DU-failure repair rate that is the reciprocal of the sum of the mean time to repair (MTTR) and half of the test interval. m represents the renewal rate. Transition rates are summarised in Table 2.

From the above Markov model and transition rates, the steady state probabilities and visit frequencies for each state can be calculated using Kolmogorov forward equations. For details on calculation, the reader is referred to Rausand and Høyland [1] and Jin et al. [34].

The above Markov model and transition rates are for equipment failures. For external failures, including physical separation, a Markov model and its transition rates can be built in a similar way. In common with LFT, the location of each component needs to be identified first. Location-RBD of the fire water supply system, including location information can be drawn, for instance, as shown in Fig. 14.

In this LRBD, the two duplicated fire pumps are in different locations so that a single external event will not result in an entire fire water supply system failure.

In common with the cooling water system in Sect. 3.1, it is assumed that an external event always disables all components in the location where the external event occurs (complete damage), and the damage of the external event cannot escalate across the boundary (perfect boundary). Similar to the Markov approach

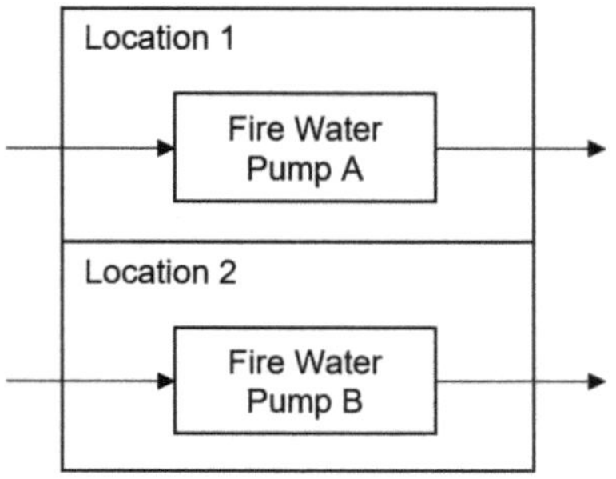

Fig. 14 LRBD of fire water supply system

Table 3 Six system states

System state	Component state	Demand
5	2 Available	Non-demand
4	1 Available, 1 Functioning	On-demand
3	1 Available, 1 Failure	Non-demand
2	1 Functioning, 1 Failure	On-demand
1	2 Failures	Non-demand
0	2 Failures	On-demand

for equipment failure, it is also assumed that (1) the two locations are identical so that the probability of occurrence of an external event in the two locations are same, (2) all transition rates are constant in time, (3) the system is restored to an "as good as new" state after the demand duration, as long as the system is not damaged by external events, and (4) the past state has no influence on the future state.

The fire water system may have six system states from State 0 to State 5, and they are given in Table 3.

Available means that every component in the location is operable and able to respond to a demand (fire). *Functioning* is a state where the components in the location are responding to the demand at that moment, and *Failure* indicates that the components in the location are damaged by external events and not able to respond to a demand.

State 5 is the initial state where (1) there is no fire, and (2) all components in both locations are operable and able to respond to a fire. State 0 is the hazardous state where (1) a fire occurs, and (2) all components in both locations are inoperable and not able to respond to the fire. If a fire occurs outside of Locations 1 and 2, then no component is damaged by the fire so that System A in Location 1 responds to the fire. The system state, thereby, moves from State 5 to State 4. After the demand duration, the System A stops to operate and the system state moves back to State 5. If a fire occurs in one of the two locations and damages all components in that location, then the fire water supply system in the other location responds to the fire. In this case, the system state moves from State 5 to State 2. The system state moves from State 5 to State 3, when external events other than fire occur in one of the two locations and damage the system in that location. If this external event damages both locations simultaneously, the system state moves from State 5 to State 1. System state can move directly from State 5 to State 0, if a single fire damages both locations so that no system is able to respond to the fire. These transitions of system states are illustrated in Fig. 15.

In this fire water supply system, the occurrence of a fire is an external event, a demand of the system, and a cause of the system failure. Let us call this primary external event (EE1), and failures due to EE1 can be called primary external failure (EF1). Other external events that can result in fire-fighting system failure can also occur, but these are not a demand of the fire water system, like explosion, flooding, sabotage, etc. We may call this secondary external event (EE2), and failures due to EE2 secondary external failure (EF2).

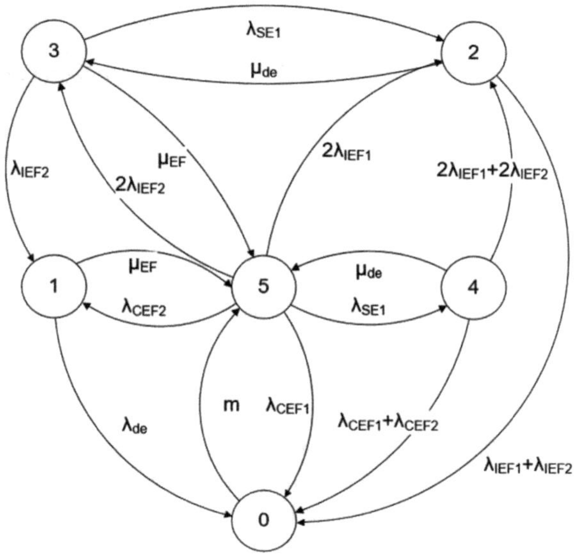

Fig. 15 Markov transition diagram for external failures

If an external event occurs outside a location, then all components in that location are not damaged and still operable after the occurrence of the external event. This can be called as survival after an external event (SE). If this external event is EE1, then the SE becomes SE1, and this applies equally to EE2 and SE2.

λ_{EEn} denotes a rate of EEn that is the sum of λ_{EFn} and λ_{SEn}.

$$\lambda_{EEn} = \lambda_{EFn} + \lambda_{SEn}\cdots \tag{3}$$

λ_{EFn} is a failure rate due to EEn, which can be obtained by

$$\lambda_{EFn} = \lambda_{EEn} \cdot \text{EFRn}\ldots \tag{4}$$

where EFR is an external failure ratio and denotes the conditional probability that an EE damages a specific location given that the EE occurs. λ_{EEn} is the rate of EEn. It is assumed that EE1 and EE2 are independent.

λ_{SEn} is a survival rate after the occurrence of EEn, which can be calculated by

$$\lambda_{SEn} = \lambda_{EEn} \cdot (1 - \text{EFRn})\ldots \tag{5}$$

λ_{IEFn} and λ_{CEFn} represents the individual EFn rate and common EFn rate, respectively. They can be obtained by

$$\lambda_{IEFn} = (1-\beta_{EFn}) \cdot \lambda_{EFn}\cdots \tag{6}$$

Table 4 Transition rates

Transition rate	Description
λ_{EEn}	EEn rate
λ_{EFn}	Failure rate due to EEn
λ_{IEFn}	Individual failure rate due to EEn
λ_{CEFn}	Common failure rate due to EEn
μ_{EF}	EF repair rate
λ_{SEn}	Survival rate after EEn
λ_{de}	Demand rate
μ_{de}	Demand duration rate
m	Renewal rate

$$\lambda_{CEFn} = \beta_{EFn} \cdot \lambda_{EFn} \cdots \tag{7}$$

where β_{EFn} is a beta factor for EFn that represents whether redundant systems are physically separated against EEn or not. If system A and B are separated by a perfect boundary against EEn so that a single EFn cannot disable the entire system, then β_{EFn} equals zero. One the other hand, if system A and B are installed in the same location so that a single EFn can disable the entire system, then β_{EFn} is 1. For an imperfect boundary, β_{EFn} may have a value from 0 to 1.

λ_{de} is the demand rate that is identical to the EE1 rate, and μ_{de} is the demand duration rate that is the reciprocal of the mean demand duration. μ_{EF} is the external failure repair rate that is the reciprocal of the mean repair time, and m is the renewal rate. Transition rates are summarised in Table 4.

The two Markov models, for equipment failure and external failure, can be combined into a single model with two assumptions: (1) equipment failure and external failure are independent, and (2) repair time (and repair rate) for DU-failure and external failure is identical. μ_{re} represents the identical repair rate, replacing

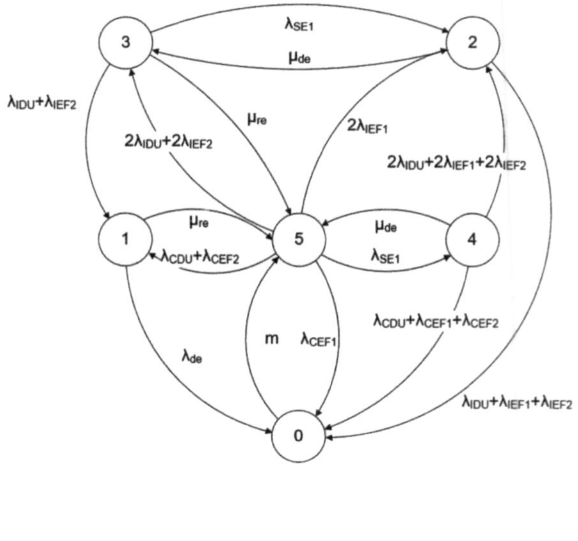

Fig. 16 A combined Markov model

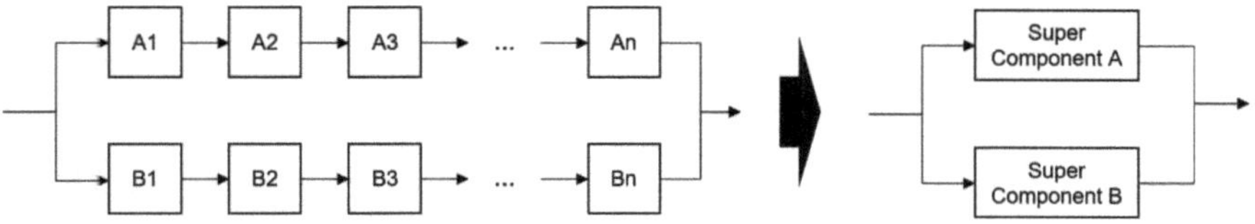

Fig. 17 Replacing series system by super component

μ_{DU} and μ_{EF}. A combined Markov model, including both equipment failure and external failure, is given in Fig. 16.

This section has explored the application of Markov method for modelling of external events and physical separation, using a fire water system as an example. Other safety systems, including their demands can also be modelled similarly.

One weakness of the Markov approach is that the number of system states increases a lot when modelling a system with large number of components. In this case, the concept of *super components* can be a solution to simplify the model. Redundant systems form to a large extent two independent series configurations, which can be replaced by two super components, as shown in Fig. 17. More details and equations for super components can be found in Rausand [2].

4 Summary

This Chapter addresses redundancy of technical systems, with a particular focus on physical separation to avoid common cause failures (CCFs), due to external events, for example, fire and flooding. Requirements to redundancy in three different industries are presented; namely the nuclear, maritime, and aviation. Physical separation can be realised by distance or by a physical barrier. There are some differences in background and application of physical separation in these three industries. Distance is preferred by the nuclear power plant and aviation industry, whereas redundant systems in maritime vessel are to be separated by a physical barrier, such as fire sub-division or watertight compartment.

The effect of redundancy on risk and reliability can be calculated by using different methods. In this chapter, fault tree analysis, including the location fault tree, is explored and demonstrated by using a cooling water supply system. This approach is not very suitable for dynamic systems, for which Markov modelling may be more feasible. A Markov method, exemplified for a fire water system, is proposed, including both internal technical failures, and external events. Even though the modelling of both technical failures and CCF may be more time consuming than a regular analysis focusing on single technical failures only, high reliability and the compliance with industry requirements can only be achieved by including both categories of events. It is believed that this Chapter can aid such types of analyses, in several different industries.

References

1. Rausand, M. and Høyland, A. (2004). *System reliability theory; models, statistical methods, and applications*, 2nd ed. Hoboken, NJ: Wiley.
2. Rausand, M. (2014). *Reliability of Safety-Critical Systems: Theory and Applications*. NJ: John Wiley & Sons.
3. IEC 60050-191. (1990). International Electrotechnical Vocabulary - Chapter 191: Dependability and quality of service, International Electrotechnical Commission.
4. Downer, J. (2011). On audits and airplanes: Redundancy and reliability-assessment in high technologies, *Accounting, Organizations and Society,* vol. 36, pp. 269-283.
5. Downer, J. (2009). When Failure is an Option: Redundancy, reliability and regulation in complex technical systems, Centre for Analysis of Risk and Regulation, London School of Economics and Political Science.
6. Rausand, M. (2011). *Risk assessment: theory, methods, and applications*: John Wiley & Sons.
7. NUREG/CR-6268. (2007). Common-cause failure database and analysis system: Event data collection, classification, and coding, US Nuclear Regulatory Commission.
8. MSC/Circ. 1023. (2002). Guidelines for Formal Safety Assessment (FSA) for use in the IMO rule-making process, Maritime Safety Committee.
9. IAEA INFCIRC/225. (2011). Nuclear security recommendations on physical protection of nuclear material and nuclear facilities, International Atomic Energy Agency.
10. ONR Guide. (2014). Diversity, redundancy, segregation and layout of mechanical plant, Office for Nuclear Regulation.
11. IAEA SSR-2/1. (2012). Safety of Nuclear Power Plants: Design, International Atomic Energy Agency.
12. IAEA. (2012). Identification of Vital Areas at Nuclear Facilities - Technical Guidance, International Atomic Energy Agency.
13. IAEA NS-G-1.7. (2004). Protection against Internal Fires and Explosions in the Design of Nuclear Power Plants, International Atomic Energy Agency.
14. MSC/Circ. 645. (1994). Guidelines for vessels with dynamic positioning systems, Maritime Safety Committee.
15. DNV. (2011). Special equipment and systems additional class - Dynamic positioning systems, Part 6, Chapter 7, Det Norske Veritas AS.
16. DNV. (2012). Special equipment and systems additional class - Redundant propulsion, Part 6, Chapter 2, Det Norske Veritas AS.
17. LR. (2011). Rules and regulations for the classification of ships - Main and auxiliary machinery, Part 5, Lloyd's Register.
18. SOLAS. (2010). International Convention for the Safety of Life at Sea, International Maritime Organization.
19. MSC/Circ. 1369. (2010). Interim explanatory notes for the assessment of passenger ship system's capabilities after a fire or flooding casualty, Maritime Safety Committee.
20. 14 CFR 25. (2011). Code of Federal Regulations - Airworthiness Standards: Transport Category Airplanes, The Office of the Federal Register National Archives and Records Administration.
21. AC 25.795-7. (2008). Advisory Circular - Survivability of Systems, Federal Aviation Administration.
22. AC 25.1701-1. (2008). Advisory Circular - Certification of Electrical Wiring Interconnection Systems on Transport Category Airplanes, Federal Aviation Administration.
23. CS-15. (2010). Certification Specifications for Large Aeroplanes, European Aviation Safety Agency.
24. ARP 4761. (1996). Certification Considerations for Highly-Integrated Or Complex Aircraft Systems, SAE International.
25. ARP 4754. (1996). Certification Considerations for Highly-Integrated Or Complex Aircraft Systems, SAE International.

26. Caldwell, R. E. and Merdgen, D. B. (1991). Zonal analysis: the final step in system safety assessment [of aircraft], in *Reliability and Maintainability Symposium, 1991. Proceedings., Annual*, pp. 277-279.
27. Gopika, V., Sanyasi Rao, V. V. S., Ghosh, A. K., and Kushwaha, H. S. (2012). PSA based vulnerability and protectability analysis for NPPs, *Annals of Nuclear Energy,* vol. 50, pp. 232-237.
28. Park, C.-K., Jung, W. S., Yang, J.-E., Kang, H. G., Gűrpinar, A., and Kim, S.-C. (2003). A PSA-based vital area identification methodology development, *Reliability Engineering & System Safety,* vol. 82, pp. 133-140.
29. Varnado, G. B. and Whitehead, D. W. (2008). Vital Area Identification for US Nuclear Regulatory Commission Nuclear Power Reactor Licensees and New Reactor Applicants, Sandia National Laboratories.
30. Misumi, Y. and Sato, Y. (1999). Estimation of average hazardous-event-frequency for allocation of safety-integrity levels, *Reliab Eng Syst Safe,* vol. 66, pp. 135-144.
31. Goble, W. M. and Cheddie, H. (2005). *Safety Instrumented Systems verification: practical probabilistic calculations*, 1st ed. USA: The Instrumentation, Systems, and Automation Society.
32. Bukowski, J. V. (2006). Incorporating process demand into models for assessment of safety system performance, in *Reliability and Maintainability Symposium, 2006. RAMS '06. Annual*, Newport Beach, CA, pp. 577-581.
33. Innal, F. (2008). Contribution to modelling safety instrumented systems and to assessing their performance critical analysis of IEC 61508 standard, Ph.D. thesis, University of Bordeaux, Bordeaux, France.
34. Jin, H., Lundteigen, M. A., and Rausand, M. (2011). Reliability performance of safety instrumented systems: A common approach for both low- and high-demand mode of operation, *Reliability Engineering & System Safety,* vol. 96, pp. 365-373.
35. Jin, H., Rausand, M., and Lundteigen, M. A. (2013). New Reliability Measure for Safety Instrumented Systems, *International Journal of Reliability, Quality and Safety Engineering,* vol. 20, p. 1350005.
36. Kim, H., Haugen, S., and Utne, I. B. (2014). Reliability Analysis Including External Failures for Low Demand Marine Systems, in *Probabilistic Safety Assessment and Management (PSAM 12)*, Hawaii.
37. Kim, H., Haugen, S., and Utne, I. B. (2015). Reliability analysis of the IMO regulation – safe return to port, *Ships and Offshore Structures,* pp. 1-10.

Preventive Maintenance of Consecutive Multi-unit Systems

Won Young Yun and Alfonsus Julanto Endharta

Abstract This chapter deals with a preventive maintenance problem for consecutive-k-out-of-n and connected-(r, s)-out-of-(m, n): F systems. The system failure probability of two types of multi-unit systems is analytically derived by utilizing the system failure paths. Dependence mechanisms between components in the systems are illustrated and the system failure paths under dependence models are shown. The expected cost rates of corrective maintenance, age preventive maintenance, and condition-based maintenance policies are estimated. The optimal maintenance policies to minimize the expected cost rates in two preventive maintenance models are obtained and three maintenance policies are compared by numerical examples.

Keywords Consecutive multi-unit system · Preventive maintenance · Expected cost rate

1 Introduction

A consecutive-k-out-of-n: F system is one type of multi-unit systems consisting of n components arranged linearly or circularly. The system fails if and only if there are at least k consecutive failed components [4, 6, 8]. This type of system structure is used the design of integrated circuits, microwave relay stations in telecommunications, oil pipeline systems, vacuum systems in accelerators, computer ring networks (k loop), and spacecraft relay stations.

The reliability of consecutive-k-out-of-n: F systems have been reviewed in several papers. The reliability of consecutive-k-out-of-n: F systems were studied first by Kontoleon [27]; however, the system name of consecutive-k-out-of-n:

W.Y. Yun · A.J. Endharta (✉)
Department of Industrial Engineering, Pusan National University, Busan, South Korea
e-mail: endharta_aj@pusan.ac.kr

W.Y. Yun
e-mail: wonyun@pusan.ac.kr

M. Ram and J.P. Davim (eds.), *Advances in Reliability and System Engineering*, Management and Industrial Engineering, DOI 10.1007/978-3-319-48875-2_2

F system originated from Chiang and Niu [8]. A mirror image of this system structure, namely consecutive-k-out-of-n: G systems have been studied and the system works if and only if there are at least k consecutive working components [24, 28, 48].

A closed form of the reliability function of a linear consecutive-2-out-of-n: F system has been developed by Chang and Niu [8]. They considered a constant failure probability p of components and obtained the exact reliability value. Another reliability function of the similar system has been proposed by Bollinger and Salvia [4] and involved the term of $N(j, k, n)$ which was interpreted as the number of binary numbers with length n containing exactly j ones with at most $k - 1$ consecutive ones. A combinatorial approach was considered for $N(j, 3, n)$ and the recursive method for the general $N(j, k, n)$ was proposed by Derman et al. [10]. A spreadsheet table was utilized in deriving $N(j, k, n)$ by Bollinger [3]. A closed form of general $N(j, k, n)$ has been also derived [21, 23, 31].

Circular structure of this system was introduced and the system reliability was provided by Derman et al. [10]. A recursive model to estimate the reliability of circular-type systems was developed by Du and Hwang [13]. A closed-form expression of the reliability of circular consecutive-k-out-of-n: F systems for $n \leq 2k + 1$ and when the components were identical with constant failure probability has been derived by Zuo and Kuo [48].

Some papers calculated the dynamic system reliability of consecutive-k-out-of-n systems in which the time to component failures is a random variable and follows a certain lifetime distribution (for example, Exponential or Weibull distributions). The expressions of the system reliability, mean time to system failure (MTTF), and the rth moment of the lifetime of consecutive-k-out-of-n: F systems can be seen in Kuo and Zuo [29].

Most existing papers related to consecutive-k-out-of-n: F systems assumed that the failures of components are independent but some papers considered dependence between component failures. Shantikumar [42] considered the linear-type system with exchangeable components, which are identical but not independent and the failure rate of operating components increases when there is one or more component failures. Papastavridis [38] considered the circular-type system with dependent components. Papastavridis and Lambiris [39] introduced another dependence mechanism called $(k - 1)$-step Markov dependence. They assumed that the reliability of component i is dependent on the states of k consecutive components besides that component i. Fu [20] considered this mechanism in a practical problem of oil transportation by pumps. Load-sharing model is a special case of Markov dependence model with $(n - 1)$-step.

A system design problem can also be considered in consecutive-k-out-of-n: F systems. In the design phase for consecutive-k-out-of-n: F systems, we should determine the system structure optimally; for example, the system parameters, such as k, n, and the component reliability, p_i for $1 \leq i \leq n$. A system design problem for a circular consecutive-k-out-of-n: F system with $(k - 1)$-step Markov dependent components has been considered in Yun et al. [46] and the optimal k and n are optimized minimizing the expected cost rate when only corrective maintenance is considered. A branch and bound algorithm was developed and used in the estimation of the system reliability for cases where the system size is small. For the cases where the system size is large, a simulation method is developed. Later, a system design problem for linear and circular consecutive-k-out-of-n: F systems with load-sharing dependent components were considered in Yun et al. [47].

Another design problem is the optimal allocation problem of given reliability values to the positions or components of the system [29, 30]. Birnbaum's importance measure was utilized in the reliability importance of component i in a linear consecutive-k-out-of-n: F system [37]. The expression for the circular-type system was also provided by Griffith and Govindarajalu [22]. For the liner-type system, the importance measure increases as the component locates nearer to the center of the system and this situation occurs for cases where the system contains only i.i.d components. The general deduction is that the optimal design is to assign the least reliable component to position 1, the next least reliable component to position n, the most reliable component to position 2, the next most reliable component to position $n-1$, and so on [12, 34]. However, for linear consecutive-k-out-of-n: F system, the mentioned optimal design cannot be generalized for all k values [35]. The optimal design is called as the invariant optimal design if the optimal solutions depend only on the ranking of the reliability values of components. Otherwise, it is called as the variant optimal design.

Another optimization problem is the maintenance problem. This problem aims to determine how and when the maintenance should be performed in order to avoid the system failures. Basic maintenance policies, such as age replacement, periodic replacement, and block replacement policies can be seen in Barlow and Proschan [1] or Nakagawa [32]. An algorithm has been developed to select the maintenance policy based on the critical component policy (CCP) for the consecutive-k-out-of-n: F systems [19]. Based on that study, the failed components are replaced if and only if the failed components are contained within the critical component sets. An age replacement model for linear and circular consecutive-k-out-of-n: F systems with load-sharing dependent components has been considered by Yun et al. [47] and a condition-based maintenance model for the linear-type systems has been considered by Endharta and Yun [15].

Salvia and Lasher [41] has introduced the multidimensional structure of the systems [29]. Two- or the three-dimensional system is a square or cubic grid of side n. Thus, the system contains n^2 or n^3 components and the system fails if and only if there is at least a square or a cube of k failed components. The more general two-dimensional system has been introduced by Boehme et al. [2] and later is called as a linear or circular connected-(r, s)-out-of-(m, n): F system. The system contains the components which are arranged into a rectangular pattern with m rows and n columns and it fails if and only if there is at least one grid of r rows and s columns comprising failed components only.

A recursive method for the reliability estimation of a linear connected-(r, s)-out-of-(m, n): F system has been constructed by Yamamoto and Miyakawa [43]. By using the method, the lower and upper bounds of the system reliability can be obtained. The component allocation and design problem showed that the existence of invariant design is very limited even in the case of superfluous consecutive systems, as well as the two-dimensional systems [25]. An age replacement model for a connected-(r, s)-out- of-(m, n): F system has been considered by Yun et al. [45], in which they developed a simulation procedure to estimate the expected cost rate, and genetic algorithm to obtain the optimal time interval. A condition-based maintenance model for one- and two-dimensional systems has been considered by Yun and Endharta [44], in which they assume a continuous monitoring and a simulation method is developed to obtain the near optimal solutions.

In this chapter, we review the maintenance problem in consecutive-k-out-of-n: F systems and connected-(r, s)-out-of-(m, n): F systems. Section 2 shows the multi-unit systems considered, such as consecutive-k-out-of-n and connected-(r, s)-out-of-(m, n): F systems. Section 3 presents the component dependence models and Sect. 4 presents the maintenance models (corrective maintenance, age-based preventive maintenance, and condition-based maintenance models). Section 5 shows the numerical examples, including comparison results of the expected cost rates of three maintenance policies. The study is concluded in Sect. 6.

2 Multi-unit Systems

In this section, consecutive-k-out-of-n and connected-(r, s)-out-of-(m, n): F systems are introduced and minimal cut sets of the systems are used to obtain the system reliability.

2.1 *Consecutive-k-Out-of-n: F Systems*

A system which fails if and only at least k of the n components fail is called a k-out-of-n: F system. A series and parallel systems are special case of this system, where the series system is a 1-out-of-n: F system and the parallel system is an n-out-of-n: F system. Another developed structure of this system is consecutive-k-out-of-n: F system. A consecutive-k-out-of-n: F system is a system which fails if and only if at least k consecutive components fail among n components.

There are two types of consecutive-k-out-of-n: F systems: linear and circular consecutive-k-out-of-n: F systems. Thus, the linear or circular consecutive-k-out-of-n: F systems consist of n components arranged linearly or circularly and fail if and only if at least k consecutive components fail. Linear and circular systems are illustrated in Fig. 1.

Definition 1 Cut sets and minimal cut sets [40]

A cut set K is a set of components which by failing causes the system to fail. A cut set is said to be minimal if it cannot be reduced without losing its status as a cut set.

The number of minimal cut sets for the system with linear form is $n - k + 1$ and the minimal cut sets are

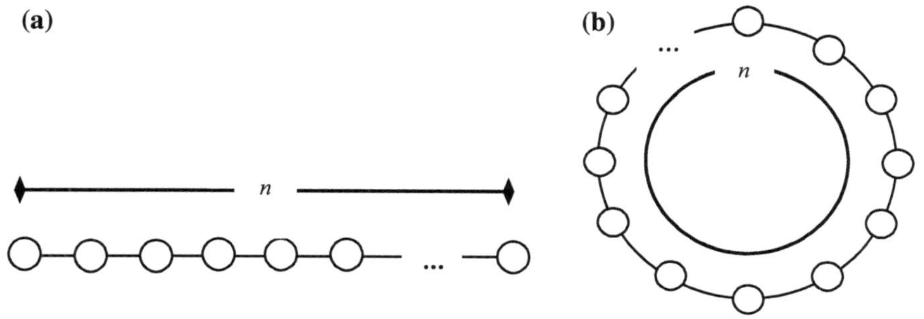

Fig. 1 **a** Linear and **b** circular consecutive-k-out-of-n: F systems

$$\begin{aligned} K_1 &= \{1,2,\ldots,k\}, K_2 = \{2,3,\ldots,k+1\},\ldots,K_{n-k+1} \\ &= \{n-k+1, n-k+2,\ldots,n\}. \end{aligned}$$

Since the component 1 and component n in the system with circular form are connected, there are n minimal cut sets for this system and they are

$$K_1 = \{1,2,\ldots,k\}, K_2 = \{2,3,\ldots,k+1\},\ldots,K_n = \{n,1,\ldots,k-1\}.$$

Based on this information, with the same number of components, n, the system with circular type tends to fail quicker than the system with linear type because the probability of system failure increases as the number of minimal cut set increases.

In order to estimate the system unreliability or the probability of the system failure, the system failure events must be known. Since the system consists of multiple components, one way to know the system failure events is to arrange the sequences of component failures in the system. Sequences from the beginning to the system failure events are the system failure paths.

The system state consists of the states of components, where the working component is 1 and failed component is 0. Thus, in the beginning, the system state is represented as a vector of ones with size n. The component fails one at a time and the sequences are called as paths. Accordingly, at the system failure event (the last step), the system state can be represented as a vector with size n, where there is at least k consecutive zeros (failed components), which are components in the minimal cut sets. The number of system failure paths is denoted as P. System failure paths are illustrated in Tables 1 and 2 for linear and circular consecutive 2-out-of-3: F systems, respectively.

Number of system failure paths P for the circular type can be reduced by assuming that the first failed component is Component 1. Thus, system failure paths in Table 2 changes into Table 3. Denote the reduced number of system failure paths in the circular-type system by p.

Table 1 System failure paths for linear consecutive 2-out of-3: F system

Path j	Step i					
	0		1		2		3
1	111	→	011	→	001		
2	111	→	011	→	010	→	000
3	111	→	101	→	001		
4	111	→	101	→	100		
5	111	→	110	→	010	→	000
6	111	→	110	→	100		

Table 2 System failure paths for circular consecutive 2-out of-3: F system

Path	Step				
	0		1		2
1	111	→	011	→	001
2	111	→	011	→	010
3	111	→	101	→	001
4	111	→	101	→	100
5	111	→	110	→	010
6	111	→	110	→	100

Table 3 Reduced number of paths for circular consecutive 2-out of-3: F system

Path	Step				
	0		1		2
1	111	→	011	→	001
2	111	→	011	→	010

It is assumed that the components are identical and the failure times of the components follow an exponential distribution with rate λ. Thus, the system failure time distribution can be derived as follows. First, a lemma is introduced.

Lemma 1 [5]

Let $Y_1, Y_2, \ldots, Y_m$ be exponentially distributed random variables with failure rates $h_1, h_2, \ldots, h_m$, and let $Z = \min(Y_1, Y_2, \ldots, Y_m)$. Then Z is also exponentially distributed with failure rate $\sum h_i$ and $\Pr\{Z = Z_i\} = h_i/\sum h_i$.

Lemma 1 provides the probability of one component fails among all working components in step i.

Lemma 1 gives the probability of selecting the component which will fail in step i among working components. The sum of failure rates of working components after ith failure and the failure rate of the ith failed component in path j are denoted as α_{ji} and β_{ji}, respectively.

Define T_j and T, respectively, as the time to complete path j and the time to system failure. The probability that the system follows path j, which is denoted as π_j, can be estimated as

$$\pi_j = \Pr\{T = T_j\} \tag{1}$$

Define X_i is the time between $(i-1)$th failure and ith failure in the system and X_{ji} as the time between $(i-1)$th failure and ith failure in path j, thus

$$\begin{aligned}\pi_j &= \Pr\{X_1 = X_{j1}, X_2 = X_{j2}, \ldots, X_{N_j} = X_{jN_j}\}\\ &= \Pr\{X_1 = X_{j1}\} \cdot \prod_{i=2}^{N_j} \Pr\{X_i = X_{ji} \mid X_1 = X_{j1}, X_2 = X_{j2}, \ldots, X_{i-1} = X_{j,i-1}\}\end{aligned} \tag{2}$$

Based on Lemma 1, we can verify that

$$\Pr\{X_i = X_{ji} \mid X_1 = X_{j1}, X_2 = X_{j2}, \ldots, X_{i-1} = X_{j,i-1}\} = \frac{\beta_{ji}}{\alpha_{ji}}.$$

Thus, the probability that the change of system states follows path j is

$$\pi_j = \prod_{i=0}^{N_j-1} \frac{\beta_{ji}}{\alpha_{ji}}. \tag{3}$$

Also based on Lemma 1, it can be seen that X_{ji} is exponentially distributed with rate α_{ji}. Thus, the distribution of T_j can be obtained through convolution of X_{ji}. Laplace transform of X_{ji} is

$$f_{X_{ji}}^{e}(s) = \frac{\alpha_{ji}}{\alpha_{ji} + s}$$

thus, Laplace transform for T_j is

$$f_j^{e}(s) = \prod_{i=0}^{N_j-1} \frac{\alpha_{ji}}{\alpha_{ji} s}$$

By using partial fraction, we can obtain

$$f_j^{e}(s) = \sum_{i=0}^{N_j-1} A_{ji} \frac{\alpha_{ji}}{\alpha_{ji} + s}$$

where $A_{ji} = \prod_{\substack{m=0\\ m \neq i}}^{N_j-1} \frac{\alpha_{jm}}{\alpha_{jm} - \alpha_{ji}}$, $i = 0, 1, \ldots, N_j - 1$.

Following the inversion of the Laplace transform, the p.d.f and c.d.f of T_j, respectively, are

$$f_j(t) = \sum_{i=0}^{N_j-1} A_{ji}\alpha_{ji}e^{-\alpha_{ji}t} \tag{4}$$

and

$$\begin{aligned} F_j(t) &= \int_0^t f_j(x)\mathrm{d}x \\ &= 1 - \sum_{i=0}^{N_j-1} A_{ji}e^{-\alpha_{ji}t} \end{aligned} \tag{5}$$

We can see that the distribution of the system failure time T is a mixture distribution of T_j. Therefore, the system failure probability at time t can be estimated as follows.

$$\begin{aligned} F(t) &= \sum_{j=1}^{P} \pi_j \cdot F_j(t) \\ &= 1 - \sum_{j=1}^{P}\sum_{i=0}^{N_j-1} \pi_j A_{ji}e^{-\alpha_{ji}t} \end{aligned} \tag{6}$$

Note that p is the reduced number of system failure paths in the systems with circular type. For the system with circular type, the probability of the system failure $F(t)$ can be estimated by

$$\begin{aligned} F(t) &= n\sum_{j=1}^{p} \pi_j \cdot F_j(t) \\ &= 1 - n\sum_{j=1}^{p}\sum_{i=0}^{N_j-1} \pi_j A_{ji}e^{-\alpha_{ji}t} \end{aligned} \tag{7}$$

Thus, the system failure probabilities of consecutive-k-out-of-n: F systems can be calculated by Eqs. 6 and 7.

2.2 *Connected-(r, s)-Out-of-(m, n): F System*

Salvia and Lasher [41] introduced a multidimensional consecutive-k-out-of- n: F system and define that the two- or the three-dimensional system is a square or cubic grid of side n. The system fails if and only if there is at least a square or a cube of k consisting of failed components only. More generalized systems are linear and circular connected-(r, s)-out-of-(m, n): F systems [2]. This system contains

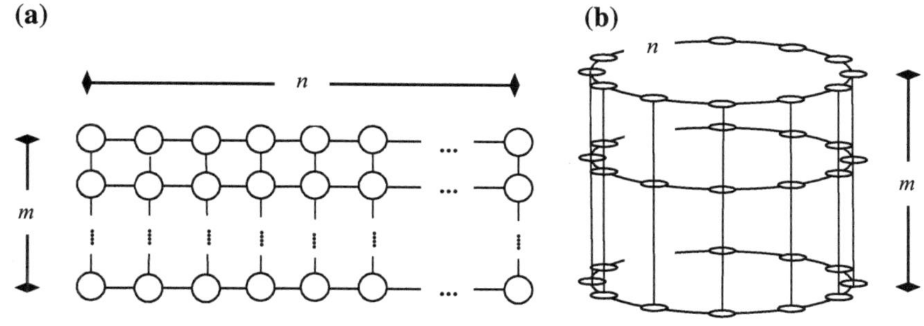

Fig. 2 **a** Linear and **b** circular connected-(r, s)-out-of-(m, n): F systems

components which are arranged into a rectangular pattern with m rows and n columns and it will fail if and only if there is at least a grid of r rows and s columns which consists of only failed components. The circular-type system forms a cylinder where the components in the first column connected to the components in the last column (see Fig. 2).

We denote the component in column a and row b by $\{a, b\}$. Thus the minimal cut sets for the linear type are

$$\begin{aligned} K_{ab} = \{&\{a,b\},\{a,b+1\},\ldots,\{a,s+b-1\},\{a+1,b\},\{a+1,b+1\},\ldots,\\ &\{a+1,s+b-1\},\ldots,\{r+a-1,b\},\{r+a-1,b+1\},\ldots,\\ &\{r+a-1,s+b-1\}\} \end{aligned}$$

for $a = 1, 2, \ldots, m - r + 1$ and $b = 1, 2, \ldots, n - s + 1$ and there are $(m - r + 1)(n - s + 1)$ minimal cut sets.

Since the components in the column 1 and column n are connected, there are $(m - r + 1)n$ minimal cut sets for the circular connected-(r, s)-out-of-(m, n): F system.

Because of the size of the system, the system failure probability for the general two-dimensional system: connected-(r, s)-out-of-(m, n): F system is difficult to obtain. Some studies derived the system reliability of two-dimensional systems with very small size (connected-(1, 2)-or-(1, 2)-out-of-(m, n): F system). In the other studies, the reliability value is estimated by simulation, approximation, or only the lower bound is obtained [25, 43–45].

3 Component Dependence Models

This section describes two component dependence mechanisms in consecutive k-out-of-n: F systems and shows the consequence in the system failure paths. Although the failures of components are dependent on each other, the system failure paths are independent on each other. Thus, it is easier to obtain the probability by exploiting the system failure paths. In this subsection, we consider the systems with identical components of which are identical and the failure rates are constant.

3.1 Systems with i.i.d Components

The failure rates of a working component in every step in each path in the system consisting of i.i.d components are identical and the value is denoted as λ. We define w_{ji} as the number of working components in step i in path j. The term α_{ji} and β_{ji} can be given as follows, respectively,

$$\alpha_{ji} = w_{ji}\lambda, \tag{8}$$

$$\beta_{ji} = \lambda. \tag{9}$$

In order to understand more easily, a small illustration is made. Table 4 shows the system failure paths and the corresponding α_{ji} and β_{ji} in step i in path j for linear consecutive-2-out-of-3: F system. Similarly, we can get the system failure paths for the circular-type system.

By substituting Eqs. (8) and (9) into Eq. (3), we obtain

Table 4 Paths for a linear consecutive-2-out-of-3: F system with i.i.d components

Path j	Step i												
	0				1				2				3
	State	α_{j0}	β_{j0}		State	α_{j1}	β_{j1}		State	α_{j2}	β_{j2}		State
1	**1**11	3λ	λ	–	0**1**1	2λ	λ	–	001				
2	**1**11	3λ	λ	–	01**1**	2λ	λ	–	0**1**0	λ	λ	–	000
3	1**1**1	3λ	λ	–	**1**01	2λ	λ	–	001				
4	1**1**1	3λ	λ	–	10**1**	2λ	λ	–	100				
5	11**1**	3λ	λ	–	**1**10	2λ	λ	–	0**1**0	λ	λ	–	000
6	11**1**	3λ	λ	–	1**1**0	2λ	λ	–	100				

$$\pi_j = \prod_{i=0}^{N_j - 1} \frac{1}{w_{ji}}. \tag{10}$$

Thus, we can estimate the system failure probability by substituting α_{ji}, β_{ji} and π_j into the system failure probability equation in the previous section.

3.2 Systems with (k − 1)-Step Markov Dependence

Another component dependence model for these systems is $(k-1)$-step Markov dependence model [39]. The reliability of component i depends on the states of $k-1$ components which are located beside that component. Load-sharing model is a special case of Markov dependence with $(n-1)$-step.

Failure rate of a component in the beginning time point denoted as λ_0 because all components are working. The failure rate of the working components becomes λ_i when there are i failed components preceding it. Table 5 shows the illustration for linear consecutive-2-out-of-3: F system with $(k-1)$-step Markov dependence model.

Thus, we can estimate the system failure probability by estimating the terms α_{ji}, β_{ji} and π_j and substituting into the system failure probability equation in the previous section.

3.3 Systems with Load-Sharing Components

If one component fails, the workload has to be shared by the remaining components. Thus, there will be an increased load share for each surviving component. In most conditions, increased workload causes a higher failure rate of the component

Table 5 Paths for a linear consecutive-2-out-of-3: F system with $(k-1)$-step Markov dependence

Path j	Step i												
	0				1				2				3
	State	α_{j0}	β_{j0}		State	α_{j1}	β_{j1}		State	α_{j2}	β_{j2}		State
1	111	$3\lambda_0$	λ_0	–	011	$\lambda_1 + \lambda_0$	λ_1	–	001				
2	111	$3\lambda_0$	λ_0	–	011	$\lambda_1 + \lambda_0$	λ_0	–	010	λ_1	λ_1	–	000
3	111	$3\lambda_0$	λ_0	–	101	$\lambda_1 + \lambda_0$	λ_0	–	001				
4	111	$3\lambda_0$	λ_0	–	101	$\lambda_1 + \lambda_0$	λ_1	–	100				
5	111	$3\lambda_0$	λ_0	–	110	$2\lambda_0$	λ_0	–	010	λ_1	λ_1	–	000
6	111	$3\lambda_0$	λ_0	–	110	$2\lambda_0$	λ_0	–	100				

Table 6 Paths for a linear consecutive-2-out-of-3: F system with load-sharing components

Path j	Step i												
	0				1				2				3
	State	α_{j0}	β_{j0}		State	α_{j1}	β_{j1}		State	α_{j2}	β_{j2}		State
1	111	$3\lambda_0$	λ_0	–	011	$2\lambda_1$	λ_1	–	001				
2	111	$3\lambda_0$	λ_0	–	011	$2\lambda_1$	λ_1	–	010	λ_2	λ_2	–	000
3	111	$3\lambda_0$	λ_0	–	101	$2\lambda_1$	λ_1	–	001				
4	111	$3\lambda_0$	λ_0	–	101	$2\lambda_1$	λ_1	–	100				
5	111	$3\lambda_0$	λ_0	–	110	$2\lambda_1$	λ_1	–	010	λ_2	λ_2	–	000
6	111	$3\lambda_0$	λ_0	–	110	$2\lambda_1$	λ_1	–	100				

and many practical studies with mechanical and computer systems showed that the failure rates of working components strongly relates to the workload. For the case of long belt conveyors used in open-cast mining, even if a roll station (for example, motor) of the belt conveyor fails, the conveyor does not stop and tt only stops (fails) when k consecutive roll stations fail [47].

Failure rate of a component in the beginning of time is λ_0 because all components are working. After the first component fails, $n - 1$ working components must carry the same workload. Thus, these $n - 1$ working components have a higher failure rate, which is λ_1. In general, when there are i failed components, the $n - i$ working components will have a failure rate λ_i (where $\lambda_0 < \lambda_1 < \ldots < \lambda_{n-k+1}$).

The failure rate of a working component where there are i failed components and given load L is denoted as $\lambda_i(L)$. There are i failed components in step i for all paths. As we defined in the previous subsection that w_{ji} is the number of working components in step i in path j, we can calculate the term α_{ji} and β_{ji}, respectively, and Table 6 shows the illustration for linear consecutive-2-out-of-3: F system with load-sharing components. In similar way, we can get the system failure paths for circular-type systems.

$$\alpha_{ji} = w_{ji}\lambda_i(L), \tag{11}$$

$$\beta_{ji} = \lambda_i(L). \tag{12}$$

By substituting Eqs. (11) and (12) into Eq. (3), we can obtain the probability that the system follows path j, π_j,

$$\pi_j = \prod_{i=0}^{N_j-1} \frac{1}{w_{ji}}. \tag{13}$$

Thus, we can estimate the system failure probability by substituting α_{ji}, β_{ji} and π_j into the system failure probability equation in the previous section.

In this subsection, we considered two dependence models but components fail one by one. However, in addition to components failing one by one, there may be other causes to result in the system failure. Common cause failures are ones of these factors. Common cause failures describe single events which cause multiple component failures [29]. These events can be external events, such as storms, floods, lightning, seismic activities, maintenance errors, other human intervention errors, and sudden environmental change, or internal events, such as the failures of other components. Studies on k-out-of-n systems with common cause failures include Chung [9], Jung [26], Chari [7] and Dhillon and Anude [11]. This dependence model can be also applied to consecutive k-out-of-n systems.

4 Maintenance Models

In this section, the maintenance problem for the consecutive-k-out-of-n: F systems are studied. The basic maintenance policies, such as corrective maintenance and age preventive maintenance policies are considered. A condition-based maintenance policy is also studied. The performance evaluation of these policies for the consecutive-k-out-of-n: F systems can be seen in detail in Endharta [14], Endharta et al. [17] and Endharta and Yun [16].

4.1 Corrective Maintenance

We assume that the system is maintained only at the system failure time. The expected time to system failure for this policy is

$$E[T] = \sum_{j=1}^{P} \pi_j \cdot E[T_j]. \tag{14}$$

Because the time to failure in path j includes the time between failures X_{ji} and the time between failures follows an exponential distribution with failure rate α_{ji}, the expected time to system failure in path j can be estimated as

$$E[T_j] = \sum_{i=0}^{N_j-1} E[X_{ji}] = \sum_{i=0}^{N_j-1} \frac{1}{\alpha_{ji}}. \tag{15}$$

Thus, the expected time to system failure ET can be written as

$$E[T] = \sum_{j=1}^{P} \sum_{i=0}^{N_j-1} \frac{\pi_j}{\alpha_{ji}}. \tag{16}$$

The expected number of failures at system failure can be estimated as

$$\begin{aligned} E[N] &= E\left[N^{\mathrm{SF}}\right] \\ &= \sum_{j=1}^{P} \pi_j \cdot N_j \end{aligned} \tag{17}$$

In this maintenance model, the cost terms which are considered are the cost spent for the corrective maintenance, c_{CM}, and the cost for component replacement, c_{R}. The expected cost rate of corrective replacement model is

$$EC_{\mathrm{CM}} = \frac{c_{\mathrm{CM}} + c_{\mathrm{R}} \cdot E[N]}{E[T]}. \tag{18}$$

Substituting Eqs. (16) and (17), the equation becomes

$$EC_{\mathrm{CM}} = \frac{c_{\mathrm{CM}} + c_{\mathrm{R}}\left(\sum_{j=1}^{P} \pi_j \cdot N_j\right)}{\sum_{j=1}^{P} \sum_{i=0}^{N_j-1} \frac{\pi_j}{\alpha_{ji}}}. \tag{19}$$

4.2 Age-Based Preventive Maintenance

In age replacement model, the system is maintained at age time T_{A} after its installation or at failure, whichever occurs first. When the system fails before $T_{\mathrm{A}}(T < T_{\mathrm{A}})$, the system is maintained correctively at the system failure. Otherwise, when there is no system failure before $T_{\mathrm{A}}(T > T_{\mathrm{A}})$, the system is maintained preventively at time T_{A}. Therefore, the expected time to renewal cycle, which is the expected time to system failure or to the preventive maintenance time, can be estimated as

$$\begin{aligned} E[T(T_{\mathrm{A}})] &= \int_0^{T_{\mathrm{A}}} t \cdot d\,\Pr\{T \le t\} + T_{\mathrm{A}} \Pr\{T > T_{\mathrm{A}}\} \\ &= \int_0^{T_{\mathrm{A}}} t \cdot d\left(\sum_{j=1}^{P} \pi_j \cdot F_j(t)\right) + T_{\mathrm{A}}\left(1 - \sum_{j=1}^{P} \pi_j \cdot F_j(t)\right) \\ &= \int_0^{T_{\mathrm{A}}} \sum_{j=1}^{P} \sum_{i=0}^{N_j-1} \pi_j A_{ji} \alpha_{ji} t \mathrm{e}^{-\alpha_{ji} t} \mathrm{d}t + T_{\mathrm{A}} \sum_{j=1}^{P} \sum_{i=0}^{N_j-1} \pi_j A_{ji} \mathrm{e}^{-\alpha_{ji} t} \end{aligned} \tag{20}$$

where $A_{jl} = \prod_{\substack{m=0 \\ m \neq l}}^{i-1} \frac{\alpha_{jm}}{\alpha_{jm} - \alpha_{jl}}$.

The failed components are replaced at maintenance time and the expected number of failed components includes the expected number of components failed at system failure and the expected number of components failed at time T_A,

$$\begin{aligned} E[N(T_A)] &= E[N^{SF}] + E[N^{T_A}] \\ &= \sum_{j=1}^{P} \pi_j N_j F_j(T_A) + \sum_{j=1}^{P} \pi_j E\left[N_j^{T_A}\right] \end{aligned} \tag{21}$$

Note that X_{ji} is the time between $(i-1)$th failure and ith failure in path j and X_{ji} follows an exponential distribution with parameter α_{ji}. Thus, we define T_{ji} as the time to ith failure in path j such that

$$T_{jm} = X_{j1} + X_{j2} + \ldots + X_{jm}.$$

The following term is necessary for the estimation, for $a \leq b$,

$$\begin{aligned} \Pr\{T_{ji} > b, T_{j,i-1} < a\} &= \Pr\{T_{j,i-1} + X_{ji} > b, T_{j,i-1} < a\} \\ &= \Pr\{X_{ji} > b - a | T_{j,i-1} < a\} \cdot \Pr\{T_{j,i-1} < a\} \\ &= \int_0^a \Pr\{X_{ji} > b - y\} \cdot d \Pr\{T_{j,i-1} < y\} \\ &= \int_0^a e^{-\alpha_{ji}(b-y)} \sum_{l=0}^{i-1} A_{jl}\alpha_{jl} e^{-\alpha_{jl} y} dy \\ &= \sum_{l=0}^{i-1} \frac{A_{jl}\alpha_{jl}}{\alpha_{ji} - \alpha_{jl}} \left(e^{-\alpha_{jl}a - \alpha_{ji}(b-a)} - e^{-\alpha_{ji}b} \right) \end{aligned} \tag{22}$$

where $A_{jl} = \prod_{\substack{m=0 \\ m \neq l}}^{i-1} \frac{\alpha_{jm}}{\alpha_{jm} - \alpha_{jl}}$.

Then, the expected number of components failed at time T_A in path j, $E[N_j^{T_A}]$, needs to be derived.

$$\begin{aligned} E[N_j^{T_A}] = 0 \cdot \Pr\{T_{j1} > T_A\} + 1 \cdot \Pr\{T_{j2} > T_A, T_{j1} < T_A\} + \ldots + (N_{j-1}) \\ \cdot \Pr\{T_{jN_j} > T_A, T_{jN_j-1} < T_A\} \end{aligned}$$

We generalize the above equation into

$$E\left[N_j^{T_A}\right] = \sum_{i=1}^{N_j-1} i \cdot \Pr\{T_{j,i+1} > T_A, T_{j,i} < T_A\} \tag{23}$$

Substituting Eq. (22) into Eq. (23), we obtain

$$E\left[N_j^{T_A}\right] = \sum_{i=1}^{N_j-1}\sum_{l=0}^{i-1}\frac{iA_{jl}\alpha_{jl}}{\alpha_{ji}-\alpha_{jl}}\left(e^{-\alpha_{ji}T_A} - e^{-\alpha_{jl}T_A}\right). \quad (24)$$

In this maintenance model, the cost terms which are considered are the cost for the corrective maintenance, c_{CM}, cost for preventive maintenance at time T_A, c_{PM}, and the cost for component replacement, c_R. The expected cost rate under age-based preventive maintenance model is

$$EC_A(T_A) = \frac{c_{CM}F(T_A) + c_{PM}(1 - F(T_A)) + c_R \cdot \left(\sum_{j=1}^{P}\pi_j N_j F_j(T_A) + \sum_{j=1}^{P}\pi_j E\left[N_j^{T_A}\right]\right)}{E[T(T_A)]}, \quad (25)$$

where $F(t) = 1 - \sum_{j=1}^{P}\sum_{i=0}^{N_j-1}\pi_j A_{ji}e^{-\alpha_{ji}t}$, $F_j(t) = 1 - \sum_{i=0}^{N_j-1}A_{ji}e^{-\alpha_{ji}t}$, $E[T(T_A)] = \int_0^{T_A}\sum_{j=1}^{P}\sum_{i=0}^{N_j-1}\pi_j A_{ji}\alpha_{ji}te^{-\alpha_{ji}t}dt + T_A\sum_{j=1}^{P}\sum_{i=0}^{N_j-1}\pi_j A_{ji}e^{-\alpha_{ji}T_A}$, and $E\left[N_j^{T_A}\right] = \sum_{i=1}^{N_j-1}\sum_{l=0}^{i-1}\frac{iA_{jl}\alpha_{jl}}{\alpha_{ji}-\alpha_{jl}}\left(e^{-\alpha_{ji}T_A} - e^{-\alpha_{jl}T_A}\right)$.

4.3 Condition-Based Maintenance

In this subsection, a condition-based maintenance policy proposed by Endharta and Yun [15] is considered. In this maintenance policy, the system is maintained when a certain system condition is satisfied. Since the system failure occurs when at least consecutive k components fail or all components in the minimal cut set fail, a condition-based maintenance policy is developed based on the number of working components in minimal cut sets and the number of minimal cut sets with specific number of working components.

a. **Condition-based maintenance with continuous monitoring**

In this model, we assume that that the system is monitored continuously and the condition about components in the minimal cut sets can be known at any time. The

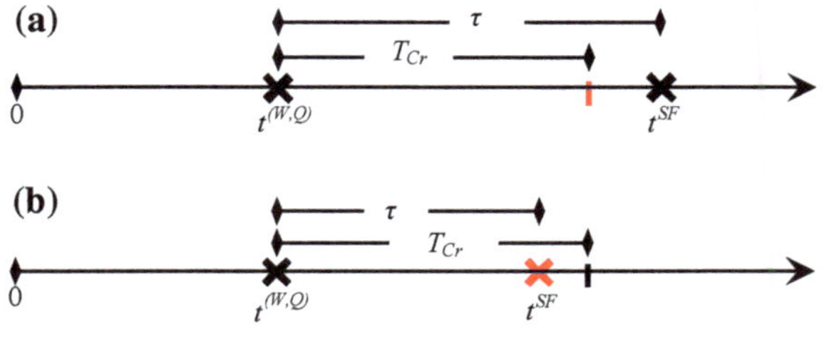

Fig. 3 Illustration of the condition-based maintenance policy. **a** PM occurs **b** CM occurs

system will be maintained preventively at a certain time point T_{Cr} after a specified condition is satisfied. The condition is that there are Q or more minimal cut sets having W or less working components. If the system fails before reaching the preventive time point, the system will be correctively maintained. The maintenance is illustrated in Fig. 3 and several terms are used. The time when the condition is satisfied is represented by $t^{(W,Q)}$, the time of system failure is t^{SF}, and the step in which the condition is satisfied is represented by $s^{(W,Q)}$.

Define a random variable $\tau = t^{SF} - t^{(W,Q)}$, representing the time difference between $t^{(W,Q)}$ and t^{SF}. If $\tau > T_{Cr}$, the renewal cycle is ended by PM and the system is maintained preventively (see Fig. 3a). Otherwise, if $\tau < T_{Cr}$, the renewal cycle is ended by CM and the system is maintained correctively (see Fig. 3b).

We know that until time $t^{(W,Q)}$ or at time to step $s^{(W,Q)}$, there is no system failure. Thus, the system failure distribution probability $F(t)$ in Eq. (7) changes into

$$F(t) = 1 - \sum_{j=1}^{P} \sum_{i=s_j^{(W,Q)}}^{N_j-1} \pi_j A_{ji} e^{-\alpha_{ji} t} \tag{26}$$

where $s_j^{(W,Q)}$ is the step where the condition is satisfied in path j,

$$\pi_j = \prod_{i=s_j^{(W,Q)}}^{N_j-1} \frac{\beta_{ji}}{\alpha_{ji}} \quad \text{and} \quad A_{ji} = \prod_{\substack{m = s_j^{(W,Q)} \\ m \neq i}}^{N_j-1} \frac{\alpha_{jm}}{\alpha_{jm} - \alpha_{ji}}.$$

The expected time of a renewal cycle includes the time to $s^{(W,Q)}$, the expected time to system failure, and the expected time to T_{Cr} as follows:

$$\begin{aligned} E[T(T_{Cr})] &= E[t^{(W,Q)}] + \int_0^{T_{Cr}} t \cdot d\Pr\{T \le t\} + T_{Cr}\Pr\{T > T_{Cr}\} \\ &= E[t^{(W,Q)}] + \int_0^{T_{Cr}} t \cdot dF(t) + T_{Cr}(1 - F(T_{Cr})) \\ &= E[t^{(W,Q)}] + \int_0^{T_{Cr}} \sum_{j=1}^{P} \sum_{i=s_j^{(W,Q)}}^{N_j-1} \pi_j A_{ji} \alpha_{ji} t e^{-\alpha_{ji} t} dt + T_{Cr} \sum_{j=1}^{P} \sum_{i=s_j^{(W,Q)}}^{N_j-1} \pi_j A_{ji} e^{-\alpha_{ji} T_{Cr}} \end{aligned} \tag{27}$$

where $E[t^{(W,Q)}] = \sum_{j=1}^{P} \sum_{i=0}^{s_j^{(W,Q)}-1} \frac{\pi_j}{\alpha_{ji}}$.

Denote $N^{(W,Q)}$ as the number of components failed at step $s^{(W,Q)}$ and $N_j^{(W,Q)}$ as the number of components failed at step $s_j^{(W,Q)}$ in path j. The components are replaced at maintenance time and the expected number of failed components includes the number of components failed at $s^{(W,Q)}$ and the expected number of components failed additionally until system failure and the expected number of components failed additionally until maintenance time T_{Cr},

$$\begin{aligned} E[N(T_{Cr})] &= E[N^{(W,Q)}] + E[N^{SF}] + E[N^{T_{Cr}}] \\ &= \sum_{j=1}^{P} \pi_j \cdot N_j^{(W,Q)} + \sum_{j=1}^{P} \pi_j N_j F_j(T_{Cr}) + \sum_{j=1}^{P} \pi_j E\left[N_j^{T_{Cr}}\right] \\ &= \sum_{j=1}^{P} \pi_j \left(N_j^{(W,Q)} + N_j F_j(T_{Cr}) + E\left[N_j^{T_{Cr}}\right]\right) \end{aligned} \quad (28)$$

where $F_j(t) = 1 - \sum_{i=s_j^{(W,Q)}}^{N_j-1} A_{ji} e^{-\alpha_{ji} t}$ and $E\left[N_j^{T_{Cr}}\right] = \sum_{i=s_j^{(W,Q)}}^{N_j-1} \sum_{m=s_j^{(W,Q)}}^{i-1} \frac{iA_{jm}\alpha_{jm}}{\alpha_{ji}-\alpha_{jm}} \left(e^{-\alpha_{ji}T_{Cr}} - e^{-\alpha_{jm}T_{Cr}}\right)$.

In this maintenance policy, we consider the cost for the corrective maintenance, c_{CM}, cost for preventive maintenance at time T_{Cr}, c_{PM}, and the cost for component replacement, c_R, in the expected cost rate. The expected cost rate under the condition-based maintenance model is

$$EC_{Cr}(T_{Cr}) = \frac{c_{CM} F(T_{Cr}) + c_{PM}(1 - F(T_{Cr})) + c_R \cdot \sum_{j=1}^{P} \pi_j \left(N_j^{(W,Q)} + N_j F_j(T_{Cr}) + E\left[N_j^{T_{Cr}}\right]\right)}{E[T(T_{Cr})]}, \quad (29)$$

where $F(t) = 1 - \sum_{j=1}^{P} \sum_{i=s_j^{(W,Q)}}^{N_j-1} \pi_j A_{ji} e^{-\alpha_{ji} t}, F_j(t) = 1 - \sum_{i=s_j^{(W,Q)}}^{N_j-1} A_{ji} e^{-\alpha_{ji} t}$, $E\left[N_j^{T_{Cr}}\right] = \sum_{i=s_j^{(W,Q)}}^{N_j-1} \sum_{l=s_j^{(W,Q)}}^{i-1} \frac{iA_{jl}\alpha_{jl}}{\alpha_{ji}-\alpha_{jl}} \left(e^{-\alpha_{ji}T_{Cr}} - e^{-\alpha_{jl}T_{Cr}}\right)$, and $E[T(T_{Cr})] = \sum_{j=1}^{P} \sum_{i=0}^{s_j^{(W,Q)}-1} \frac{\pi_j}{\alpha_{ji}} + \int_0^{T_{Cr}} \sum_{j=1}^{P} \sum_{i=s_j^{(W,Q)}}^{N_j-1} \pi_j A_{ji} \alpha_{ji} t e^{-\alpha_{ji} t} dt + T_{Cr} \sum_{j=1}^{P} \sum_{i=s_j^{(W,Q)}}^{N_j-1} \pi_j A_{ji} e^{-\alpha_{ji} T_{Cr}}$.

Endharta and Yun [15] studied an special case with $W = 1$ and $Q = 1$, in which the system will be maintained preventively at certain time point after there is at least one minimal cut set having only one working component and proposed the equation to estimate the expected cost rate by utilizing the system failure paths.

b. **Condition-based maintenance with periodic inspection**

In practical situation, it may be impossible to know the system status, that is, whether components in the system fail or not. Thus, in this subsection, we consider inspection problem to know the system condition together. Endharta and Yun [15] considered a condition-based maintenance policy with periodic inspection.

We inspect the system periodically and if there are Q or more minimal cut sets having W or less working components, the system will be maintained immediately at inspection times. The condition can be represented as $x(n_W) \geq Q$, where $x(n_W)$ represents the number of minimal cut sets consisting of n_W working components

($n_W = 1, 2, \ldots, W$). In this policy, the decision variables are the minimum number of minimal cut sets, Q, the minimum number of the working components W in the minimal cut set, and the inspection interval T_I. Endharta and Yun [15] used a simulation method to obtain the near-optimal decision parameters which minimize the expected cost rate. The expected cost rate includes the system failure cost, preventive maintenance cost, replacement cost per component, and inspection cost.

5 Numerical Examples

In this subsection, we are going to compare three maintenance policies numerically. Several examples were studied and we selected some parts of existing numerical results. Tables 7 and 8 are parts of the numerical results from Endharta [14], Endharta et al. [17] and Endharta and Yun [16].

First, we consider linear consecutive-k-out-of-n: F systems with i.i.d components. Table 7 shows the expected cost rates for three maintenance policies. If the optimal interval in the age replacement model equals to infinite, the system should be maintained at system failures (corrective maintenance). Otherwise, if it equals to 0, the system should be maintained immediately at the time when the condition is satisfied.

Based on Table 7, the expected cost rates of consecutive-3-out-of-4 and 3-out-of-7: F systems are same, which means that three policies give same performance. However, because the optimal interval time is infinite, corrective maintenance policy is better than others for these particular systems. For consecutive-5-out-of-6 and 5-out-of-8: F systems, the expected cost rates of condition-based replacement policy are the smallest. Thus, for these particular systems, condition-based replacement policy should be applied and when there is one working component left in the minimal cut set, the system should be maintained immediately by replacing all failed components.

The example for the system with load-sharing dependence mechanism is shown in Table 8. Suppose that the relationship between the load and the component failure rate is as follows:

$$\lambda_i = a\left(\frac{L}{n-i}\right)^b \quad \text{where } a > 0 \text{ and } b > 0.$$

Filus [18] considered this function to determine the load size which maximizes the average asymptotic gain per unit time. Meeker and Escobar [36] also used this function in the accelerated life testing (ALT). For illustration, we use the following example with $L = 10$. In the numerical result for linear consecutive-5-out-of-8: F systems with load-sharing components (see Table 8), the optimal solution cannot be obtained due to the computational problem.

Table 7 Expected cost rates with various system parameters for linear consecutive-k-out-of-n: F systems with i.i.d components where $c_{CM} = 2$, $c_{PM} = 1$, $c_R = 0.01$ and $\lambda = 0.01$

k	n	Corrective replacement			Age replacement				Condition-based replacement			
		$E[N]$	$E[T]$	EC_{CM}	$E[N]$	$E[T]$	EC_A	T_A^*	$E[N]$	$E[T]$	EC_{Cr}	T_{Cr}^*
3	4	3.50	158.333	0.0129	3.50	158.333	0.0129	∞	3.50	158.333	0.0129	∞
	7	4.54	97.619	0.0210	4.54	97.619	0.0210	∞	4.54	97.619	0.0210	∞
5	6	5.67	211.667	0.0097	5.54	199.366	0.0097	346.2	4.40	115.000	0.0091	0
	8	6.86	178.333	0.0116	6.86	178.333	0.0116	∞	5.11	94.881	0.0111	0

Table 8 Expected cost rates with various system parameters for linear consecutive-k-out-of-n: F systems with load-sharing components where $L = 10$, $c_{CM} = 2$, $c_{PM} = 1$, $c_R = 0.01$, $a = 0.01$ and $b = 0.5$

k	n	Corrective replacement			Age replacement				Condition-based replacement			
		$E[N]$	$E[T]$	EC_{CM}	$E[N]$	$E[T]$	EC_A	T_A^*	$E[N]$	$E[T]$	EC_{Cr}	T_{Cr}^*
3	4	3.50	22.8446	0.0891	3.39	21.8807	0.0889	42.3	2.17	11.9520	0.0855	0
	7	4.54	20.7643	0.0985	4.13	18.4381	0.0971	26.9	3.57	15.3824	0.0945	6.7
5	6	5.67	33.0659	0.0622	4.90	26.7673	0.0595	32.3	4.40	22.1565	0.0471	0
	8	6.86	33.5291	0.0617	NA	NA	NA	NA	5.11	21.7638	0.0483	0

Based on Table 8, condition-based maintenance policy outperforms the others because the expected cost rates are the smallest. For consecutive-3-out-of-4, 5-out-of-6, and 5-out-of-8: F systems, the system should be maintained immediately by replacing all failed components when there are one working component in the minimal cut set. For consecutive-3-out-of-7: F system, the system should be maintained by replacing all failed components at 6.7 time unit after there is one working component in the minimal cut set. If the system fails before reaching that time point, the system is maintained correctively by replacing all failed components.

For more numerical studies, refer to Endharta [14], Endharta et al. [17] and Endharta and Yun [16].

6 Conclusion and Future Works

This chapter considered the optimal maintenance problem for consecutive multi-unit systems, such as consecutive-k-out-of-n: F systems. Systems with i.i.d components, load-sharing dependence and $(k-1)$-step Markov dependence mechanisms are analyzed. Three maintenance policies: corrective maintenance, age preventive maintenance and condition-based maintenance policies were considered. The expected cost rates were obtained analytically by utilizing the system failure paths. The cost terms involved are the system failure cost, preventive maintenance cost, and component replacement cost.

The condition-based maintenance policy is based on the number of working components in minimal cut sets. When the system can be monitored continuously and the component condition can be known at any time, the system is maintained preventively at a certain time point after there are at least Q minimal cut sets with at most W working components. If the system fails before reaching the preventive maintenance time point, the system is correctively maintained by replacing all failed components. The closed-form equation for estimating the expected cost rate is obtained.

However, the system usually cannot be known without inspection. Thus, we considered a condition-based maintenance with periodic inspection and the system is maintained immediately at inspection times if there are at least Q minimal cut sets with at most W working components. We used simulation to estimate the expected cost rate and some metaheuristics to obtain the near optimal decision parameters: W, Q and inspection interval.

We compared the three maintenance policies numerically and knew the condition-based replacement policy outperforms other replacement policies based on the expected cost rate.

For further studies, the following topics may be promising ones in consecutive multi-unit systems;

1. Maintenance problem for consecutive-k-out-of-n: F systems with nonidentical components: We can consider various maintenance policies for the system with non-identical components.
2. Maintenance problem for consecutive-k-out-of-n: F systems with finite spare components: We consider spare parts provisioning problem and maintenance problem together.
3. Maintenance problem for consecutive-k-out-of-n: F systems with limited maintenance duration: When we maintain the system, we should select the failed components for replacement and sometimes we cannot replace all the failed components at maintenance times.
4. Maintenance problem for consecutive-k-out-of-n: F systems with non-periodic inspection intervals: Since the system fails significantly quicker as the time goes, non-periodic inspection intervals might be considered.
5. Dependence models in two and three dimensional cases?
6. Maintenance problem for toroidal-type systems (refer [33]: Toroidal-type system is a circular connected-(r, s)-out-of-(m, n): F system where the components in row 1 are connected to those in row m; thus, this system forms a torus.

Acknowledgments This research was supported by Basic Science Research Program through the National Research Foundation of Korea (NRF) funded by the Ministry of Education, Science and Technology (NRF-2013R1A1A2060066).

References

1. Barlow, R.E. and Proschan, F. (1965). *Mathematical Theory of Reliability*. New York: John Wiley & Sons, Inc.
2. Boehme, T.K., Kossow, A., and Preuss, W. (1992). A generalization of consecutive-k-out-of-n: F system. *IEEE Transactions on Reliability*, R-41(3), 451-457.
3. Bollinger, R.C. (1982). Direct computation for consecutive-k-out-of-n: F systems. *IEEE Transactions on Reliability*, R-31(5), 444-446.
4. Bollinger, R.C. and Salvia, A.A. (1982). Consecutive-k-out-of-n: F networks. *IEEE Transactions on Reliability*, R-31(1), 53-55.
5. Bollinger, R.C. and Salvia, A.A. (1985). Consecutive-k-out-of-n: F system with sequential failures. *IEEE Transactions on Reliability*, R-34, 43-45.
6. Chao, M.T. and Lin, G.D. (1984). Economical design of large consecutive-k-out-of-n: F systems. *IEEE Transactions on Reliability*, R-33(5), 411-413.
7. Chari, A.A. (1994). Optimal redundancy of k-out-of-n: G system with two kinds of CCFS. *Microelectronics and Reliability*, 34(6), 1137-1139.
8. Chiang, D.T. and Niu, S.C. (1981). Reliability of consecutive-k-out-of-n: F system. *IEEE Transactions on Reliability*, R-30(1), 87-89.
9. Chung, W.K. (1990). Reliability analysis of a k-out-of-n: G redundant system with multiple critical errors. *Microelectronics and Reliability*, 30(5), 907-910.
10. Derman, C., Lieberman, G.J, and Ross. S.M. (1982). On the consecutive-k-out-of-n: F system. *IEEE Transactions on Reliability*, R-31(1), 57-63.

11. Dhillon, B.S. and Anude, O.C. (1995). Common-cause failure analysis of a *k*-out-of-*n*: G system with non-repairable units. *International Journal of System Science*, 26(10), 2029-2042.
12. Du, D.Z. and Hwang, F.K. (1986). Optimal consecutive-0-out-of-*n* systems. *Mathematics of Operations Research*, 11(1), 187-191.
13. Du, D.Z. and Hwang, F.K. (1988). A direct algorithm for computing reliability of a consecutive-*k* cycle. *IEEE Transactions on Reliability*, R-37(1), 70-72.
14. Endharta, A.J. (2016). *Optimal preventive maintenance for consecutive multicomponent systems* (Unpublished doctoral thesis). Pusan National University, Busan, South Korea.
15. Endharta, A.J. and Yun, W.Y. (2015). Condition-based maintenance policy for linear consecutive-*k*-out-of-*n*: F system. *Communications in Statistics – Simulation and Computation*, DOI:10.1080/03610918.2015.1073305.
16. Endharta, A.J. and Yun, W.Y. (2016). A preventive maintenance of circular consecutive-*k*-out-of-*n*: F systems. *International Journal of Quality and Reliability Management* (Manuscript submitted for publication).
17. Endharta, A.J., Yun, W.Y. and Yamamoto, H. (2015). Preventive maintenance policy for linear consecutive-*k*-out-of-*n*: F system. *Journal of the Operations Research Society of Japan* (Manuscript submitted for publication).
18. Filus, J. (1986). A problem in reliability optimization. *Journal of the Operational Research Society*, 37, 407-412.
19. Flynn, J. and Chung, C.S. (2004). A heuristic algorithm for determining replacement policies in consecutive-*k*-out-of-*n* systems. *Computers & Operations Research*, 31, 1335-1348.
20. Fu, J.C. (1986). Reliability of consecutive-*k*-out-of-*n*: F systems with (*k*-1)-step Markov dependence. *IEEE Transactions on Reliability*, R-35(5), 600-606.
21. Goulden, I.P. (1987). Generating functions and reliabilities for consecutive-*k*-out-of-*n*: F systems. *Utilitas Mathematic*, 30(1), 141-147.
22. Griffith, W.S. and Govindarajalu, Z. (1985). Consecutive-*k*-out-of-*n* failure systems: Reliability and availability, component importance, and multi state extensions. *American Journal of Mathematical and Management Sciences*, 5(1,2), 125-160.
23. Hwang, F.K. (1986). Simplified reliabilities for consecutive-*k*-out-of-*n*: F systems. *SIAM Journal on Algebraic and Discrete Methods*, 7(2), 258-264.
24. Hwang, F.K. (1989). Invariant permutations for consecutive-*k*-out-of-*n* cycles. *IEEE Transactions on Reliability*, R-38(1), 65-67.
25. Hwang, F.K. and Shi, D. (1987). Redundant consecutive-*k*-out-of-*n*: F systems. *Operations Research Letters*, 6(6), 293-296.
26. Jung, K.H. (1992). MTBF and MSFT for a *k*-out-of-*n*: G system with two types of forced outages. *Reliability Engineering and System Safety*, 35(2), 117-125.
27. Kontoleon, J.M. (1980). Reliability determination of *r*-successive-out-of-*n*: F system. *IEEE Transactions on Reliability*, R-29, 600-602.
28. Kuo, W., Zhang, W., and Zuo, M. (1990). A consecutive-*k*-out-of-*n*: G system: The mirror image of a consecutive-*k*-out-of-*n*: F system. *IEEE Transactions on Reliability*, R-39(2), 244-253.
29. Kuo, W. and Zuo, M.J. (2003). *Optimal Reliability Modeling: Principles and Applications*. Hoboken, NJ: John Wiley & Sons, Ltd.
30. Kuo, W. and Zhu, X. (2012). *Importance Measures in Reliability, Risk, and Optimization: Principles and Applications*. UK: John Wiley & Sons, Ltd.
31. Lambiris, M. and Papastavridis, S.G. (1985). Exact reliability formulas for linear and circular consecutive-*k*-out-of-*n*: F systems. *IEEE Transactions on Reliability*, R-34(2), 104-106.
32. Nakagawa, T. (2005). *Maintenance Theory of Reliability*. USA: Springer.
33. Nakamura, T., Yamamoto, H., Shinzato, T., Akiba, T. and Xiao, X. *Reliability of toroidal connected-(1,2)-or-(2,1)-out-of-(3,n): F lattice system*. Presented at 7th Asia-Pacific International Symposium on Advanced Reliability and Maintenance Modelling (APARM 2016), Seoul, Korea, 24-26 August.

34. Malon, D.M. (1984). Optimal consecutive-2-out-of-*n*: F component sequencing. *IEEE Transactions on Reliability*, R-33(5), 414-418.
35. Malon, D.M. (1985). Optimal consecutive-*k*-out-of-*n*: F component sequencing. *IEEE Transactions on Reliability*, R-34(1), 46-49.
36. Meeker, W.Q. and Escobar, L.A. (1998). *Statistical Methods for Reliability Data*. New York: John Wiley & Sons, Inc.
37. Papastavridis, S.G. (1987). The most important component in a consecutive-*k*-out-of-*n*: F system. *IEEE Transactions on Reliability*, R-36(0), 066-068.
38. Papastavridis, S.G. (1989). Lifetime distribution of circular consecutive-*k*-out-of-*n*: F systems with exchangeable lifetimes. *IEEE Transactions on Reliability*, R-38(4), 460-461.
39. Papastavridis, S.G. and Lambiris, M. (1987). Reliability of a consecutive-*k*-out-of-*n*: F system for Markov-dependent components. *IEEE Transactions on Reliability*, R-36(1), 78-79.
40. Rausand, M. and Hoyland, A. (2004). *System Reliability Theory: Models, Statistical Methods, and Applications*. New Jersey: John Wiley & Sons, Inc.
41. Salvia, A.A. and Lasher, W.C. (1990). 2-dimensional consecutive-*k*-out-of-*n*: F models. *IEEE Transactions on Reliability*, R-39(3), 380-385.
42. Shanthikumar, J.G. (1985). Lifetime distribution of consecutive-*k*-out-of-*n*: F systems with exchangeable lifetimes. *IEEE Transactions on Reliability*, R-34(5), 480-483.
43. Yamamoto, H. and Miyakawa, M. (1995). Reliability of a linear connected-(*r*,*s*)-out-of-(*m*,*n*): F lattice system. *IEEE Transactions on Reliability*, R-44(2), 333-336.
44. Yun, W.Y. and Endharta, A.J. (2016). A preventive replacement policy based on system critical condition. *Proceedings of the Institution of Mechanical Engineers, Part O: Journal of Risk and Reliability*, 230(1), 93-100.
45. Yun, W.Y., Kim, G.R., and Jeong, C.H. (2004). A maintenance design of connected-(*r*,*s*)-out-of-(*m*,*n*): F system using Genetic Algorithm. *Journal of Korean Institute of Industrial Engineers*, 30(3), 250-260.
46. Yun, W.Y., Kim, G.R., and Yamamoto, H. (2007). Economic design of a circular consecutive-*k*-out-of-*n*: F system with (*k*-1)-step Markov dependence. *Reliability Engineering and System Safety*, 90, 464-478.
47. Yun, W.Y., Kim, G.R., and Yamamoto, Y. (2012). Economic design of a load-sharing consecutive-*k*-out-of-*n*: F system. *IIE Transactions*, 44, 55-67.
48. Zuo, M. and Kuo, W. (1990). Design and performance analysis of consecutive-*k*-out-of-*n* structure. *Naval Research Logistics*, 37, 223-230.

Connectivity-Based Survivability Analysis with Border Effects for Wireless Ad Hoc Network

Zhipeng Yi, Tadashi Dohi and Hiroyuki Okamura

Abstract Taking account of border effects in communication network areas is one of the most important problems to quantify accurately the performance/dependability of wireless ad hoc networks (WAHNs), because the assumption on uniformity of network node density is often unrealistic to describe the actual communication areas. This problem appears in both modeling the node behavior of WAHNs and quantification of their network survivability. In this article, we focus on the border effects in WAHNs and reformulate the network survivability models based on a semi-Markov process, where two kinds of communication network areas are considered; square area and circular area. Based on some geometric ideas, we improve the quantitative network survivability measures for three stochastic models by taking account of the border effects, and revisit the existing lower and upper bounds of connectivity-based network survivability. While some analytical formulas on the quantitative network survivability have been proposed, they have not been validated yet by comparing with the exact value of network survivability in a comprehensive way. We develop a simulation model in two communication areas. It is shown through simulation experiments that the analytical solutions often fail the exact network survivability measurement in some parametric circumstances.

Keywords Network survivability · Network connectivity · WAHN · Semi-Markov model · Dos attack · Border effects · Simulation

1 Introduction

Network survivability is defined as an attribute that network is continually available even though a communication failure occurs, and is regarded as the most fundamental issue to design resilient networks. Since unstructured networks such as P2P

Z. Yi · T. Dohi (✉) · H. Okamura
Department of Information Engineering, Hiroshima University, 1-4-1 Kagamiyama, Higashi-Hiroshima 739-8527, Japan
e-mail: dohi@rel.hiroshima-u.ac.jp

M. Ram and J.P. Davim (eds.), *Advances in Reliability and System Engineering*, Management and Industrial Engineering, DOI 10.1007/978-3-319-48875-2_3

network and wireless ad hoc network (WAHN) can change dynamically their configurations, the survivability requirement for unstructured networks is becoming much more popular than static networks. Network survivability is defined by various authors [1–4]. Chen et al. [5], Cloth and Haverkort [6], Heegaard and Trivedi [7], Liu et al. [8], Liu and Trivedi [9] consider the survivability of virtual connections in telecommunication networks and define the quantitative survivability measures related to performance metrics like the loss probability and the delay distribution of non-lost packets. Their survivability measures are analytically tractable and depend on the performance modeling under consideration, where the transition behavior has to be described by a continuous-time Markov chain (CTMC) or a stochastic Petri net. More recently, Zheng et al. [10] conduct a survivability analysis for virtual machine-based intrusion tolerant systems, and take the similar approach to the works [1–4].

On the other hand, Xing and Wang [11, 12] perceive the survivability of a WAHN as the probabilistic *k*-connectivity, and provide a quantitative analysis on impacts of both node misbehavior and failure. They approximately derive the lower and upper bounds of network survivability based on *k*-connectivity, which implies that every node pair in a network can communicate with at least *k* neighbors. On the probabilistic *k*-connectivity, significant research works are done in [13, 14] to build the node degree distribution models. Unfortunately, the resulting upper and lower bounds are not always tight to characterize the connectivity-based network survivability, so that a refined measure of network survivability should be defined for analysis. Roughly speaking, the connectivity-based network survivability analysis in [11, 12] focuses on the path behavior and can be regarded as a myopic approach to quantify the survivability attribute. The performance-based network survivability analysis in [1–4] is, on the other hand, a black-box approach to describe the whole state changes. Although both approaches possess advantage and disadvantage, in this article we concern only the former case.

The authors in [15] develop somewhat different stochastic models from Xing and Wang [11, 12] by introducing different degree distributions. More specifically, they propose binomial and negative binomial models in addition to the familiar Poisson model, under the assumption that mobile nodes are uniformly distributed. Okamura et al. [16] extend the semi-Markov model [15] to a Markov regenerative process model and deal with a generalized stochastic model to describe a power-ware MANET. However, it should be noted that the above network survivability models are based on multiple unrealistic assumptions to treat an ideal communication network environment. One of them is ignorance of border effects arising to represent network connectivity. In typical WAHNs such as sensor networks, it is common to assume that each node is uniformly distributed in a given communication network area. Since the shape of communication network area is arbitrary in real world, it is not always guaranteed that the node degree is identical even in the border of network area. In other words, border effects in communication network area tend to decrease both the communication coverage and the node degree, which reflect the whole network availability. Laranjeira and Rodrigues [17] show that the relative average node degree for nodes in borders is independent of the node transmission range and of the overall network node density in a square communication network area. Bettsetetter [18] also gives a mathematical formula to calculate the average node degree for nodes in borders for a circular communication network area. However, these works just concern to investigate the connectivity of a wireless sensor network, but not the quantitative survivability evaluation for WAHNs. This fact motivates us to reformulate the existing network survivability models [11, 12, 15] for WAHNs with border effects.

In this article, we resolve the above two problems, and propose refined measures for network survivability taking account of the expected number of active nodes and border effects in WAHNs. More specifically, we represent an approximate form of connectivity-based network survivability with the expected number of active nodes, instead of its upper and lower bounds [11, 12, 16], and consider border effects in both two types of communication network areas; square area [17] and circular area [18]. Next, we develop a simulation model to quantify the network survivability accurately. In past, several simulation models have been proposed in the literature to quantify network connectivity or to detect survival routes in WAHNs (see Caro et al. [19] and Guo [20]). To our best knowledge, an accurate simulation model to quantify the network survivability based on connectivity [11, 12, 15, 16] has not been proposed yet. It is definitely needed to check the resulting quantitative survivability based on the analytical approaches by comparing with the simulation solution. It is indeed significant to point out that the analytical solutions in the literature [11, 12, 15, 16] have not been validated in comparison with the simulation solution because of its complexity. This point is gained from our earlier paper [21].

The remaining part of this article is organized as follows. In Sect. 2, we define the state of each node in stochastic models to describe a WAHN, and refer to an isolation problem under the Blackhole attack, which is equivalent to the well-known DoS attack. Based on the familiar semi-Markov analysis [11, 12, 15], the transition behavior of the network node is analyzed. Section 3 is devoted to the network survivability analysis, where the node isolation, network connectivity, and network survivability are defined. Here, we formulate a problem on network connectivity in the presence of misbehaving nodes, and present two node distribution models [15], in addition to the seminal Poisson model [11, 12]. In Sect. 4, we refine the network survivability models by taking account of border effects in two kinds of communication network areas; square area and circular area. Based on some geometric ideas, we improve the quantitative network survivability measures for three stochastic models. Section 5 concerns about simulation algorithms to calculate the exact network survivability in both the types of communication areas. Numerical examples are given in Sect. 6 where we conduct a Monte Carlo simulation on the node degree and investigate the impact of border effects. We compare our refined network survivability models with the existing ones without border effects in both steady-state and transient survivability analyses, and refer to the limitation. Finally, the article is concluded with some remarks in Sect. 7.

2 Preliminary

2.1 State of Node

Since nodes in WAHNs cooperate with the routing processes to maintain network connectivity, each of node is designed as it behaves autonomously, but its discipline to require, send and receive the route information is defined as a strict protocol. At the same time, it is also important to decide the protocol in order to prevent propagation of the erroneous route information caused by malicious attacks. Xing and Wang [11, 12] and Yi and Dohi [15] consider a WAHN that suffers such a malicious attack, whose node states are defined as follows:

- *Cooperative state* (C): Initialized state of a node, which responds to route discoveries and forwards data packets to others.
- *Selfish state* (S): State of a node, which may not forward control packets or data packets to others, but responds to only route discoveries for its own purpose from the reason of low power.
- *Jellyfish state* (J): State of a node, which launches Jellyfish DoS attack.
- *Black hole state* (B): State of a node, which launches Black hole DoS attack.
- *Failed state* (F): State of a node, which can no longer perform basic functions such as initiation or response of route discoveries.

For common DoS attacks, the node in Jellyfish attack receives route requests and route replies. The main mechanism of Jellyfish state is to delay packets without any reason. On the other hand, the node in Black hole attack can respond a node with a fake message immediately by declaring as it is in the optimal path or as it is only one-hop away to other nodes.

Based on the node classification above, we consider a semi-Markov model to describe the stochastic behavior of a node by combining states with respect to the wellness. Suppose that a node may change its behavior under the following assumptions:

- A cooperative node may become a failed node due to energy exhaustion or misconfiguration. It is apt to become a malicious node when it launches DoS attack.
- A malicious node cannot become a cooperative node again, but may become a failed node.
- A node in failed state may become a cooperative node again after it is repaired and responds to routing requests to others.
- A failed node can become cooperative again if it is recovered and responds to routing operations.

2.2 Semi-Markov Node Model

Similar to [11, 12, 15], let $S = \{C, S, J, B, F\}$ be a state space, and describe the node behavior transition by a stochastic process, $\{Z(t), t \geq 0\}$, associated with space S. Let X_n denote the state at transition time t_n. Define

$$\begin{aligned} &\Pr(X_{n+1} = x_{n+1} | X_0 = x_0, \ldots, X_n = x_n) \\ &= \Pr(X_{n+1} = x_{n+1} | X_n = x_n), \end{aligned} \tag{1}$$

where $x_i \in S$ for $0 \leq i \leq n+1$. From Eq. (1), the stochastic process $\{X_n, n = 0, 1, 2, \ldots\}$ constitutes a continuous-time Markov chain (CTMC) with state space S, if all the transition times are exponentially distributed. However, since the transition time from one state to another state is subject to random behavior of a node, it is not realistic to characterize all the transition times by only exponentially distributed random variables. For instance, if a node is more inclined to fail due to energy consumption as time passes, and the less residual energy is left, then the more likely a node changes its behavior to selfish. This implies that the future action of a node may depend on how long it has been in the current state and that transition intervals may have arbitrary probability distributions.

From the above reason it is common to assume a semi-Markov process (SMP) for $\{Z(t), t \geq 0\}$ to describe the node behavior transitions, which is defined by

$$Z(t) = X_n, \quad \forall t_n \leq t \leq t_{n+1}. \tag{2}$$

Letting $T_n = t_{n+1} - t_n$ be the sojourn time between the n-th and $(n+1)$-st transitions, we define the associated semi-Markov kernel $\mathbf{Q} = (Q_{ij}(t))$ by

$$Q_{ij}(t) = \Pr(X_{n+1} = j, T_n \leq t | X_n = i) = p_{ij} F_{ij}(t), \tag{3}$$

where $p_{ij} = \lim_{t \to \infty} Q_{ij}(t) = \Pr(X_{n+1} = j | X_n = i)$ is the transition probability between states i and j $(i, j = c, s, j, b, f)$ corresponding to $S = \{C, S, J, B, F\}$, and $F_{ij}(t) = \Pr(T_n < t | X_{n+1} = j, X_n = i)$ is the transition time distribution from state i to j.

Figure 1 illustrates the transition diagram of the homogeneous SMP, $\{Z(t), t \geq 0\}$, under consideration, which is somewhat different from the SMP in [11, 12], because it is somewhat simplified by eliminating redundant states. Let $1/\lambda_{ij}$ denote the mean transition time from state i to state j, and define the Laplace–Stieltjes transform (LST) by $q_{ij}(s) = \int_0^\infty \exp\{-st\} \mathrm{d}Q_{ij}(t)$. From the familiar SMP analysis technique, it is immediate to see that

$$q_{cs}(s) = \int_0^\infty \exp\{-st\} \overline{F}_{cj}(t) \overline{F}_{cb}(t) \overline{F}_{cf}(t) \mathrm{d}F_{cs}(t) \tag{4}$$

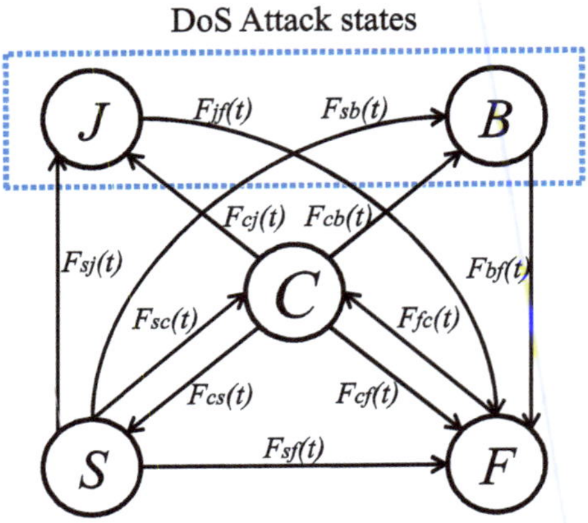

Fig. 1 Semi-Markov transition diagram for node behavior

$$q_{cj}(s) = \int_0^\infty \exp\{-st\}\overline{F}_{cs}(t)\overline{F}_{cb}(t)\overline{F}_{cf}(t)\mathrm{d}F_{cj}(t) \tag{5}$$

$$q_{cb}(s) = \int_0^\infty \exp\{-st\}\overline{F}_{cs}(t)\overline{F}_{cj}(t)\overline{F}_{cf}(t)\mathrm{d}F_{cb}(t) \tag{6}$$

$$q_{cf}(s) = \int_0^\infty \exp\{-st\}\overline{F}_{cs}(t)\overline{F}_{cj}(t)\overline{F}_{cb}(t)\mathrm{d}F_{cf}(t) \tag{7}$$

$$q_{sc}(s) = \int_0^\infty \exp\{-st\}\overline{F}_{sj}(t)\overline{F}_{sb}(t)\overline{F}_{sf}(t)\mathrm{d}F_{sc}(t) \tag{8}$$

$$q_{sj}(s) = \int_0^\infty \exp\{-st\}\overline{F}_{sc}(t)\overline{F}_{sb}(t)\overline{F}_{sf}(t)\mathrm{d}F_{sj}(t) \tag{9}$$

$$q_{sb}(s) = \int_0^\infty \exp\{-st\}\overline{F}_{sc}(t)\overline{F}_{sj}(t)\overline{F}_{sf}(t)\mathrm{d}F_{sb}(t) \tag{10}$$

$$q_{sf}(s) = \int_0^\infty \exp\{-st\}\overline{F}_{sc}(t)\overline{F}_{sj}(t)\overline{F}_{sb}(t)\mathrm{d}F_{sf}(t) \tag{11}$$

$$q_{jf}(s) = \int_0^{\infty} \exp\{-st\} \mathrm{d}F_{jf}(t) \tag{12}$$

$$q_{bf}(s) = \int_0^{\infty} \exp\{-st\} \mathrm{d}F_{bf}(t) \tag{13}$$

$$q_{fc}(s) = \int_0^{\infty} \exp\{-st\} \mathrm{d}F_{fc}(t), \tag{14}$$

where in general $\overline{\psi}(\cdot) = 1 - \psi(\cdot)$. Define the recurrent time distribution from state C to state C and its LST by $H_{cc}(t)$ and $h_{cc}(s)$, respectively. Then, from the one-step transition probabilities from Eqs. (4)–(14), we have

$$\begin{aligned} h_{cc}(s) &= \int_0^{\infty} \exp\{-st\} \mathrm{d}H_{cc}(t) \\ &= q_{cs}(s)q_{sc}(s) + q_{cs}(s)q_{sj}(s)q_{jf}(s)q_{fc}(s) \\ &\quad + q_{cs}(s)q_{sb}(s)q_{bf}(s)q_{fc}(s) + q_{cs}(s)q_{sf}(s)q_{fc}(s) \\ &\quad + q_{cj}(s)q_{jf}(s)q_{fc}(s) + q_{cb}(s)q_{bf}(s)q_{fc}(s) \\ &\quad + q_{cf}(s)q_{fc}(s). \end{aligned} \tag{15}$$

Let $P_{ci}(t)$ denote the transition probability from the initial state C to respective states $i \in \{c, s, j, b, f\}$ corresponding to $S = \{C, S, J, B, F\}$. Then, the LSTs of the transition probability, $p_{ci} = \int_0^{\infty} \exp\{-st\} \mathrm{d}P_{ci}(t)$, are given by

$$p_{cc}(s) = \{\overline{q}_{cs}(s) - q_{cj}(s) - q_{cb}(s) - q_{cf}(s)\}/\overline{h}_{cc}(s) \tag{16}$$

$$p_{cs}(s) = q_{cs}(s)\{\overline{q}_{sc}(s) - q_{sj}(s) - q_{sb}(s) - q_{sf}(s)\}/\overline{h}_{cc}(s) \tag{17}$$

$$p_{cj}(s) = \{q_{cm}(s) + q_{cs}(s)q_{sj}(s)\}\overline{q}_{mf}(s)/\overline{h}_{cc}(s) \tag{18}$$

$$p_{cb}(s) = \{q_{cm}(s) + q_{cs}(s)q_{sb}(s)\}\overline{q}_{mf}(s)/\overline{h}_{cc}(s) \tag{19}$$

$$\begin{aligned} p_{cf}(s) = \{&q_{cf}(s) + q_{cs}(s)q_{sf}(s) + q_{cs}(s)q_{sj}(s)q_{jf}(s) \\ &+ q_{cs}(s)q_{sb}(s)q_{bf}(s) + q_{cj}(s)q_{jf}(s) \\ &+ q_{cb}(s)q_{bf}(s)\}\overline{q}_{fc}(s)/\overline{h}_{cc}(s). \end{aligned} \tag{20}$$

From Eqs. (16)–(20), the transient solutions, $P_{ci}(t)$, $i \in \{c, s, j, b, f\}$, which mean the probability that the state travels in another state i at time t, can be derived numerically, by means of the Laplace inversion technique (e.g. see [22]). As a special case, it is easy to derive the steady-state probability $P_i = \lim_{t\to\infty} P_{ci}(t), i \in \{c, s, b, j, f\}$ corresponding to S. Based on the LSTs, $p_{ci}(s)$, we calculate $P_i = \lim_{t\to\infty} P_{ci}(t) = \lim_{s\to 0} p_{ci}(s)$ from Eqs. (16)–(20).

3 Quantitative Network Survivability Measure

3.1 Node Isolation and Connectivity

An immediate effect of node misbehaviors and failures in WAHNs is the node isolation problem [12]. It is a direct cause for network partitioning, and eventually affects network survivability. The node isolation problem is caused by four types of neighbors; Failed, Selfish, Jellyfish, and Blackhole nodes. For an example, we suppose in Fig. 2 that the node u has four neighbors when it initiates a route discovery to another node v. Then it must go through by its neighbors $x_i (i = 1, 2, 3, 4)$. If all neighbors of u are Failed, Selfish or Jellyfish nodes, then u can no longer communicate with the other nodes. In this case, we find that u is isolated by Failed and Selfish neighbors. On the other hand, if one of neighbors is Blackhole (i.e., x_2 in Fig. 3), it gives u a faked one-hope path, and makes u always choose it. In this case, we find that u is isolated by the Blackhole neighbor.

We define the node degree $D_{(u)}$ for node u by the maximum number of neighbors [13], and let $D_{(i,u)}$ be the number of node u's neighbors at state

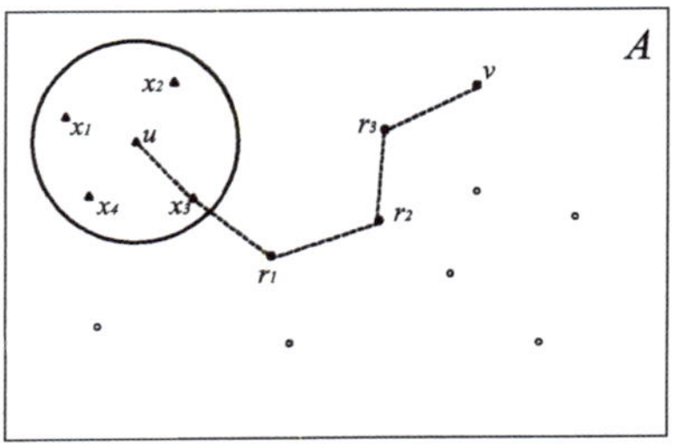

Fig. 2 Isolation by failed, selfish, or jellyfish neighbors

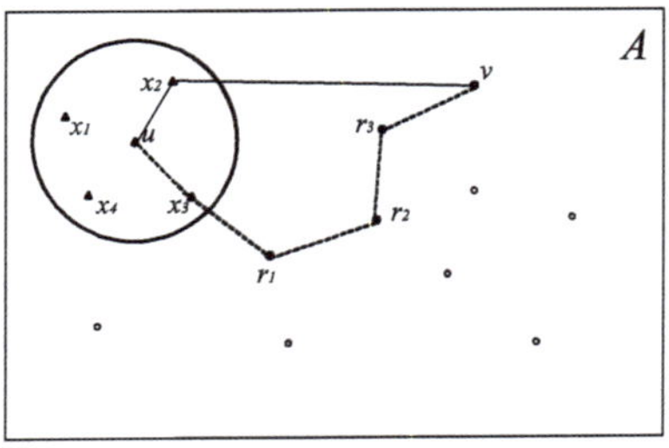

Fig. 3 Isolation by Blackhole neighbors

$i \in \{c, s, j, b, f\}$. Then the isolation problem in our model can be formulated as follows: Given a node u with degree d, *i.e.*, $D_{(u)} = d$, if $D_{(s,u)} + D_{(f,u)} + D_{(j,u)} = d$ or $D_{(b,u)} \geq 1$, then the cooperative degree is zero, i.e., $D_{(c,u)} = 0$, and u is isolated from the network, so it holds that

$$\begin{aligned}\Pr(D_{(c,u)} = 0|D_{(u)} = d) &= \Pr(D_{(b,u)} \geq 1|D_{(u)} = d)\\ &+ \Pr(D_{(s,u)} + D_{(f,u)} + D_{(j,c)} = d|D_{(u)} = d)\\ &= 1 - (1 - P_b)^d + (1 - P_c - P_b)^d,\end{aligned} \tag{21}$$

where P_c is the steady-state probability of a node in a cooperative state and P_b is the steady-state probability of a node launching Blackhole attacks. In the transient case, the steady-state probabilities P_c and P_b are replaced by $P_{cc}(t)$ and $P_{cb}(t)$, respectively.

In this article, a node is said to be *k-connected* to a network if its associated cooperative degree is given by $k\ (\geq 1)$. Given node u with degree d, i.e., $D_{(u)} = d$, u is said to be k-connected to the network if the cooperative degree is k, i.e., $D_{(c,u)} = k$, which holds only if u has no Blackhole neighbor and has exactly k cooperative neighbors, i.e., $D_{(b,u)} = 0$ and $D_{(c,u)} = k$, respectively. Then, from the statistical independence of all nodes, it is straightforward to see that

$$\begin{aligned}&\Pr\big(D_{(c,u)} = k|D_{(u)} = d\big)\\ &= \Pr\big(D_{(c,u)} = k, D_{(b,u)} = 0|D_{(u)} = d\big)\\ &= \Pr\big(D_{(c,u)} = k, D_{(b,c)} = 0, D_{(s,u)} + D_{(f,u)} + D_{(J,u)} = d - k\big)\\ &= \binom{d}{k}(P_c)^k(1 - P_c - P_b)^{d-k}.\end{aligned} \tag{22}$$

3.2 Network Survivability

In the seminal paper [12], the network survivability is defined as the probability that WAHN is a k-vertex-connected graph. Strictly speaking, it is difficult to validate the vertex-connectivity in the graph whose configuration dynamically changes such as WAHNs. Therefore, Xing and Wang [12] derive approximately low and upper bounds of network survivability when the number of nodes is sufficiently large by considering the connectivity of a node in a WAHN. The upper and lower bounds of connectivity-based network survivability are given by

$$\mathrm{SVB}_U = \big(1 - \Pr(D_{(c,u)} < k)\big)^{N_D}, \tag{23}$$

$$\mathrm{SVB}_L = \max\big(0, 1 - E[N_a](\Pr(D_{(c,u)} < k))\big), \tag{24}$$

respectively, where u is an arbitrary node index in the active network. In Eq. (24), $E[N_a] = \lfloor N(1 - P_f) \rfloor$ is the expected number of active nodes in the network, where $\lfloor * \rfloor$ is the maximum integer less than $*$, P_f is the steady-state probability that a node is failed, and N denotes the total number of nodes. In Eq. (23), N_{D} is the number of points whose transmission ranges is mutually disjoint over the WAHN area. Let A and r be the area of WAHN and the node transmission radius, respectively. The number of disjoint points is given by $N_{\mathrm{D}} = \lfloor N/(\lambda \pi r^2) \rfloor$, where $\lambda = N/A$ is the node density.

In this article, we follow the same definition of network survivability as reference [12], but consider the expected network survivability instead of the low and upper bounds. Getting help from the graph theory, the expected network survivability is approximately given by the probability that expected active node in the network is k-connected

$$\mathrm{SVB} \approx \left\{1 - \Pr(D_{(c,u)} < k)\right\}^{E[N_a]}. \tag{25}$$

By the well-known total probability law, we have

$$\Pr\big(D_{(c,u)} < k\big) = \sum_{d=k}^{\infty} \Pr\big(D_{(c,u)} < k | D_{(u)} = d\big) \Pr\big(D_{(u)} = d\big), \tag{26}$$

so that we need to find the explicit forms of $\Pr(D_{(c,u)} < k | D_{(u)} = d)$ and $\Pr(D_{(u)} = d)$. From Eqs. (21) and (22), it is easy to obtain

$$\begin{aligned}
&\Pr\big(D_{(c,u)} < k | D_{(u)} = d\big) \\
&= \Pr\big(D_{(c,u)} = 0 | D_{(u)} = d\big) + \sum_{m=1}^{k-1} \Pr\big(D_{(c,u)} = m | D_{(u)} = d\big) \\
&= 1 - (1 - P_b)^d + \sum_{m=0}^{k-1} \binom{d}{m} P_c^m (1 - P_c - P_b)^{d-m} \\
&= 1 - (1 - P_b)^d + \sum_{m=0}^{k-1} B_m(d, P_c, 1 - P_c - P_b),
\end{aligned} \tag{27}$$

where B_m denotes the multinomial probability mass function. Replacing P_b and P_c by $P_{cb}(t)$ and $P_{cc}(t)$ respectively, we obtain the transient network survivability at an arbitrary time t. Since the node distribution $\Pr(D_{(u)} = d)$ strongly depends on the model property, we introduce three specific stochastic models [15] as follows.

(i) *Poisson Model* [11, 12]

Suppose that N nodes in a WHAN are uniformly distributed over a two-dimensional square with area A. The node transmission radius, denoted by r, is assumed to be identical for all nodes. To derive the node degree distribution $\Pr(D_{(u)} = d)$, we divide the area into N small grids virtually, so that the grid size has the same order as the physical size of a node. Consider the case where the network area is sufficiently larger than the physical node size. Then, the probability that a node occupies a specific grid, denoted by p, is very small. With large N and small p, the node distribution can be modeled by the Poisson distribution

$$\Pr\left(D_{(u)} = d\right) = \frac{\mu^d}{d!}e^{-\mu}, \tag{28}$$

where $\mu = \rho\pi r^2$, and $\rho = E[N_a]/A$ is the node density depending on the underlying model. Finally, substituting Eqs. (26)–(28) into Eq. (25) yields

$$\mathrm{SVB} \approx \left\{e^{-\mu P_b}\left[1 - \frac{\Gamma(k, \mu P_c)}{\Gamma(k)}\right]\right\}^{E[N_a]}, \tag{29}$$

where $\Gamma(x) = (x-1)!$ and $\Gamma(h, x) = (h-1)!e^{-x}\sum_{l=0}^{h-1} x^l/l!$ are the complete and incomplete gamma functions, respectively.

(ii) *Binomial Model* [15]

It is evident that the Poisson model just focuses on an ideal situation of node behavior. In other words, it is not always easy to measure the physical parameters such as r and A in practice. Let p denote the probability that each node is assigned into a communicate network area of a node. For the expected number of activate nodes $E[N_a]$, we describe the node distribution by the binomial distribution:

$$\Pr(D_{(u)} = d) = \binom{E[N_a]}{d} p^d(1-p)^{E[N_a]-d} = B_d(E[N_a], p), \tag{30}$$

where B_d is the binomial probability mass function. Substituting Eq. (30) into Eq. (25) yields an alternative formula of the network survivability

$$\mathrm{SVB} \approx \left\{\sum_{k=0}^{E[N_a]} B_d(E[N_a], p)\left[(1-P_b)^d - \sum_{m=0}^{k-1} B_m(d, P_c, 1-P_c-P_b)\right]\right\}^{E[N_a]}. \tag{31}$$

Even though each node is assigned into a communication network area of a node with probability $p = \pi r^2/A$, then the corresponding binomial model results a different survivability measure from the Poisson model.

(iii) *Negative Binomial Model* [15]

The negative binomial model comes from a mixed Poisson distribution instead of Poisson distribution. Let $f(\mu)$ be the distribution of parameter μ in the Poisson model. This implicitly assumes that the parameter μ includes uncertainty, and that the node distributions for all disjoint areas have different Poisson parameters. Then the node distribution can be represented by the following mixed Poisson distribution:

$$P(D_{(u)} = d) = \int_0^\infty e^{-\mu} \frac{\mu^d}{d!} f(\mu) \mathrm{d}\mu. \tag{32}$$

For the sake of analytical simplicity, let $f(\mu)$ be the gamma probability density function with mean $\pi r^2 N(1 - P_f)/A$ and coefficient of variation c. Then we have

$$P(D_{(u)} = d) = \frac{\Gamma(a+d)}{d!\Gamma(a)} \left(\frac{b}{1+b}\right)^a \left(\frac{1}{1+b}\right)^d = \lambda_d(a, b), \tag{33}$$

where $a = \lfloor 1/c^2 \rfloor$ and $b = \lfloor A/(\pi r^2 N(1 - P_f)c^2) \rfloor$. It should be noted that Eq. (33) corresponds to the negative binomial probability mass function with mean $\pi r^2 N(1 - P_f)/A$, and that the variance is greater than that in the Poisson model. In other words, it can represent the overdispersion or underdispersion property dissimilar to the Poisson model. From Eq. (33), we can obtain an alternative representation of the network survivability with an additional model parameter c.

$$\mathrm{SVB} \approx \left\{ \sum_{k=0}^{E[N_a]} \lambda_d(a, b) \left[(1 - P_b)^d - \sum_{m=0}^{k-1} B_m(d, P_c, 1 - P_c - P_b) \right] \right\}^{E[N_a]}. \tag{34}$$

4 Border Effects of Communication Network Area

The results on network survivability presented in Sect. 3 are based on the assumption that network area A has a node density $\rho = E[N_a]/A$. This strong assumption means that the expected number of neighbors of a node in the network has the exactly same value as $\rho \pi r^2$. In other words, such an assumption is not realistic in the real world communication network circumstance. It is well known that the border effect tends to decrease both the communication coverage and the node degree, which reflect the whole network availability. Laranjeira and Rodrigues [17] show that the relative average node degree for nodes in borders is independent of the node transmission range and of the overall network node density in a square communication network area. Bettsetetter [18] calculates the average node degree for nodes in borders for a circular communication network area. In the remaining part of this section, we consider the above two kinds of areas, and apply both results by Laranjeira and Rodrigues [17] and Bettsetetter [18] to the network survivability quantification.

4.1 Square Border Effect

Given a square area of side L in Fig. 4, the borders correspond to region B and C. We call the rectangular region B *the lateral border*, the square region C *the corne border*, and the square region I *the inner region*, respectively. In Fig. 4, for a node v located in the inner region of the network area, the expected effective number of neighbors, E_I, is given by $\rho\pi r^2$. This result is from the fact that the effective coverage area of a point in the inner region I is precisely equal to $\rho\pi r^2$. However, the effective coverage area for a point in the border region B (node u_1) or C (node u_2) is less than πr^2 as shown in the shadow areas of Fig. 4. Consequently, the expected effective number of neighbors of nodes located in the border areas will be smaller than $\rho\pi r^2$. Since the connectivity properties of the network depend on the expected effective number of neighbors of nodes, it is needed to obtain the expected effective number of neighbors of nodes in these regions (I, B, and C), in order to understand the connectivity properties in the network.

In Fig. 5, the effective coverage area for a node located at a point $P(x, y)$, in relation to the origin $O(0,0)$ and the lateral border B, is defined as $\mathrm{EA_B}$. Let $\angle P_1PP_4 = \alpha$. Then, it can be seen that $\mathrm{EA_B}$ is equal to the area of the portion of a circle, which corresponds to the area through the points P, P_1, P_2 and P_3. Then we have

$$\mathrm{EA_B}(x, y) = r^2[(\pi - \alpha) + (\sin\alpha)(\cos\alpha)], \tag{35}$$

where $\alpha = \cos^{-1}(x/r)$. After a few algebraic manipulations, we get

$$\mathrm{EA_B}(x, y) = r^2\left[\left(\pi - \cos^{-1}\left(\frac{x}{r}\right)\right) + \frac{x}{r}\sqrt{1 - \left(\frac{x}{r}\right)^2}\right]. \tag{36}$$

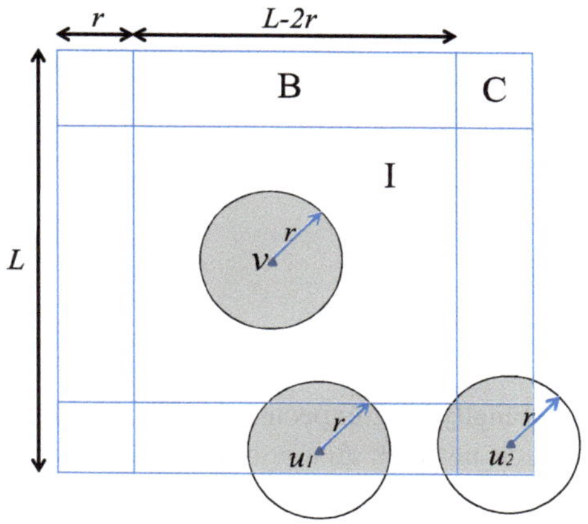

Fig. 4 Border effects in square area

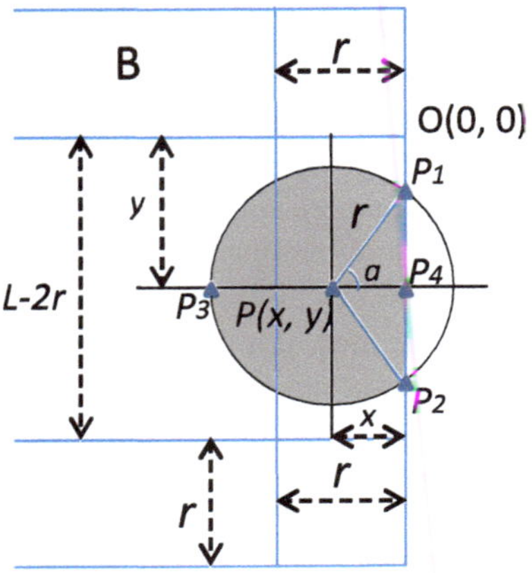

Fig. 5 Effective coverage area for a node in the lateral border

Next we derive the average effective coverage area ϕ_{lat}, for nodes in the lateral border B by integrating EA_B in Eq. (36) over the entire lateral border and diving the result by its total area $r(L-2r)$

$$\phi_{\text{lat}} = \frac{1}{r(L-2r)} \int_0^{L-2r} \left(\int_0^r \text{EA}_\text{B}(x,y)\text{d}x \right) \text{d}y. \tag{37}$$

Substituting Eq. (35) into Eq. (37), we get

$$\phi_{\text{lat}} = r^2\left(\pi - \frac{2}{3}\right) \approx 0.787793\pi r^2. \tag{38}$$

Then, the expected effective node degree E_{lat} for nodes in the lateral border region B is given by

$$E_{\text{lat}} = \rho\phi_{\text{lat}} = 0.787793\rho\pi r^2. \tag{39}$$

In the similar way, the average effective coverage area ϕ_{cor} for nodes in the corner border C and the expected effective node degree E_{cor} for nodes in the lateral corner region C are obtained as [17]

$$\phi_{\text{cor}} \approx 0.615336\pi r^2 \tag{40}$$

and

$$E_{\text{cor}} = \rho\phi_{\text{cor}} = 0.615336\rho\pi r^2. \tag{41}$$

Finally, the expected effective node degree μ_s for nodes in square communication network area is obtained as the weighted average with E_I, E_{lat} and E_{cor}

$$\mu_s = \frac{E_I A_I + E_{\text{lat}} A_B + E_{\text{cor}} A_C}{A}, \tag{42}$$

where $A_I = (L - 2r)^2$, $A_B = 4r(L - 2r)$, $A_C = 4r^2$ and $A = L^2$. Substituting the area and expected node degree values in Eq. (42) and simplifying the resulting expression, we have

$$\mu = \mu_s = \frac{\rho \pi r^2 \sigma}{L^2}, \tag{43}$$

where $\sigma = (L - 2r)^2 + 3.07492r(L - 2r) + 2.461344r^2$.

For the binomial model, we need to find the probability that a node is assigned into the expected coverage of node in square communicate area (p_s). It can be obtained as

$$p = p_s = \frac{\pi r^2 \sigma}{L^4}. \tag{44}$$

On the other hand, the parameter b in the negative binomial model with square border effect, b_s, can be derived by

$$b = b_s = 1/\left(\pi r^2 N_a c^2 \sigma\right) \tag{45}$$

for a given coefficient of variation c.

4.2 Circular Border Effect

For the circular area A with radius r_0, define the origin O in A and use the coordinates r for a node in Fig. 6. Nodes in the circular communication network area that are located at least r away from the border, are called *the center nodes*, which are shown as a node v in Fig. 6. They have a coverage area equal to πr^2 and an expected node degree $E[N_a]\pi r^2/A$. On the other hand, nodes located closer than r to the border are called *the border nodes* (node u in Fig. 6), which have a smaller coverage area, leading to a smaller expected node degree. The expected node degree of both center nodes and border nodes can be obtained by

$$\mu = \begin{cases} N_a \hat{r}_1^2 & 0 \le \hat{r} \le 1 - \hat{r}_1 \\ \left(\frac{n}{\pi}\right)\left[\hat{r}_0^2 \arccos\frac{\hat{r}^2 + \hat{r}_1^2 - 1}{2\hat{r}\hat{r}_1} + \arccos\frac{\hat{r}^2 - \hat{r}_1^2 + 1}{2\hat{r}} - \frac{1}{2}\xi\right] & 1 - \hat{r}_1 < \hat{r} \le 1, \end{cases} \tag{46}$$

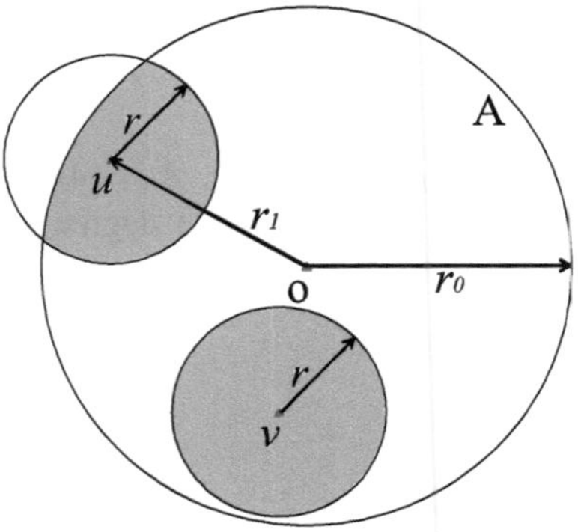

Fig. 6 Border effects in circular area

where $\xi = (\hat{r}+\hat{r}_1+1)(-\hat{r}+\hat{r}_1+1)(\hat{r}-\hat{r}_1+1)(\hat{r}+\hat{r}_1-1)$ and $\hat{r} = r/r_0$ and $\hat{r}_1 = r_1/r_0$.

Bettsettter [18] obtains the expected node degree of a node $\mu = \mu_c$ in circular communicate area:

$$\mu_c = \frac{N_a}{2\pi}\left[4(1-\hat{r}^2)\arcsin\frac{\hat{r}}{2} + 2\hat{r}^2\pi - (2\hat{r}+\hat{r}^3)\sqrt{1-\frac{\hat{r}^2}{4}}\right], \tag{47}$$

which can be simplified by using Taylor series as

$$\mu_c \approx N_a\hat{r}^2\left(1-\frac{4\hat{r}}{3\pi}\right). \tag{48}$$

Then, $p = p_c$ for the binomial model and $b = b_c$ for the negative binomial model can be given in the following:

$$p_c = \frac{\tau}{2\pi}, \tag{49}$$

$$b_c = \frac{1}{2\pi\hat{r}^2N_ac^2\tau}, \tag{50}$$

where $\tau = 4(1-\hat{r}^2)\arcsin(\hat{r}/2) + 2\hat{r}^2\pi - (2\hat{r}+\hat{r}^3)\sqrt{1-\hat{r}^2/4}$. By replacing the square border effect parameters μ, p and b in Eqs. (29), (31) and (34) by μ_s, p_s and b_s in Eqs. (43)–(45), we obtain the connectivity-based network survivability formulae with square border effects. Also, using μ_c, p_c and b_c in Eqs. (48)–(50), we calculate the connectivity-based network survivability in circular communication network area. We summarize refined network survivability formulae by taking account of border effects in Table 1.

Table 1 Summary of expected network survivability formulae with/without border effects

Border Effects	Poisson	Binomial	Negative Binomial
Ignorance	$\left\{e^{-\mu P_b}\left[1-\frac{\Gamma(k,\mu P_c)}{\Gamma(k)}\right]\right\}^{E[N_a]}$	$\{\sum_{k=0}^{n} B_d(n,p)[*]\}^{E[N_a]}$	$\{\sum_{k=0}^{n} \lambda_d(a,b)[*]\}^{E[N_a]}$
Square	$\left\{e^{-\mu_s P_b}[1-\frac{\Gamma(k,\mu_s P_c)}{\Gamma(k)}]\right\}^{E[N_a]}$	$\{\sum_{k=0}^{n} B_d(n,p_s)[*]\}^{E[N_a]}$	$\{\sum_{k=0}^{n} \lambda_d(a,b_s)[*]\}^{E[N_a]}$
Circular	$\left\{e^{-\mu_c P_b}[1-\frac{\Gamma(k,\mu_c P_c)}{\Gamma(k)}]\right\}^{E[N_a]}$	$\{\sum_{k=0}^{n} B_d(n,p_c)[*]\}^{E[N_a]}$	$\{\sum_{k=0}^{n} \lambda_d(a,b_c)[*]\}^{E[N_a]}$
$*$	$(1-P_b)^d - \sum_{m=0}^{k-1} B_m(d,P_c,1-P_c-P_b)$		

5 Simulation Algorithms

For our SMP modulated network survivability model, it is needed to quantify the network survivability throughout Monte Calro simulation, because the analytical solutions (upper and lower bounds) may not be validated without knowing the exact solution. Unfortunately, the simulation approach mentioned in Xing and Wang [12] is oversimplified and does not seem to catch up impacts of both node misbehavior and node failure accurately. Once the shape of communication area, such as a square area, is given, a fixed number of nodes are uniformly distributed to the area. A commonly used technique for a square communication area is to locate points randomly with abscissa x and ordinate y for each node (x, y) in a Cartesian coordinate system. We generate the random numbers for x and y, which are sampled from the continuous uniform distribution with lower and upper endpoints at 0 and 1, respectively, where the side length of communication area is given by L. In our simulation, we never take account of effects of the speed and destination of moving nodes, to simplify the simulation procedure, although this is because we employ the SMP modulated network survivability model. Figure 7 presents the pseudo code to give the node location for a square area. Here, we suppose that a circle area can be approximated by the sum of infinitesimal triangles, BCD, where the point C is at the origin, and the points B and D are located on the circumference. Since the sum of two triangles BCD is equivalent to the area of a parallelogram BCDE, then we can apply the similar algorithm to Fig. 7. Figure 8 is the pseudocode to give the node location for circular areas. For a circular communication area with radius R, we randomly choose two points on BC and CD. Let a and b be the distances between

Fig. 7 A node location algorithm for square areas

```
Set Node and Node_i to empty;
Set N to the total number of node in the WAHN;
Set L to the side length of square area;
For (i = 0; i ≤ N; i++) {
Set x and y to 0;
Randomly generate (x, y) (x, y ∈ [0, L])
for Node_i (i ∈ [0, N]);
Add Node_i to Node;}
/*Node_i is the i-th element of Node*/
```

Fig. 8 A node location algorithm for circular areas

```
Set Node and Node_i to empty;
Set N to the total number of node in the WAHN;
Set R to the radius of circular area;
For (i = 0; i ≤ N; i++) {
Set α, a, b, z, x and y to 0;
Randomly generate α (α ∈ [0, 2π));
Randomly generate a and b (a, b ∈ [0, R]);
if a + b > R then z = 2R − (a + b),
otherwise z = a + b;
calculate (x, y) by:
x = z cos α;
y = z sin α;
for Node_i (i ∈ [0, N]);
Add Node_i to Node;}
/*Node_i is the i-th element of Node*/
```

origin and these chosen points on BC and CD. Then let z equal to $2R - (a+b)$ if $a+b > R$, otherwise $a+b$. We can select one of triangles BCD by picking an angle $\alpha \in [0, 2\pi)$, so the random points in a circular area with abscissa x and ordinate y can be calculated as $(x, y) = (z \cos \alpha, z \sin \alpha)$.

In this way, when each node is located over a two-dimensional area, the next step is to modulate each node state by an SMP. If we focus on the steady-state behavior of WAHNs, then the steady-state probability $P_i, i \in \{c, s, b, j, f\}$ in Eqs. (16)–(20) can be used. In our simulation experiments, we generate the node state with probability P_i uniformly, where the unit length, $\sum_{i=1}^{5} P_i = 1$, is divided into five portions proportional to P_i. Let $N_i, i \in \{c, s, j, s, f\}$ be the number of nodes in state i in the network. If $N_B \neq 0$, then the minimum cooperative degree of network M, is given by $\theta(\mathcal{M}) = 0$, otherwise, divide the active nodes into two state groups; Cooperative nodes and Selfish/Jellyfish nodes, and calculate the $\theta(\mathcal{M})$ of an arbitrary node in a WAHN $\mathcal{M}$. For a given transmission radius r and the number of node N, we generate the node location 100 times and make 100 state transitions for each node. Finally, we execute 10,000 simulation runs to represent the node location and state for a fixed size of networks, say, N. Then the connectivity-based network survivability in our simulation experiments is calculated by

$$SVB_k(\mathcal{M}) = \frac{\sum_{i=1}^{10,000} I_{\{A_i\}}}{10,000}, \tag{51}$$

where A_i indicates the event $\theta(\mathcal{M}) \geq k$ at i-th simulation and $I_{\{A\}}$ is the indicator function to output 1 or 0 for occurrence of the event A.

Figures 9 and 10 illustrate simulated examples of network topology in square and circular areas, respectively, where small points denote Cooperative nodes and can be used to transmission, and larger points denote Jellyfish and Selfish nodes which initiate the transmission. Counting the number of cooperative neighbors for

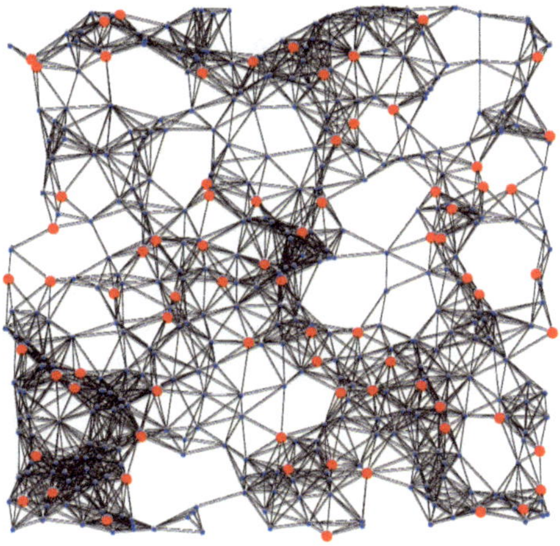

Fig. 9 Network topology used in simulation (square area)

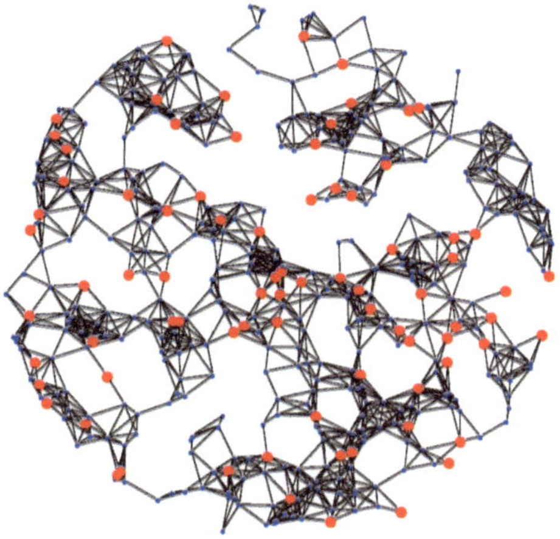

Fig. 10 Network topology used in simulation (circular area)

all active nodes, we can find the minimum cooperative degree $\theta(\mathcal{M})$. An algorithm to change the state of each node and to find the minimum cooperative degree is given in Fig. 11. Let *Node* and $Node_i$ be the sequence of node in a WAHN and i-th element of *Node*. Also let *NodeC*, *NodeSJ* and *NodeB* denote the subsets of sequence *Node* with Cooperative node, Selfish/Jellyfish node and Blackhole node, respectively. For each $Node_i$, we choose a value v randomly from 0 to 1, and identify the state j $(\in C, S, J, B, F)$. If the subset *NodeB* is not empty, then the minimum cooperative degree $\theta(\mathcal{M})$ equals to 0, otherwise, we need to count the number of cooperative neighbors of each node in subsets *NodeC* and *NodeSJ*. For a given transmission radius r, we calculate the distance of each element between

Fig. 11 Algorithm to find the minimum cooperative degree

```
Pc = 0.7299,   Ps = 0.1629,   Pj = 6.696e-6,
Pb = 7.44e-7,   Pf = 0.1072.
Set P_C, P_S, P_J, P_B, P_F to the steay state
probability of each state;
Set NodeC, NodeSJ, NodeB, θ(M_a) to empty;
For (i = 0; i ≤ N; i++){
Set v to 0;
Randomly generate j (j ∈ [0, 1]);
if 0 ≤ v < P_C then add Node_i to NodeC;
if P_C ≤ v < P_C + P_S + P_J
then add Node_i to NodeSJ;
if P_C + P_S + P_J ≤ v < P_C + P_S + P_J + P_B
then add Node_i to NodeB;}
/*NodeC, NodeSJ, NodeB are the set of nodes
in the states C, S and J, B, respectively*/
if NodeB is not empty then add 0 to θ(M_a),
else Set NoC, NoSJ to the number of elements
of NodeC and NodeSJ, respectively;
Set r to the transmission radius;
Set degree to empty;
For (i = 0; i ≤ NoSJ; i++){
Set count = 0;/*count the number of
cooperative neighbor*/
For (j = 0; j ≤ NoC; j++){
if distance between NodeSJ_i and
NodeC_j is no greater than r,
count++;}
Add count to degree;};
For (i = 0; i ≤ NoC; i++){
Set count = 0;
For (j = 0; j ≤ NoC; j++){
if i ≠ j and distance between NodeC_i and
NodeC_j is no greater than r,
count++;}
Add count to degree;};
Add min{degree} to θ(M_a);/*min{degree}
is the smallest element of degree*/
```

NodeSJ and *NodeC*. Besides, we also calculate the distance of each element in *NodeC*. If the distance of a node pair is not greater than *r*, then they are considered as neighbors. After counting the number of cooperative neighbors for all node in *NodeC* and *NodeSJ*, we can find the minimum cooperative degree $\theta(\mathcal{M})$.

To our best knowledge, the simulator developed in this article is a unique tool to quantify the connectivity-based network survivability with higher accuracy. However, as well known, the computation cost to seek the survivability measure is troublesome and very expensive. In other words, it is quite hard to simulate the node behavior and calculate the network survivability in online procedure. Hence, the analytical solution is still valuable to measure the connectivity-based network survivability in real WAHNs.

6 Numerical Examples

6.1 Comparison of Steady-State Network Survivability

In this section, we investigate border effects in the connectivity-based network survivability quantification with three stochastic models. We setup the following model parameters [12]:

$$\begin{aligned}
&\lambda_{c,s} = 1/240.0[1/\mathrm{s}], \quad \lambda_{c,j} = 3/2.0\mathrm{e}+7[1/\mathrm{s}],\\
&\lambda_{c,b} = 1/6.0\mathrm{e}+7[1/\mathrm{s}], \quad \lambda_{c,f} = 1/500.0[1/\mathrm{s}],\\
&\lambda_{s,c} = 1/60.0[1/\mathrm{s}], \quad \lambda_{s,j} = 3/2.0\mathrm{e}+7[1/\mathrm{s}],\\
&\lambda_{s,b} = 1/6.0\mathrm{e}+7[1/\mathrm{s}], \quad \lambda_{s,f} = 1/500.0[1/\mathrm{s}],\\
&\lambda_{j,f} = 1/50.0[1/\mathrm{s}], \quad \lambda_{b,f} = 1/50.0[1/\mathrm{s}],\\
&\lambda_{j,f} = 1/60.0[1/\mathrm{s}],
\end{aligned}$$

where λ_{ij} are transition rates from state i to state j in the exponential distributions. Under the above model parameters, the node probabilities in the steady state are given by

$$\begin{aligned}
&P_c = 0.7299, \quad P_s = 0.1629, \quad P_j = 6.696\mathrm{e}{-6},\\
&P_b = 7.44\mathrm{e}-7, \quad P_f = 0.1072.
\end{aligned}$$

We also assume the network parameters as follows: $A = 1000\,(\mathrm{m}) \times 1000\,(\mathrm{m})$: the area of WAHN. $N = 500$: the number of mobile nodes.

First, we compare the lower and upper bounds of connectivity-based network survivability [11, 12] with our expected network survivability in Eq. (29), (31) or (34). We set the transition radius r from 80 to 130, and connectivity requirement k from 1 to 3. The comparative results are shown in Table 2. From this table, we can see that the difference between lower and upper bounds of network survivability is very large for specific values of r and k. For example, when $r = 100$ and $k = 3$, the lower and upper bounds of network survivability in the Poisson model are equal to 0.0000 and 0.9296, respectively. On the other hand, the expected network survivability always takes a value between lower and upper bounds. This result shows us that our expected network survivability measure is more useful than the bounds for quantification of network survivability. Since it is known in [17, 18] that the number of neighbors of a node located in border areas is smaller than that located in the inner area, i.e., the border effect is effective, we compare the expected number of neighbors in both types of network area in Fig. 12. From this result, it can be seen that as transmission radius increases, the gap between two cases with and without border effects becomes remarkable. Especially, it is found that the node in square area affects the border effect more than that in circular area.

To evaluate the accuracy of our analytical results on border effects, we conduct a Monte Carlo simulation, where two types of network communication areas; square

Table 2 Comparison of lower and upper bounds with expected network survivability

r	k	Poisson			Binomial			Negative binomial		
		SVB_L	SVB	SVB_U	SVB_L	SVB	SVB_U	SVB_L	SVB	SVB_U
80	1	0.3595	0.5268	0.9311	0.3807	0.5381	0.9333	0.3278	0.5103	0.927
	2	0.0000	0.0079	0.5830	0.0000	0.0088	0.5902	0.0000	0.0066	0.5716
	3	0.0000	0.0000	0.1219	0.0000	0.0088	0.1256	0.0000	0.0000	0.1154
90	1	0.8844	0.8908	0.9899	0.8908	0.8965	0.9904	0.8753	0.8827	0.9891
	2	0.0000	0.3519	0.9122	0.0020	0.3682	0.9158	0.0000	0.3295	0.9069
	3	0.0000	0.0073	0.6487	0.0000	0.0086	0.6576	0.0000	0.0058	0.6354
100	1	0.9793	0.9796	0.9985	0.9808	0.9810	0.9986	0.9773	0.9776	0.9984
	2	0.8156	0.8316	0.9869	0.8289	0.8427	0.9879	0.7971	0.8163	0.9856
	3	0.0000	0.3593	0.9296	0.0367	0.3812	0.9336	0.0000	0.3304	0.9241
110	1	0.9925	0.9925	0.9996	0.9928	0.9928	0.9996	0.9921	0.9922	0.9995
	2	0.9694	0.9699	0.9982	0.9723	0.9727	0.9984	0.9654	0.9660	0.9980
	3	0.8264	0.8406	0.9898	0.8426	0.8543	0.9908	0.8041	0.8220	0.9885
120	1	0.9931	0.9931	0.9997	0.9932	0.9932	0.9997	0.9931	0.9931	0.9997
	2	0.9905	0.9906	0.9995	0.9910	0.9910	0.9996	0.9898	0.9899	0.9995
	3	0.9713	0.9717	0.9986	0.9745	0.9748	0.9987	0.9667	0.9673	0.9984
130	1	0.9921	0.9921	0.9997	0.9921	0.9922	0.9997	0.9921	0.9921	0.9997
	2	0.9919	0.9919	0.9997	0.9920	0.9920	0.9997	0.9918	0.9918	0.9997
	3	0.9898	0.9899	0.9996	0.9903	0.9904	0.9996	0.9891	0.9892	0.9995

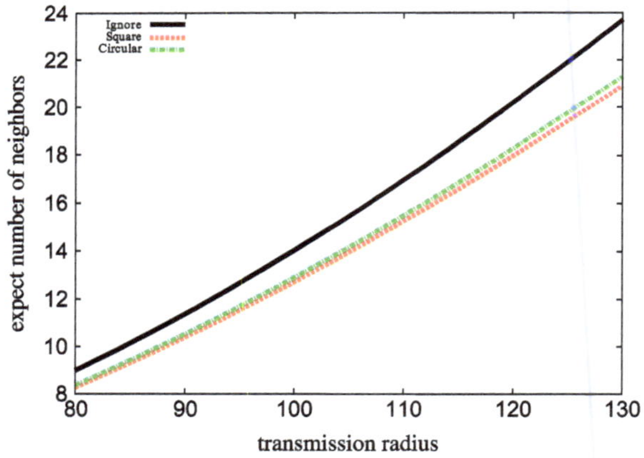

Fig. 12 Effects of r on the expected node degree

area and circular area, are assumed identically to be $A = 1{,}000{,}000$ (m^2). The expected active number of nodes, $E[N_a]$, is given by 446 with different communication radius r, which ranges from 80 to 130. We make r increase by 5. The random generation of active nodes is made 50 times for one radius. In each simulation, 20 nodes are randomly chosen, and the number of neighbors is calculated for each node. Finally we get 1000 values of number of neighbors for each r. From

Table 3 Simulation and analytical results on node degree

r	Number of neighbors				
	Ignorance	Square/a	Square/s	Circular/a	Circular/s
80	8.9751	8.3288	8.1390	8.4350	8.5840
85	10.1321	9.3582	9.3510	9.4842	9.6990
90	11.3592	10.4421	10.5950	10.5901	10.6790
95	12.6563	11.5797	11.6510	11.7519	11.7150
100	14.0237	12.7700	12.8280	12.9687	13.0730
105	15.4611	14.0124	13.9660	14.2398	14.3040
110	16.9686	15.3059	15.2910	15.5645	15.6250
115	18.5463	16.6497	16.6480	16.9418	17.1330
120	20.1941	18.0429	17.9260	18.3711	18.3100
125	21.9120	19.4848	19.6190	19.8515	19.7780
130	23.7000	20.9746	21.1300	21.3823	21.2570

this result, we calculate the average number of neighbors for a node in both square and circular communication areas. Table 3 compares the simulation results with analytical ones in terms of the number of neighbors, where 'ignorance' denotes the case without border effects in [11, 12, 15], 'Square/a' ('Circular/a') is the number of neighbors based on the analytical approach in Eq. (43) [Eq. (48)], and 'Square/s' ('Circular/s') is the simulation result in square (circular) area. In the comparison, we can see that the analytical results taking account of border effects get closer to the simulation results. However, the ignorance of border effects leads to an underestimation of the number of neighbors.

Table 4 presents the dependence of connectivity k and the number of nodes N on the steady-state network survivability among three stochastic models with and without border effects. Poisson model (Poisson), binomial model (Binomial) and negative binomial model (Negative Binomial) are compared in cases without border effects, which are denoted by Ignorance, Square, and Circular in the table. From these results, it is shown that the network survivability is reduced fiercely as k increases when the number of nodes N is relatively small. In Fig. 13, we show the dependence of r and k on the steady-state network survivability in the Poisson model. From this figure, we find that the transition radius rather affects the steady-state network survivability, if each node has a relatively large r which is greater than 120 (m). In this case even for the 3-connected network, the steady-state network survivability becomes 0.8 and tends to take a lower value. On the other hand, if n is sufficiently large and p is sufficiently small under $\mu = np$, from the small number's law, the binomial distribution can be well approximated by the Poisson distribution. This asymptotic inference can be confirmed in Fig. 14. So, three stochastic models provide almost similar performance in terms of connectivity-based network survivability in such a case.

Table 4 Steady-state network survivability with three stochastic models

N		Poisson			Binomial			Negative binomial		
	k	Ignorance	Square	Circular	Ignorance	Square	Circular	Ignorance	Square	Circular
500	1	0.9787	0.9549	0.9602	0.9809	0.9594	0.9642	0.9780	0.9545	0.9592
	2	0.8249	0.6473	0.6824	0.8416	0.6735	0.7072	0.8193	0.6476	0.6780
	3	0.3435	0.1052	0.1350	0.3786	0.1257	0.1587	0.3373	0.1092	0.1351
600	1	0.9909	0.9865	0.9876	0.9912	0.9873	0.9883	0.9905	0.9856	0.9868
	2	0.9611	0.9078	0.9200	0.9643	0.9146	0.9260	0.9569	0.8986	0.9121
	3	0.7971	0.5700	0.6147	0.8120	0.5916	0.6355	0.7781	0.5415	0.5884
700	1	0.9905	0.9904	0.9905	0.9906	0.9905	0.9906	0.9905	0.9902	0.9903
	2	0.9853	0.9732	0.9762	0.9859	0.9749	0.9777	0.9843	0.9708	0.9743
	3	0.9482	0.8680	0.8866	0.9524	0.8772	0.8948	0.9421	0.8556	0.8761
800	1	0.9881	0.9890	0.9889	0.9881	0.9890	0.9889	0.9881	0.9889	0.9888
	2	0.9872	0.9855	0.9860	0.9874	0.9859	0.9864	0.9870	0.9849	0.9856
	3	0.9799	0.9597	0.9649	0.9809	0.9624	0.9672	0.9786	0.9561	0.9620
900	1	0.9850	0.9863	0.9861	0.9850	0.9863	0.9861	0.9850	0.9863	0.9861
	2	0.9849	0.9856	0.9856	0.9849	0.9857	0.9857	0.9848	0.9855	0.9855
	3	0.9835	0.9799	0.9810	0.9838	0.9807	0.9816	0.9832	0.9790	0.9802

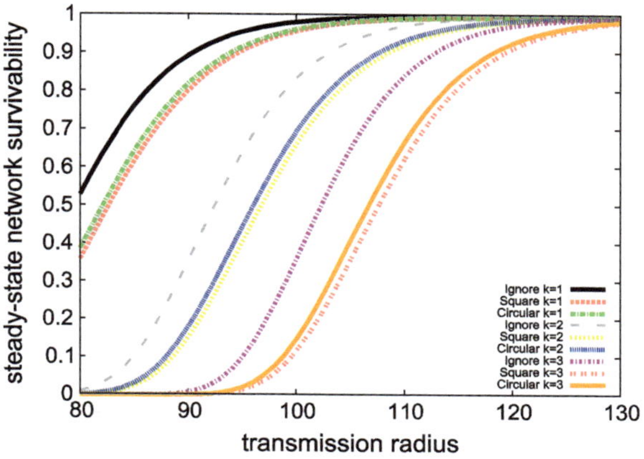

Fig. 13 Effects of k on the steady-state network survivability

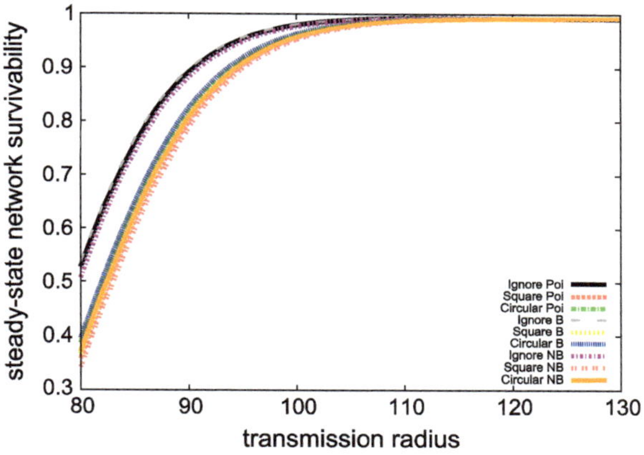

Fig. 14 Comparison of three stochastic models with two types of border effect

6.2 *Transient Analysis of Network Survivability*

Next we concern the transient network survivability at arbitrary time t. For the numerical inversion of Laplace–Stieltjes transform, we apply the well-known Abate's algorithm [22]. Although we omit to show here for brevity, it can be numerically checked that the transient probability $P_{cc}(t)$ decreases first and approaches to the steady-state solution as time goes on. The other probabilities $P_{cs}(t)$, $P_{cj}(t)$, $P_{cb}(t)$, and $P_{cf}(t)$ increase in the first phase, but converge to their associated saturation levels asymptotically. Reminding these asymptotic properties on transition probabilities, we set $N = 500$ and $r = 100$, and consider the transient network survivability of three stochastic models with and without border effects in Table 5. The network survivability with or without border effects has almost the

Table 5 Transient network survivability with three stochastic models

t	k	Poisson			Binomial			Negative binomial		
		Ignorance	Square	Circular	Ignorance	Square	Circular	Ignorance	Square	Circular
0	1	0.9999	0.9997	0.9998	0.9999	0.9997	0.9998	0.9999	0.9996	0.9997
	2	0.9987	0.9953	0.9962	0.9990	0.9960	0.9968	0.9984	0.9944	0.9954
	3	0.9895	0.9645	0.9707	0.9910	0.9688	0.9744	0.9871	0.9591	0.9653
40	1	0.9950	0.9915	0.9923	0.9953	0.9921	0.9929	0.9947	0.9906	0.9906
	2	0.9719	0.9289	0.9388	0.9746	0.9350	0.9442	0.9680	0.9204	0.9204
	3	0.8395	0.6425	0.6825	0.8527	0.6642	0.7032	0.8208	0.6131	0.6131
80	1	0.9893	0.9785	0.9810	0.9899	0.9799	0.9822	0.9883	0.9762	0.9791
	2	0.9186	0.8150	0.8371	0.9241	0.8252	0.8466	0.9091	0.7979	0.8220
	3	0.6076	0.3221	0.3696	0.6253	0.3397	0.3882	0.5768	0.2926	0.3401
120	1	0.9845	0.9678	0.9716	0.9855	0.9700	0.9735	0.9830	0.9646	0.9689
	2	0.8752	0.7336	0.7627	0.8834	0.7475	0.7758	0.8624	0.7130	0.7442
	3	0.4682	0.1919	0.2321	0.4888	0.2072	0.2492	0.4368	0.1699	0.2087
160	1	0.9819	0.9619	0.9664	0.9830	0.9643	0.9686	0.9800	0.9580	0.9631
	2	0.8515	0.6924	0.7245	0.8608	0.7073	0.7386	0.8368	0.6698	0.7040
	3	0.4057	0.1456	0.1809	0.4259	0.1586	0.1958	0.3742	0.1267	0.1602

similar initial values (0.9999), and the differences between them will be remarkable as time elapses.

Because three stochastic models show the quite similar tendency, hereafter we focus on only the Poisson model with k-connectivity ($k = 1, 2, 3, 4$) to investigate the impact on the transient network survivability. From Fig. 15, it is seen that the Poisson model has higher transient network survivability when k is lower. Also, when the connectivity level becomes higher, the transient network survivability gets closer to 0 with time t elapsing. Finally we compare the Poisson model with and without border effects in terms of the transient network survivability. Figure 16 illustrates the transient network survivability when $k = 1$. It is shown that if the

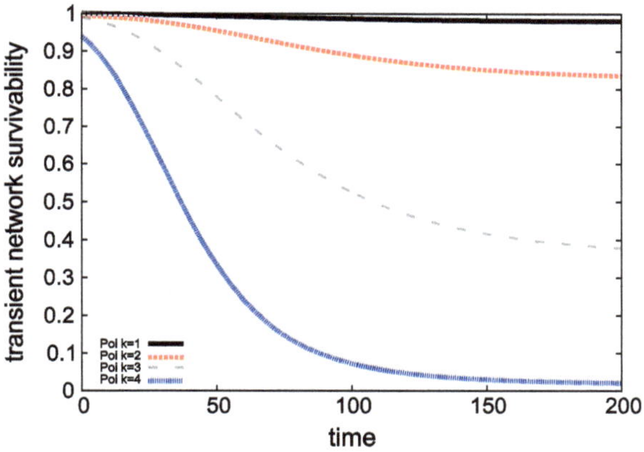

Fig. 15 Transient network survivability with varying k

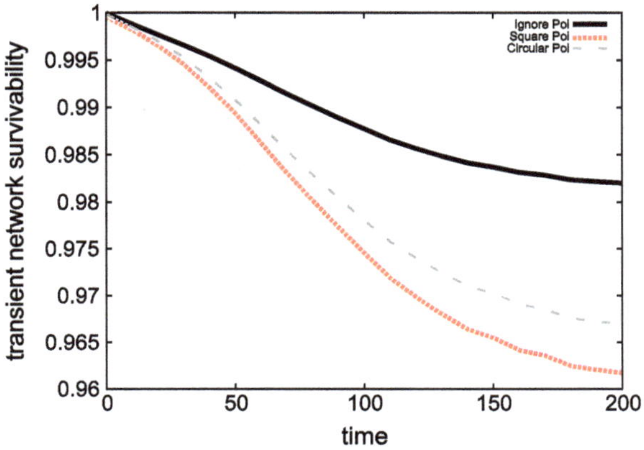

Fig. 16 Transient network survivability with border effects

border effects are taken into consideration, the transient network survivability drops down as the operation time goes on. However, the transient solution without border effects still keeps higher levels in the same situation. This fact implies that the ignorance of border effects leads to an underestimation of network survivability. Since such an optimistic assessment of network survivability may result a risk through the network operation, it is recommended to take account of border effects in the connectivity-based network survivability assessment in WAHNs.

6.3 Accuracy and Limitation

We have already shown in Table 3 that our analytical models with border effects provided relatively nice performance comparing with the simulation solutions in terms of the number of neighbors. Here, we compare the network survivability in analytical models with the simulation results. As analytical solutions, we calculate the expected network survivability and its associated bounds in two cases with/without border effects. Suppose that the number of nodes N equals to 500 for varying transition radius r from 80 to 130 by 5. Since the network survivability depends on the connectivity level k, we set $k = 1, 2, 3$ in the simulation experiments. In Tables 6 and 7 the steady-state network survivability for varying node transmission radius r in square and circular areas are calculated. It can be seen that the simulation result is two-sided bounded in a few cases with $(k, r) = (1, 125), (1, 130), (2, 80), (2, 85), (2, 90), (3, 80), (3, 85), (3, 90), (3, 95), (3, 100)$ in Table 6. Looking at the expected network survivability, it takes rather different values from the simulation results. In Tables 8 and 9, we compare our analytical solutions with the simulated ones for varying N. In the combination of $(k, N) = (1, 800),\ (1, 900), (1, 1000), (2, 1000), (3, 500)$, it is shown that the simulation result is two-sided bounded, but the expected network survivability does not always take the closed values to simulation results. When 'Ignorance' is compared with 'Square' or 'Circular', the latter can takes the closer value than the former, but fail to get the satisfactory approximate performance. In other words, our analytical models taking account of border effects still fail to evaluate the accurate network survivability except in very few cases. This negative observation implies that there is no satisfactory analytical model to assess the connectivity-based network survivability, and is a challenging issue to develop a more sophisticated stochastic model to evaluate it. The lesson learned from the comparative study in this article will motivate to investigate the other stochastic modeling approach for the purpose.

Table 6 Steady-state network survivability for varying node transmission radius r in square area

$k = 1$		Ignorance		Square	
r	Simulation	Expected	Bounds	Expected	Bounds
80	0.3522	0.5268	[0.3595, 0.9311]	0.3557	$[0.0000, 0.8920]^*$
85	0.5365	0.7459	[0.7227, 0.9730]	0.6054	$[0.5149, 0.9532]^*$
90	0.7071	0.8908	[0.8844, 0.9899]	0.7891	[0.7780, 0.9806]
95	0.7964	0.9547	[0.9524, 0.9962]	0.8952	[0.9009, 0.9922]
100	0.8867	0.9796	[0.9793, 0.9985]	0.9481	[0.9558, 0.9969]
105	0.9451	0.9928	[0.9893, 0.9993]	0.9721	[0.9792, 0.9987]
110	0.9532	0.9925	[0.9925, 0.9996]	0.9821	[0.9886, 0.9993]
115	0.9632	0.9973	[0.9933, 0.9996]	0.9859	[0.9921, 0.9996]
120	0.9854	0.9931	[0.9931, 0.9997]	0.9872	[0.9932, 0.9997]
125	0.9943	0.9931	$[0.9927, 0.9997]^*$	0.9873	$[0.9932, 0.9997]^*$
130	0.9963	0.9921	$[0.9921, 0.9997]^*$	0.9871	$[0.9929, 0.9997]^*$
$k = 2$		Ignorance		Square	
r	Simulation	Expected	Bounds	Expected	Bounds
80	0.0019	0.0079	$[0.0000, 0.5830]^*$	0.0014	$[0.0000, 0.4436]^*$
85	0.0279	0.1073	$[0.0000, 0.7962]^*$	0.0283	$[0.0000, 0.6877]^*$
90	0.1335	0.3519	$[0.0000, 0.9122]^*$	0.1577	$[0.0000, 0.8466]^*$
95	0.2744	0.6249	[0.5515, 0.9652]	0.3997	$[0.0959, 0.9310]^*$
100	0.4286	0.8316	[0.8156, 0.9869]	0.6453	[0.5835, 0.9707]
105	0.6579	0.9271	[0.9260, 0.9952]	0.8158	[0.8143, 0.9881]
110	0.7492	0.9699	[0.9694, 0.9982]	0.9098	[0.9185, 0.9952]
115	0.8078	0.9882	[0.9853, 0.9992]	0.9549	[0.9635, 0.9980]
120	0.8767	0.9906	[0.9905, 0.9995]	0.9745	[0.9820, 0.9991]
125	0.9378	0.9958	[0.9919, 0.9996]	0.9824	[0.9890, 0.9995]
130	0.9606	0.9919	[0.9919, 0.9997]	0.9852	[0.9914, 0.9996]
$k = 3$		Ignorance		Square	
r	Simulation	Expected	Bounds	Expected	Bounds
80	0.0000	0.0000	$[0.0000, 0.1219]^*$	0.0000	$[0.0000, 0.0498]^*$
85	0.0000	0.0002	$[0.0000, 0.3762]^*$	0.0000	$[0.0000, 0.2213]^*$
90	0.0012	0.0073	$[0.0000, 0.6487]^*$	0.0008	$[0.0000, 0.4811]^*$
95	0.0121	0.1108	$[0.0000, 0.8334]^*$	0.0195	$[0.0000, 0.7104]^*$
100	0.0538	0.3593	$[0.0000, 0.9296]^*$	0.1265	$[0.0000, 0.8574]^*$
105	0.2062	0.6336	[0.5693, 0.9725]	0.3545	$[0.0000, 0.9355]^*$
110	0.3367	0.8406	[0.8264, 0.9898]	0.6074	[0.5270, 0.9725]
115	0.4664	0.9408	[0.9313, 0.9963]	0.7927	[0.7898, 0.9887]
120	0.5940	0.9717	[0.9713, 0.9986]	0.8981	[0.9082, 0.9955]
125	0.7363	0.9882	[0.9854, 0.9993]	0.9492	[0.9590, 0.9981]
130	0.8174	0.9899	[0.9898, 0.9996]	0.9717	[0.9797, 0.9991]

* The case in which the upper and lower bounds include the exact (simulation) value

Table 7 Steady-state network survivability for varying node transmission radius r in circular area

$k = 1$		Ignorance		Circular	
r	Simulation	Expected	Bounds	Expected	Bounds
80	0.3852	0.5268	[0.3595, 0.9311]*	0.3831	[0.0516, 0.8996]*
85	0.6175	0.7459	[0.7227, 0.9730]	0.6315	[0.5572, 0.9572]*
90	0.7735	0.8908	[0.8844, 0.9899]	0.8072	[0.8004, 0.9826]
95	0.8437	0.9547	[0.9524, 0.9962]	0.9057	[0.9121, 0.9931]
100	0.9301	0.9796	[0.9793, 0.9985]	0.9535	[0.9611, 0.9972]
105	0.9542	0.9928	[0.9893, 0.9993]	0.9746	[0.9816, 0.9988]
110	0.9793	0.9925	[0.9925, 0.9996]	0.9832	[0.9896, 0.9994]
115	0.9920	0.9973	[0.9933, 0.9996]	0.9863	[0.9925, 0.9996]
120	0.9937	0.9931	[0.9931, 0.9997]*	0.9873	[0.9932, 0.9997]*
125	0.9970	0.9931	[0.9927, 0.9997]*	0.9873	[0.9932, 0.9997]*
130	0.9989	0.9921	[0.9921, 0.9997]*	0.9870	[0.9928, 0.9997]*
$k = 2$		Ignorance		Circular	
r	Simulation	Expected	Bounds	Expected	Bounds
80	0.0052	0.0079	[0.0000, 0.5830]*	0.0020	[0.0000, 0.4676]*
85	0.0422	0.1073	[0.0000, 0.7962]*	0.0367	[0.0000, 0.7079]*
90	0.1755	0.3519	[0.0000, 0.9122]*	0.1850	[0.0000, 0.8596]*
95	0.3392	0.6249	[0.5515, 0.9652]	0.4385	[0.1916, 0.9381]*
100	0.5507	0.8316	[0.8156, 0.9869]	0.6789	[0.6341, 0.9742]
105	0.6825	0.9271	[0.9260, 0.9952]	0.8375	[0.8396, 0.9897]
110	0.8191	0.9699	[0.9694, 0.9982]	0.9216	[0.9306, 0.9959]
115	0.9169	0.9882	[0.9853, 0.9992]	0.9606	[0.9690, 0.9983]
120	0.9497	0.9906	[0.9905, 0.9995]	0.9771	[0.9842, 0.9992]
125	0.9706	0.9858	[0.9919, 0.9996]	0.9834	[0.9899, 0.9995]
130	0.9824	0.9919	[0.9919, 0.9997]	0.9856	[0.9917, 0.9996]
$k = 3$		Ignorance		Circular	
r	Simulation	Expected	Bounds	Expected	Bounds
80	0.0000	0.0000	[0.0000, 0.1219]*	0.0000	[0.0000, 0.0590]*
85	0.0000	0.0002	[0.0000, 0.3762]*	0.0000	[0.0000, 0.2451]*
90	0.0019	0.0073	[0.0000, 0.6487]*	0.0013	[0.0000, 0.5104]*
95	0.0223	0.1108	[0.0000, 0.8334]*	0.0276	[0.0000, 0.7339]*
100	0.1035	0.3593	[0.0000, 0.9296]*	0.1567	[0.0000, 0.8722]*
105	0.2500	0.6336	[0.5693, 0.9725]	0.4016	[0.1016, 0.9435]*
110	0.4206	0.8406	[0.8264, 0.9898]	0.6503	[0.5954, 0.9764]
115	0.6730	0.9408	[0.9313, 0.9963]	0.8212	[0.8236, 0.9905]
120	0.7696	0.9717	[0.9713, 0.9986]	0.9136	[0.9240, 0.9962]
125	0.8414	0.9882	[0.9854, 0.9993]	0.9568	[0.9660, 0.9985]
130	0.8967	0.9899	[0.9898, 0.9996]	0.9750	[0.9826, 0.9993]

* The case in which the upper and lower bounds include the exact (simulation) value

Table 8 Steady-state network survivability for varying number of node N in square area

$k = 1$		Ignorance		Square	
N	Simulation	Expected	Bounds	Expected	Bounds
500	0.9065	0.9787	[0.9793, 0.9985]	0.9549	[0.9611, 0.9972]
600	0.9715	0.9909	[0.9908, 0.9995]	0.9865	[0.9876, 0.9993]
700	0.9853	0.9905	[0.9905, 0.9995]	0.9904	[0.9905, 0.9995]
800	0.9960	0.9881	$[0.9880, 0.9995]^*$	0.9890	$[0.9888, 0.9995]^*$
900	0.9987	0.9850	$[0.9849, 0.9994]^*$	0.9863	$[0.9860, 0.9994]^*$
1000	0.9988	0.9815	$[0.9814, 0.9993]^*$	0.9832	$[0.9828, 0.9994]^*$
$k = 2$		Ignorance		Square	
N	Simulation	Expected	Bounds	Expected	Bounds
500	0.5231	0.8249	[0.8156, 0.9869]	0.6473	[0.6341, 0.9742]
600	0.8055	0.9611	[0.9604, 0.9976]	0.9078	[0.9166, 0.9951]
700	0.9027	0.9853	[0.9851, 0.9992]	0.9732	[0.9759, 0.9988]
800	0.9569	0.9872	[0.9871, 0.9994]	0.9855	[0.9859, 0.9994]
900	0.9872	0.9849	$[0.9848, 0.9994]^*$	0.9856	$[0.9855, 0.9994]^*$
1000	0.9960	0.9815	$[0.9813, 0.9993]^*$	0.9830	$[0.9827, 0.9994]^*$
$k = 3$		Ignorance		Square	
N	Simulation	Expected	Bounds	Expected	Bounds
500	0.0910	0.3435	$[0.0000, 0.9296]^*$	0.1052	$[0.0000, 0.8722]^*$
600	0.4031	0.7971	[0.7734, 0.9866]	0.5700	[0.5138, 0.9715]
700	0.6545	0.9482	[0.9468, 0.9973]	0.8680	[0.8796, 0.9939]
800	0.8189	0.9799	[0.9797, 0.9991]	0.9597	[0.9643, 0.9984]
900	0.9351	0.9835	[0.9834, 0.9993]	0.9799	[0.9808, 0.9992]
1000	0.9758	0.9813	[0.9811, 0.9993]	0.9818	[0.9817, 0.9993]

* The case in which the upper and lower bounds include the exact (simulation) value

Table 9 Steady-state network survivability for varying number of node N in circular area

$k = 1$		Ignorance		Circular	
N	Simulation	Expected	Bounds	Expected	Bounds
500	0.8599	0.9787	[0.9793, 0.9985]	0.9602	[0.9558, 0.9969]
600	0.9536	0.9909	[0.9908, 0.9995]	0.9876	[0.9865, 0.9992]
700	0.9775	0.9905	[0.9905, 0.9995]	0.9905	[0.9903, 0.9995]
800	0.9912	0.9881	$[0.9880, 0.9995]^*$	0.9889	$[0.9889, 0.9995]^*$
900	0.9974	0.9850	$[0.9849, 0.9994]^*$	0.9861	$[0.9862, 0.9995]^*$
1000	0.9988	0.9815	$[0.9814, 0.9993]^*$	0.9829	$[0.9830, 0.9994]^*$
$k = 2$		Ignorance		Circular	
N	Simulation	Expected	Bounds	Expected	Bounds
500	0.3888	0.8249	[0.8156, 0.9869]	0.6824	[0.5835, 0.9707]
600	0.7211	0.9611	[0.9604, 0.9976]	0.9200	[0.9034, 0.9943]
700	0.8404	0.9853	[0.9851, 0.9992]	0.9762	[0.9728, 0.9986]
800	0.9148	0.9872	[0.9871, 0.9994]	0.9860	[0.9854, 0.9993]

(continued)

Table 9 (continued)

$k = 2$		Ignorance		Circular	
N	Simulation	Expected	Bounds	Expected	Bounds
900	0.9739	0.9849	[0.9848, 0.9994]	0.9856	[0.9855, 0.9994]
1000	0.9862	0.9815	[0.9813, 0.9993]*	0.9828	[0.9829, 0.9994]*
$k = 3$		Ignorance		Circular	
N	Simulation	Expected	Bounds	Expected	Bounds
500	0.0464	0.3435	[0.0000, 0.9296]*	0.1350	[0.0000, 0.8574]*
600	0.3154	0.7971	[0.7734, 0.9866]	0.6147	[0.4386, 0.9672]
700	0.5176	0.9482	[0.9468, 0.9973]	0.8866	[0.8584, 0.9928]
800	0.6979	0.9799	[0.9797, 0.9991]	0.9649	[0.9589, 0.9982]
900	0.8770	0.9835	[0.9834, 0.9993]	0.9810	[0.9797, 0.9992]
1000	0.9263	0.9813	[0.9811, 0.9993]	0.9819	[0.9817, 0.9993]

* The case in which the upper and lower bounds include the exact (simulation) value

7 Concluding Remarks

In this article, we have refined the network survivability models by taking account of border effects in both square and circular areas. Based on the definition of border effects in communication areas, we have calculated the expected coverage of node in WAHNs which results the expected node degree, and have formulated the network survivability with border effects. Also we have developed a simulation model to quantify the network survivability. In numerical experiments, we have calculated the expected node degree and the connectivity-based network survivability measures in both analytical and simulation models, and shown that the border effects were significant to evaluate the number of neighbors accurately. We have also compared the steady-state network survivability and the transient network survivability in three stochastic models, and shown numerically that the network survivability was reduced fiercely as k increased when N was small and that the connectivity-based network survivability without border effects was higher than that without border effects. However, in the comparison of connectivity-based network survivability, we have shown that both the analytical bounds and the expected network survivability did poorly worked except in a few cases, although it has been shown that the border effects were still useful to evaluate the number of neighbors accurately when the transmission radius r changed.

In this article the network survivability has been defined by the minimum cooperative degree, but it is worth mentioning that it can be considered as an approximate measure. In future, we will develop a comprehensive network survivability model and investigate whether the approximate method for network survivability itself can work well in several random network environments.

References

1. R. Ellison, D. Fisher, R. Linger, H. Lipson, T. Longstaff and N. Mead, "Survival network systems: an emerging discipline," *Technical Report of SEI/CMU*, CMU/SEI-97-TR-013 (1997).
2. J. C. Knight and K. J. Sullivan, "On the definition of survivability," *Technical Report of Dept. of Computer Science/University of Virginia*, CS-TR-33-00 (2000).
3. S. C. Liew and K. W. Lu, "A framework for characterizing disaster-based network survivability," *IEEE Transactions on Selected Areas in Communications*, vol. 12, no. 1, pp. 52–58 (1994).
4. A. Zolfaghari and F. J. Kaudel, "Framework for network survivability performance," *IEEE Transactions on Selected Areas in Communications*, vol. 12, no. 1, pp. 46–51 (1994).
5. D. Chen, S. Garg and K. S. Trivedi, "Network survivability performance evaluation: a quantitative approach with applications in wireless ad-hoc networks," *Proceeding of ACM International Conference on Modeling, Analysis and Simulation of Wireless and Mobile Systems (MSWiM-2002)*, pp. 61–68, ACM (2002).
6. L. Cloth and B. R. Haverkort, "Model checking for survivability," *Proceeding of the 2nd IEEE Conference on Quantitative Evaluation of Systems (QEST-2005)*, pp. 145–154, IEEE CPS (2005).
7. P. E. Heegaard and K. S. Trivedi, "Netrwork survaivability modeling," *Computer Networks*, vol. 53, pp. 1215–1234 (2009).
8. Y. Liu, V. B. Mendiratta and K. S. Trivedi, "Survaivability analysis of telephone access network," *Proceeding of the 15th IEEE International Symposium on Software Reliability Engineering (ISSRE-2004)*, pp. 367–378, IEEE CPS (2004).
9. Y. Liu and K. S. Trivedi, "Survaivability quantification: the analytical modeling approach," *International Journal of Performability Engineering*, vol. 2, no. 1, pp. 29-44 (2006).
10. J. Zheng, H. Okamura and T. Dohi, "Survivability analysis of VM-based intrusion tolerant systems," *IEICE Transactions on Information & Systems (D)*, vol. E98-D, no. 12, pp. 2082-2090 (2015).
11. F. Xing and W. Wang, "Modeling and analysis of connectivity in mobile ad hoc networks with misbehaving nodes," *Proceeding of IEEE Conference on Communications (ICC-2006)*, pp. 1879–1884, IEEE CPS (2006).
12. F. Xing and W. Wang, "On the survivability of wireless ad hoc networks with node misbehaviors and failures," *IEEE Transactions on Dependable and Secure Computing*, vol. 7, no. 3, pp. 284–299 (2010).
13. C. Bettstetter, "On the minimum node degree and connectivity of a wireless multihop network," *Proceeding of ACM International Symposium on Mobile Ad Hoc Networking and Computing (MobiHoc-2002)*, pp. 80–91, ACM (2002).
14. C. Bettstetter, J. Klinglmayr and S. Lettner, "On the degree distribution of *k*-connected random networks," *Proceeding of IEEE Conference on Communications (ICC-2010)*, pp. 1–6, IEEE CPS (2010).
15. Z. Yi, T. Dohi, "Survivability analysis for a wireless ad hoc network based on semi-Markov model," *IEICE Transactions on Information & Systems*, vol. E95-D, no. 12, pp. 2844–2851 (2012).
16. H. Okamura, Z. Yi, T. Dohi: Network survivability modeling and analysis for a power-aware MANETs by Markov regenerative processes, *Telecommunication Systems Journal*, vol. 60, pp. 471–484 (2015).
17. L. Laranjeira and G. N. Rodrigues, "Border effect analysis for reliability assurance and continuous connectivity of wireless sensor networks in the presence of sensor failures," *IEEE Transactions on Wireless Communications*, vol. 13, no. 8, pp. 4232–4246 (2014).
18. [18] C. Bettstetter, "On the connectivity of ad hoc networks," *The Computer Journal*, vol. 47, no. 4, pp. 432–447 (2004).

19. G. D. Caro, F. Ducatelle and L. M. Gambardella, "AntHocNet: An adaptive nature-inspired algorithm for routing in mobile ad hoc networks," *European Transactions on Telecommunications*, vol. 16, no. 5, pp. 443–455 (2005).
20. L. Guo, "A new and improved algorithm for dynamic survivable routing in optical WDM networks," *Computer Communications*, vol. 30, no. 6, pp. 1419–1423 (2007).
21. Z. Yi, T. Dohi and H. Okamura, "Survivability quantification of wireless ad hoc network taking account of border effects," *Proceedings of the 21st IEEE Pacific Rim International Symposium on Dependable Computing (PRDC 2015)*, pp. 149–158 (2015).
22. P. Valko and J. Abate, "Numerical inversion of laplace transform with multiple precision using the complex domain," http://library.wolfram.com/MathSource/5026/.

Mobile Ad Hoc Network Reliability: An Imperative Research Challenge

Sanjay K. Chaturvedi and N. Padmavathy

Abstract A rapid development in the areas of sensor and wireless networks has motivated today's researchers to model and assess the performance of these dynamic networks on various counts resulting into the proliferation of interest and researches in the area of mobile ad hoc networks (MANET). MANET is a network that exchanges information among the entities that are potentially mobile without any pre-defined infrastructure based communication support. These networks are of practical importance in applications like environmental monitoring, health care, military, location tracking, disaster recovery and many more. However, the design and analysis of a reliable MANET introduces a formidable challenge since the required knowledge encompasses a whole range of topics viz., network complexity, routing, scalability, heterogeneity, protocol issues, clustering, reliability, bandwidth management, mobility management etc. Therefore, performance modelling and evaluation to ensure their successful deployment and exploitation in practice are becoming quite a formidable task. The contribution towards reliability modelling and its assessment of such systems are still scarce. In this chapter, an extensive study pertaining to the reliability modelling and assessment under no capacity constraints and the effect of propagation models (hybrid model and fading model) on reliability are presented.

Keywords Mobile ad hoc network · Geometric random graphs · Propagation models · Monte Carlo simulation · Network reliability

---

S.K. Chaturvedi (✉)
Reliability Engineering Centre, Indian Institute of Technology Kharagpur,
Kharagpur, West Bengal, India
e-mail: skcrec@hijli.iitkgp.ernet.in

N. Padmavathy
Dept. of Electronics and Communication Engineering,
Vishnu Institute of Technology, Bhimavaram, Andhra Pradesh, India

M. Ram and J.P. Davim (eds.), *Advances in Reliability and System Engineering*,
Management and Industrial Engineering, DOI 10.1007/978-3-319-48875-2_4

1 Introduction

Historically, the wireless media had been the means of communication since ages (e.g., use of drum communication; use of smoke signals for long distance communication; use of carrier pigeons for data communication; signal torches and many more) even before the invention of telephone by Alexander Graham Bell in 1876, and the advent of the wired networks of US defense-"US DARPA or ARPA" in early 1960s with its subsequent variants thereafter.

There is a phenomenal technological growth and advances in the past two decades, both in hardware and software, in the direction of providing sophisticated uninterrupted services by making the devices more portable, cheaper, user-friendly, and more importantly reliable. In recent times, Mobile Ad hoc network (MANET), an emerging technology, allows devices to establish communication on the concept of *anytime-anywhere* without the aid of a dedicated centralized infrastructure with enormous *power-on-the-palm* through wielding a variety of wireless mobile devices.

The communication using wireless media has been witnessing a phenomenal growth in recent past with a tremendous growth in mobile users and myriad amount of innovative applications. The present mobile technology have made the human lives more convenient and influential by providing access with the global community, thus enabling the users to communicate in the absence of ubiquitous and clumsy wires or cables seen some years back.

In general, all these devices are equipped with one or more wireless networking interfaces like Wi-Fi and/or Bluetooth using which, one can communicate with others through wireless media. These devices can connect with each other either directly (*single-hop*-when a mobile node and its neighbor(s) are within its vicinity) or through intermediate users (*multi-hop*-when there exists no direct link between communicating nodes) in a decentralized fashion giving rise to an arbitrary, scalable, and dynamic network configuration. Moreover, its multi-hoping feature allows and extends the capability of MN even well-beyond its line-of-sight based applications.

So far, among many intriguing issues with such infrastructureless networks, reliability analysis is one of the challenging issues of interest, first, because of its ever changing topology, variant properties of mobile nodes and links, and its resilience for a specific purpose viz., military scenario, commercial scenarios, etc. Second, the failure of these networks has significant impact on the services offered as the networks are application oriented. MANET reliability is an area of great research to devise better and efficient reliability metrics and/or reliability evaluation techniques to study their performance and feasibility of deployment.

This chapter discusses the performance study of the MANET on different terrains or environments that includes reliability evaluation of MANET using two different types of propagation models—(i) hybrid model and (ii) fading models, i.e., modeling the link existence based on TR-FRG and fading models. As a scenario, we consider a user-centric ad hoc network (infantry—soldiers on foot; armored vehicles carrying soldiers equipped with wireless communication devices) which consists of highly mobile battery/regenerative powered nodes without any infrastructure support. We assume that the time-to-failure of MN is governed by Weibull distribution. We further assume that the nodes failure is statistically independent and their repair during a mission is not feasible. In other words, these MN remain failed from the time it has failed till the end of the mission duration, if the MN fails in-between the mission duration. The MN movement follows *Random Way Point model* with their respective uniformly distributed velocity and direction.

The subsequent sections cover basics of MANET. A brief survey of the literature on various aspects of MANET and network reliability is provided in Sects. 2 and 3, respectively. In Sects. 4 and 5, we cover the performance study of MANET scenario (stated above) on different terrains or environments. We conclude this chapter in Sect. 6.

1.1 Mobile Ad Hoc Networks: Terminology, Classifications, and Characteristics

MANET or peer-to-peer network is a collection of solely independent transceivers, called Mobile Nodes (MN), communicating among them either directly/indirectly to sustain connectivity by forming an arbitrary topology in a decentralized infrastructure manner. The MN may be server or client. If a node works as a server, then it should have attributes like high memory size, high computational power, high battery capacity, and high storage size whereas the remaining with lesser capabilities are the clients. Each MN of the network can act as an independent source, destination or router and generates/routes the data. Every MN forwards each other's information to enable sharing among other MN. The network configuration depends on the factors such as positions of the MN, their coverage zones, and their transmission power. Moreover, the creation and termination of links among the communicating MN depends on the mobility and is commonly referred to as MANET's self-forming aspect. An unambiguous ad hoc network at some arbitrary point in time is shown in Fig. 1.

A MANET can be deployed almost in no time in a defined geographical region alone or as a wireless extension to an existing wired/infrastructure network. Its scalability and quick deployment feature are some of the driving force for their deployment in emergency situations like natural or man-made disasters; search and rescue; location tracking; disaster recovery military conflicts; emergency grounds; commercial applications; home networking, etc., the list is almost endless. There are numerous examples in the literature where the system supporting such

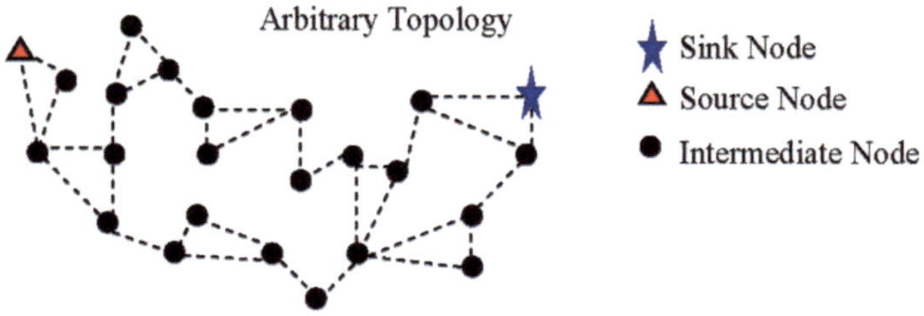

Fig. 1 An arbitrary network

applications is a versatile network, with its base structure being installed either in a perpetual or an impromptu way. It is expected that, the increasing demand from military, national security, and commercial customers will further necessitate a large-scale deployment of such ad hoc networks.

Some of the other prominent features of such networks are mobility and flexibility with rapidly changing network topology; deployed for a specific application of short durations (mission time); infrastructurelessness; self-creating networks operating at high frequency and high BER with low power consumption and low bandwidth [1–4].

1.2 MANET Broad Classification

Based on the performance characteristics, the MANET can broadly be put into as homogeneous and heterogeneous networks [1]. Figure 2a, b depict the homogeneous and heterogeneous network, respectively.

In homogeneous networks, the devices with identical specifications (i.e., transmission range, battery capacity, memory size, etc.) within their transmission range take part in forming and establishing communication among themselves. The heterogeneous networks provide communication among mobile devices of varied nature (due to the employment of different software/hardware configurations, memory size, computational power, battery capacity and physical size, and so on), which result into formation of asymmetric links. Design and analysis of such networks are still a challenging issue before the researchers as network needs to adapt to the dynamically changing power and channel conditions of different devices [5–11].

1.3 Failures in MANET

An assortment of reasons viz., low transmission range, out of coverage area, climatic effects, physical obstructions, and limited battery life can lead to MANET failures. Other possible sources of failures could be: the source and/or terminal MN gets isolated

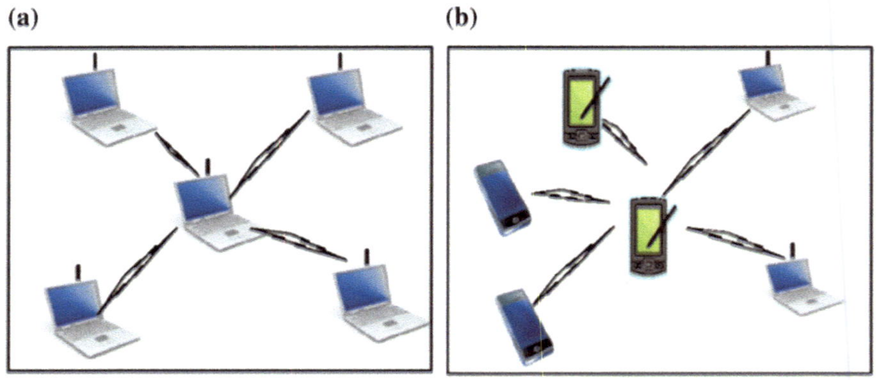

Fig. 2 **a** Homogeneous network and **b** heterogeneous network

or experiences some technical glitches in them; intermediate MN serving as router becomes unavailable due to the above reasons. Even if it does not happen, the fast movements of MN might make a MN to move out of the coverage area, thereby, snapping a connection. Similarly, a link failure can occur when two MN fall out of their vicinity or the fast movements of MNs, noise, rapidly changing network configurations and/or congestion of wireless links. Similarly, connectivity might not establish among MN due to hindrances created by some obstacle, viz., buildings, hills, trees, etc. The other reasons could be signal fading, and excessive interference, etc.

The MN and link failures are shown in Fig. 3a, b, respectively. In Fig. 3a, u_1 is the source MN, u_2 is the terminal MN and the remaining MN are intermediate ones. The single-hop/multi-hop communication between the source and terminal MN does not exist since u_2 lies outside the transmission range of all other MN. There may be a situation where u_1 or u_2 may be completely isolated from all other nodes of the network, thus, losing out the transmission or reception of signals. Besides, when no direct communication is possible, and source and terminal MN are to communicate through the intermediate node(s) (say, MNu_3).

In summary, all types of failure can lead to path or connectivity failure, thus, affecting the MANET reliability adversely.

1.4 *Attributes and Requirements*

The unique attributes and requirements of MANET are presented in great detail in the MANET literature [7–11]. Here, we briefly provide some attributes and challenges of MANET from their operation, communication, and design perspectives. They are

- *Bandwidth Management*: A process of quantifying and streaming the information flow on a network link that guarantees quality communication to its maximum capacity. Due to improper bandwidth management and network congestion, the network performance degrades. The effect of high bit error rates leading to fluctuating link capacity might be more profound in multi-hop networks.
- *Heterogeneity*: A network formed by connecting devices with different capabilities (like mobile phones, mobile electronic devices) must be able to exchange messages with one another in a consistent and unified way. This way of

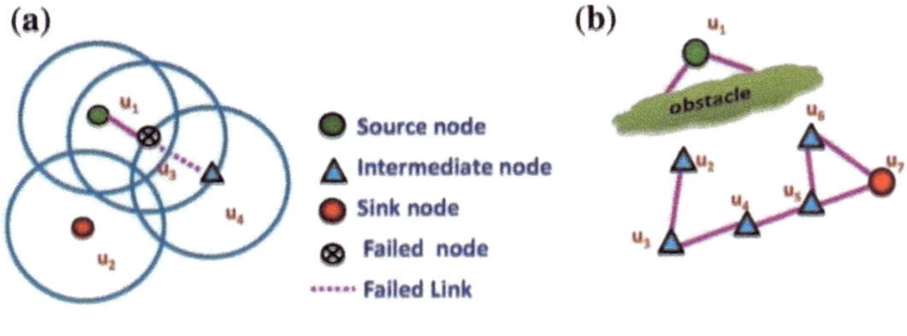

Fig. 3 **a** Node failure, **b** link failure

connecting different devices with some dedicated potential roles is heterogeneity. Device specifications like memory, storage space, and processor speed vary significantly among them which makes it difficult to implement a uniform programming interface on all nodes.

- *Infrastructureless*: A group of wireless devices communicating with each other using wireless media or radio frequency connectivity without the aid of any centralized controller.
- *Interoperability*: It is a fundamental property that ensures sharing of authorized information (audio/video/data) signals wirelessly in real time at any instant with the available resources. Interoperability means the flawless connectivity of mobile appliances all the way through the transmission and reception process with real-time flows of information and analysis.
- *Mobility Management*: The nodes in the network are rapidly repositioned and/or move in a deployed area. MN mobility results in changing network topologies. The aim of mobility management is to change attitudes and the travel behavior with the ultimate goal to create a new mobility culture in effectively delivering wireless services to the moving mobile users.
- *Multi-hop Routing*: It is a type of multi-hop network communication where the routing between nodes is relayed with the help of intermediate nodes. Reasons for multi-hop routing may be due to the absence of direct path between the designated MN pairs, obstacles in the route, and/or energy consumption.
- *Power Management*: Resource limited devices such as laptops, PDAs, etc., are battery-driven devices that have very limited power supply. The limited power supply will affect the CPU processing time, memory usage, signal processing, and communication related functions like connectivity which in turn limits the services supported applications by each node. Power management helps in reducing the overall energy consumption, prolonging battery life, reducing noise, reducing operating costs for energy and cooling.
- *Reliability*: In general, the wireless medium is significantly less reliable than the wired medium since the channel is unprotected from signals from other sources. In fact lack of fixed infrastructure, flexibility, and the mobility characteristics of nodes make it difficult to achieve a very high reliable performance.
- *Scalability*: It is true that performance of a network system suffers with increasing network size; therefore, sustaining performance with the increasing complexity is a real challenge. However a dynamic network; the absence of infrastructure; limited radio range affects scalability.
- *Security*: This is one of the primary difficulties confronted by engineers of these systems, namely ad hoc networks, today on account of communication over radio channel, noisy environment, decentralization, hubs reassociation, and constrained assets.
- *Self-organization*: The MANET is deployed and managed independently of a preexisting infrastructure. Individual MN in the network are responsible for dynamically discovering other MN they can communicate with. These networks determine their own configuration and their ability to create a dynamic configuration is self-creating network.

Designing of an ad hoc network, operating in harsh terrain to exchange messages more efficiently and reliably, has become a major challenge. Besides, the advances in wireless communications technology are required to overcome the inherent limitations of broadcast radio networks.

1.5 MANET Scenario Metrics

A list of MANET scenario metrics for, namely, node movement/dynamic topology based on the average velocity of the nodes, network size, terrain, transmission range, node density, pause time, traffic patterns, and number of links are highlighted in [12]. These metrics describe the network environment and define its scenario.

MN velocity defines the speed with which MN traverse within a specified geographical region. *Network size* defines the number of MN participating in a specified geographical region. This parameter has considerable effect on the network connectivity, thereby, on reliability. *Network coverage area/service area* (also known as Terrain) is the boundary within which the nodes move in and/or out. It affects the arrangement of the nodes and the hop count between the (*source*, *terminal*) pair. The *transmission range* is the range, with respect to the transmitting/receiving MN within which a communication can be successfully established. When MN wants to communicate with another MN, outside its vicinity, a multi-hop routing with the involvement of some intermediate MN is the solution. Besides, a survey and examination of MANET mobility models proposed in the recent research literature can be seen in [13, 14].

2 Ad Hoc Network Research: A Brief Survey

2.1 MANET Modeling

The *Graph theory* is a natural framework for the mathematical representation of several complex engineering systems and one can find a plethora of applications in a wide range of arena to solve problems in engineering, physical and social, biological sciences, and natural sciences [15, 16]. According to this theory, any system can be represented as a graph composed of nodes (or vertices) and links (or edges). A network can be directed (undirected) as well as sparse (complete). The network components are prone to failures, attacks, and errors. The graph is known as probabilistic graph if these entities (nodes/edges) have certain probability associated with respect to their operational success/failure.

Traditionally, any infrastructure-based engineering system (communication systems, water distribution systems, transportation systems, homeland security systems, environment monitoring systems, mission critical military systems, etc., have been modeled as probabilistic graph taking connectivity as the major criterion for defining network reliability [16]. However, the especial characteristics of MANET preclude the direct application of tools and techniques employed for modeling and reliability analysis of infrastructure-based systems. For instance, *Erdős and Rényi Graph* (ERG) Model selects pairs of nodes from a set of nodes of the graph with equal probability and connects them with a predefined probability thereafter [17]. Although, ERG have been found suitable for modeling several complex systems, yet, found to be a poor choice for modeling real-world systems like wireless sensor networks and ad hoc networks [18–20].

2.2 Geometric Random Graph Model

The fundamental assumption in random graph is that the link existence (non-existence) between a node-pair is entirely independent of the presence (absence) of any other link in the network. This may not be true in the case of MANET as the active nodes within the coverage area of each others are certainly connected. In other words, the existence of links is a function of the geometric distance between the nodes and the transmission range of the nodes. The *Geometric Random Graph* (*GRG*) also termed as Waxman Model (spatial graphs) is an improved version of ERG model. A random geometric graph can be obtained by distributing n points (nodes) uniformly at unit disk and connecting the node-pairs "iff" their geometric distance (Euclidean, Manhattan) is nearly the radius of the nodes, where the distance depends on the chosen metric. This graph can be a viable solution for modeling real-time networks like MANET and sensor networks [21]. Therefore, modeling networks using GRG is a more realistic alternative to the classical random graph models of *Erdős–Rényi* [20, 21].

2.3 Evolving Graph Model

Evolving graph model (EGM) models [22] approach appear to be another preferable approach to model some dynamic systems such as social networks, communication networks like MANET, transportation networks and many more. In general, any dynamic system/network, the topology structure is highly complex when MN adds into the network over time or sometimes less complex when MN

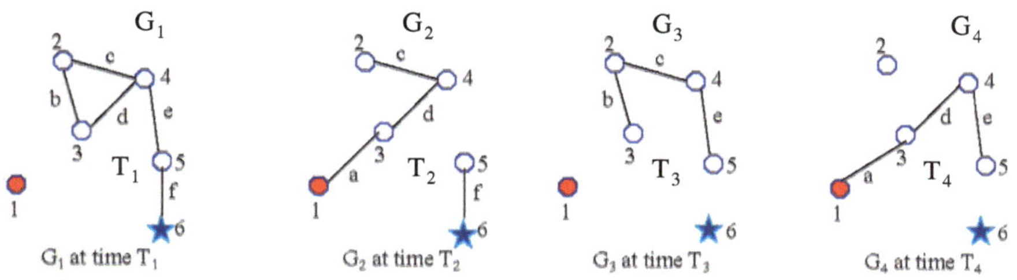

Fig. 4 The evolution of 6-node MANET over 4 time-slots

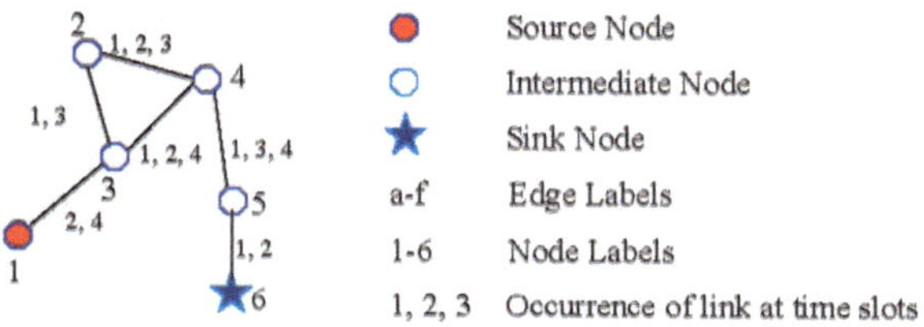

Fig. 5 Evolving graph corresponding to Fig. 4

leaves the network. When a MN leaves the network, existing links break and similarly when a MN adds, links are created. This frequent change in configurations or topologies with time is very well captured using EGM.

An evolving graph *G*(*V*, *E*) models time-dependent networks as a sequence of static graphs, $G_t(V_t, E_t)$ such that their topology change with time, say, $S_G = G_1, G_2, G_3, \ldots, G_T; T \in n$. Then the evolving graph is represented as $G = (S_G, G)$. A node n_i is active at time slot t_i, if this node n_i is connected to at least another node n_j at the same time slot. Moreover, the path between node n_i at time slot t_1 and node n_j at time slot t_n is sequence of nodes $\{(n_i, t_1), (n_{i+1}, t_2), \ldots, (n_j, t_n)\} \subseteq t_1 \leq t_2, \leq \cdots \leq t_n$ and $((n_n, t_k), (n_{n+1}, t_{k+1})) \in E_{t_k}$ if $t_k = t_{k+1}$ otherwise, $n_n = n_{n+1}$; is a temporal path. Thus, if a temporal path exists between the nodes (n_i, t_m) and node (n_j, t_n), it may be concluded that both these nodes are temporally connected.

At a given time instant (index point), each network topology is represented as a subgraph. From Fig. 4, it can be seen that connectivity "*b*" is present at time instants T_1 and T_3, respectively (see subgraphs G_1 and G_3). These graphs clearly depict the route between the designated nodes (source and sink). Multi-paths (say 1–3–2–4–5–6; 1–3–4–5–6) are prevalent (see Fig. 5) and finally route selection depends on the minimum number of hops taken for information exchange between the designated nodes.

Several researchers have demonstrated that EGM based system modeling permits to tackle the issues like connectedness, congestion control, and routing in impromptu systems [23, 24]. It seems that the EGM could be a reasonable and alternate way to focus on the reliability issue with some originality.

2.4 Node and Link Failures Models in MANET

2.4.1 Node Failure in MANET

Many researchers have concentrated on studying the impact of node failures on issues like network partitioning, packet delays, congestion, and routing.

Dimitar et al. [25] emphasized that the node failures alters the topology of MANET, thus, brings in severe changes in the network performance. Ye et al. [26] showed that the probability of successful reliable path between a random node-pair using a routing protocol increases with small number of reliable nodes. According to Shastri et al. [27] signal attenuation, high bit error rate are certain factor that can cause performance degradations. These performance degradations have significant impact on the node failure, which is more prevalent in wireless ad hoc environments due to run-out battery time. Tekiner and Ashish [28] also showed that the consequences of congestion, low energy levels, and unpredictable nature of MANET are prevalent issues for failure of node leading to connectivity loss. Shivashankar et al. [29] developed a compensation algorithm using Network Simulator model to prevent network partitioning and due to limited bandwidth, limited residual battery power and poor throughput the MNs failed. Zadin and Thomas [30] focused on the effect of node failures on connectivity and its stability using node protection algorithm to improve the use of MANET application. Fei et al. [31] proposed a classification method of node failures to show that node failures have great impact on the network connectivity. It has also been shown that the impact of node failures shall be small as the network size increases. Merkel et al. [32] introduced an algorithm that mitigated the effect of mobility in terms of hop count based on distance estimation.

A node reliability model was proposed by Zhao et al. [33] to show that node mobility has more impact on terminal reliability and link reliability, and necessary measures should be taken to balance the issues related with both deployment and protocol design of MANET. Migov and Vladimir [34] used an undirected probabilistic graph which considers imperfect nodes (because of scuffing or intrusions) with perfect links to model to show its effects on the reliability of ad hoc network. However, in reliability engineering, treating time-to-failure of a node as a random variable, the node failure models can be modeled by any suitable statistical distribution.

2.4.2 Link Model in MANET

In support of the growing advancements, a good modeling of the radio channel has always been a most important issue in the design of any wireless networks, and MANET is not an exception. The Link reliability models have been used as a universal measure to examine the MANETs performance [25]. Generally, radio wave propagation models have been used to determine the mean path loss and suitable fading margins wherein the characteristics of a radio propagation model for a given environment can be represented by a mathematical expression, diagrams, and/or algorithms. Usually, statistical methods or site-specific propagation models have been used in modeling the radio channels to represent the realistic scenarios [35, 36].

Several researchers [25, 37–43] have contributed to develop efficient working models on link reliability, which help in achieving a reliable communication between the specified set of nodes. Dong and Banerjee [38] proposed an energy efficient link reliability model that minimizes the energy requirement for total transmission through either hop-to-hop retransmission or end-to-end retransmission. A model developed by Khandani et al. [39] considered a transmitter (source) node that has the lowest power. This power is calculated as a function of distance (d_{ij}) between the source and the destination node. In this model, the receiver node receives message with certainty provided that the information is transferred by a low power transmitter. Moreover, these authors have also developed another model as a function of MN range, distance (d_{ij}) channel fade state for perfect reception of information. Wei and Guo [41] proposed a routing protocol using a statistical-based conditional probability model to determine a reliable route. The theoretical approach [42] was borrowed and implemented [43] to determine the optimum link of multi-hop wireless broadcast networks.

Besides, most of the proposed algorithms /technique applied on ad hoc networks operate in path loss environment. These techniques have been adopted for connectivity analysis, improving the link reliability, protocol, and routing studies [44, 45].

3 Network Reliability

Network systems are ubiquitous in communication networks, power line networks, computer networks, video conferencing, grid computing systems, and traffic networks necessitating the network reliability as an important criterion. The increasing complexity and scale of systems imply that reliability issues will not only continue to be a challenge but also require more accurate models, metrics, and efficient techniques for viable solutions [46].

Qualitatively yet another metric coined as *all-operational* terminal reliability [48] is also being used and is defined as the possibility that all the active nodes of a network must communicate with every other node. This measure is very useful as it deals with the network whose nodes become disconnected as a result of failure of the relay (intermediate) nodes. The network reliability could be defined as the ability of the network to provide network services in an uninterrupted manner. The purpose of reliability analysis is to quantify this ability with the impact of component failures, identify the weakness in the network, and suggest possible remedies to overcome such weaknesses. At the basic quantitative level, k-terminal reliability of a given network can be defined as the *probability* that there exists at least one feasible and *minimal* path between a specified set of k-out of-N nodes of the network under specified conditions of use [47]. When $k = 2$, then it is known as *2-terminal (terminal-pair /s-t) reliability*, which is the probability that a specified pair of nodes (the source, *s* and the terminal, *t*), with known success/failure probabilities, remain connected by a path of operating nodes and links. When $k = N$, then it is known as *all-terminal reliability*, which is the probability that all nodes are directly (or indirectly) connected.

These above-stated measures have been in vogue for infrastructure-based network systems, and have also been extended to the MANET.

3.1 Network Reliability Evaluation Approaches

In the past, the network reliability design, evaluations, and other related issues for infrastructure-based networks have been dealt with either using the analytical technique and/or the simulation method. These algorithms have been extensively based on graph theory, probability laws, and Boolean algebra such as state enumeration method, reduction and factoring (or decomposition) method, transformation methods, direct methods, approximation method, and in recent times by employing Binary Decision Diagrams (BDD) [47]. The most common assumptions in these techniques have been the perfectness or imperfectness of links and/or nodes and statistical independence of failures.

Although, these common assumptions can also be extended to MANET, yet, the peculiar characteristics and attributes of random and dynamic topologies; fast node movements; group formation, etc., preclude the direct application of existing analytical methods of infrastructure-based network due to analytical complexity, and computational cost to develop a close form type of solution.

An Enhanced Ordered Binary Decision Diagram (EOBDD) algorithm is also proposed to efficiently evaluate the reliability of the wireless networks based on the considerations of common cause failure and large network size in [49]. Mo et al. [50] proposed an efficient truncated BDD-based reliability evaluation methods to calculate the reliability of large-scale distributed systems (peer-to-peer network/ad hoc network and WSN). A BDD method has also been proposed to evaluate the reliability for wireless sensor networks with different network configuration characteristics that includes connectivity, average hop count, average node degree, and clustering coefficient [51]. Reference [52] illustrated a novel approach of integrating coverage area and connectivity for reliability analysis of WSN using computationally efficient ROBDD approach. These works have shown that by using BDD for reliability studies is still a promising area of research.

An approach proposed in [53] uses fuzzy logic concepts in enhancing the performance and evaluating the reliability of ad hoc network. A fuzzy logic technique for gossip-based reliable broadcasting in MANET was proposed in [54]. It demonstrates that using gossiping approach, the import design issues like unreliable links, multi-hop connectivity, dynamic topology, and adaptability can be tackled. However, fuzzy logic concept has mostly been used in determining route reliability, stability, route selection and hop count, and reliable multicasting [55–58]. Besides, simulators like NS-2, NS-3, OMNet++, SSFNet, NCTUns, J-Sim, ShoX, QualNet, OPNET, Shawn, DIANEmu, GloMosim, GTNets, Jane, NAB; have been in use to study the MN moving patterns, unstable topology, and connectivity in a MANET [59].

A scalable approach to model and simulate the reliability of the infrastructure-oriented wireless networks has been proposed in [60]. Their results depict the impact of number of subscribers, network size on the expected number of failures for different component/link reliability conditions. However, use of either only a simulation approach or combination of analytical–simulation approaches [61–64] can be applied to evaluate reliability. MCS approach has been used to characterize MANET for approximating 2TR [65–67].

4 Reliability: Wireless Mobile Ad Hoc Networks

For the purpose of designing a reliable network, most of the researchers have considered the link existence purely binary with no limitation on the flow of the size of information carried by the links, i.e., the information carrying capacity of the links are assumed to be quite large. Besides, under certain practical situations, the fact that the nodes are perfectly reliable may not always be true leading to inaccurate estimation of reliability metrics and decision about the network capability thereof. Therefore, the network reliability of ad hoc network is measured by considering both nodes and links are unreliable and other properties like changing topology, limited transmission, node movements, limited life, etc.

Recently, this area has drawn the attention of several researchers. A symbolic reliability expression was derived using an algorithm which can handle imperfect nodes and dynamic network connectivity [68]. The $2TR_m$ of the grid structured ad hoc network is computed and their work mainly focuses on the component-wise reliability of fixed networks. A combination of random graph theory, percolation theory, and linear algebra for analyzing the mobile sensor multi-hop networks was used in [69]. They applied a probabilistic version of adjacency matrix to analyze the connectivity of such networks without addressing the concern of mobility and movements of MN.

In fact, the authors of [70] have considered the existence of the link as a probabilistic event with respect to the nodes status and provided a basic framework to deal with MANET reliability. Its further extension by considering the mobility to determine the $2TR_m$ by employing MCS can be found in [65, 66]. As algorithm in [64] simulates the status of (s, t) pair as well, the reliability estimate turns out to be a conservative estimate. The example scenario is restated here for the sake of brevity.

Scenario: *A network composed of 18 dismounted infantry (soldiers on foot) equipped with identical man-portable radios. Each radio is capable of* *communicating with other radios and is required to operate for duration of 72 h. Each MN has a transmission range of 3 miles; each with a reliability that is described by Weibull distribution with parameters θ = 1000 h and β = 1.5. The soldiers move randomly in a square coverage area of 64 square miles with a maximum and minimum velocity of 6 and 3 miles per hour respectively, in a random fashion.*

Ref. [66] improves the approach of [65] and has also provided effects due to various parameters viz., varying network size (NS), varying transmission range (TR) and varying network coverage area (NCA), on 2-terminal reliability. The approach proposed in [66] also helps in taking the decision on deployment of MANET with respect to the network size, transmission range of MN, etc. For instance, with respect to a hypothetical scenario taken in [65, 66], the summary of the outcome is

- For a given coverage area, one can determine the *optimal transmission range of MN* where further increase in transmission range of MN would have no significant effect on reliability as shown in Fig. 6a. The results also confirm the fact that with increase in the MN transmission range increases the reliability as more nodes get connected with each other (Fig. 6b).
- Suitability of a fixed number of available MN to cover a certain area of operation. For a fixed number of available MN (18 in number) with their average transmission range (3 miles), the change in $2TR_m$ with the varying coverage area is shown in Fig. 7, indicates a sharp decline in reliability and the unsuitability of transmitters for a simulation boundary of even 64 miles^2.
- The variation of $2TR_m$ with network size is shown in Fig. 8 to decide upon the number of MN requirements for a targeted reliability to cover a 64 square mile in a critical operation clearly indicating that to achieve a reliability of more than 0.9, about 50 MN are required. The significance is that as the network size increases, the link consistency increases leading to the availability of large number of paths and hence high reliability.

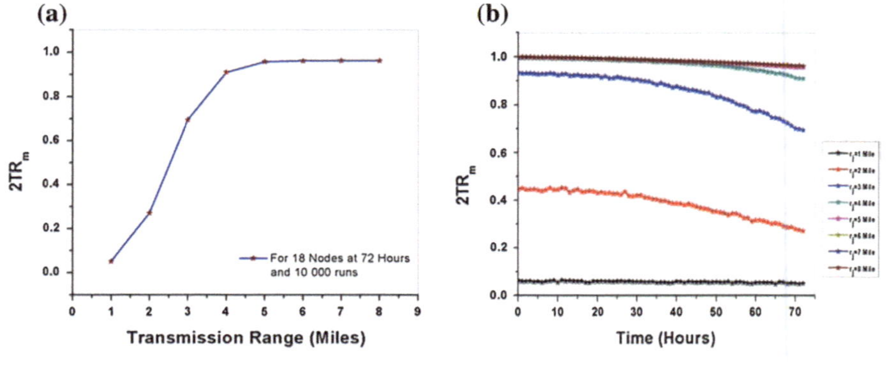

Fig. 6 Effect of transmission range on **a** $2TR_m$ **b** $2TR_m$ with a mission duration

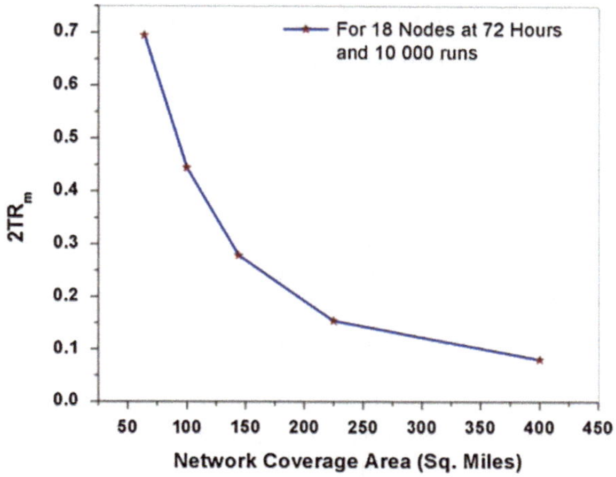

Fig. 7 Effect of NCA on $2TR_m$

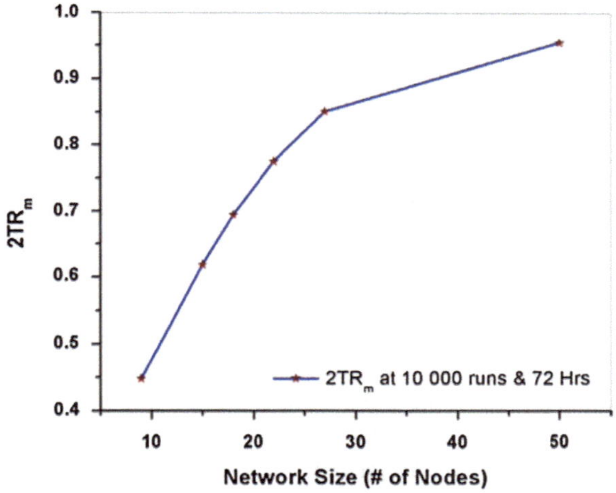

Fig. 8 Effect of network size on $2TR_m$

Summarily, if the reliability of MN is poor, then even by deploying a large number of nodes with long transmission range or less coverage area, the target reliability of MANET would be constrained by reliabilities of designated communicating nodes. Similar outcomes have been resulted once the proposed algorithm is applied to study the effect on *all-terminal reliability* (ATR_m) and *all operational* where results have shown that $2TR_m > AoTR_m > ATR_m$.. Besides, the pattern obtained for $AoTR_m$ is found to be almost the same as that attained for ATR_m. For more details, readers can refer [71]. However, both the approaches in [65, 66] did not give any attention to signal degradation or loss.

5 Reliability: Wireless Ad Hoc Network Under Different Environment Scenarios

It is well known in communication engineering that the communication signal strength is greatly affected by the signal frequency, incidence angle, and nature of the medium. Therefore, it is worthwhile to briefly discuss communication exchange via radio waves /electromagnetic /acoustic waves over space using the media.

The wireless radio channel is the most challenging issue related with achieving reliable communication as the channel is susceptible to noise, interference, channel impediments (attenuation, distortion, etc.) and mobility of users [72]. Practically, the radio wave propagation through the environment gets reflected, scattered, or diffracted because of obstacles and other interfering objects. Further, the strength of the signal decreases as the distance between the transceiver increases. The variation in signal strength with the distance between the mobile nodes may be small within line-of-sight (LOS) or severe in non-line-of sight (NLOS) regions, i.e., obstructed by buildings, mountains, and foliage.

There are propagation models that focus on the average signal strength at a given distance from the transmitter (path loss models) or on the randomness in the signal strength at close proximity to a particular location (fading models) and are classified based on the distance of separation between the mobile users as large-scale propagation models and small-scale or fading models [73]. The large-scale propagation models are characterized as models that can predict the average signal strength over large T-R separation distances (say several 100 s to 1000 s of meters) while the small-scale propagation models are characterized over very short distances (a few wavelength).

A successful communication in wireless media is related with the perfect reception of the transmitted signal by the receiver node at the destination. But the transmitted radio signals attenuate as they traverse through space and is called as *terrestrial losses*. These losses occur because the radio wave is affected by the topography of the terrain, man-made obstacles (buildings), and natural obstacles (trees, hills, and mountains) [74]. The presence of these obstacles along the path of the radio signal introduces many random variations like scatter, reflection, diffraction. Figures 9 and 10 depicts the propagation mechanisms for both indoor and outdoor environment applications.

Owing to the *reflection*, *diffraction,* and *scattering* of the radio waves the transmitted wave more often reach the receiver by more than one path, resulting in a phenomenon known as multipath fading [36]. *Reflections* occur when the radio signals impinge on obstructions. Reflections are commonly caused by the surface of the earth and from buildings, walls, and other such obstructions, and are dominant in indoor applications. *Diffraction* is a mechanism where the signals are incident along the edges of the buildings, walls, or other larger objects and is most prominent when the receiver is in non-line-of-sight (NLOS) of the transmitter

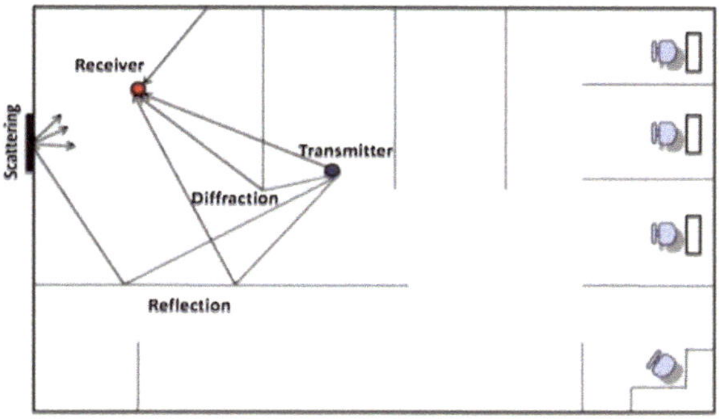

Fig. 9 Radio propagation mechanisms in indoor environment

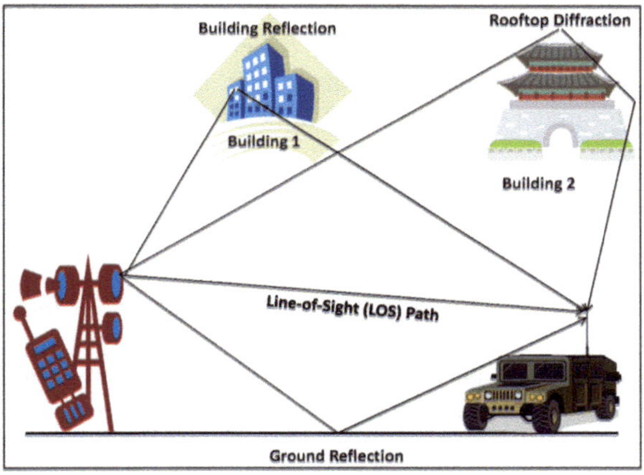

Fig. 10 Radio propagation mechanisms in outdoor environment

(outdoor environments). This phenomena cause to arise the secondary wave from the obstructing surface. In mobile radio communication, the signal scattering is mainly due to foliage. *Scattering* is a mechanism wherein the signal spread out in all directions in form of spherical waves because of obstacles with rough surfaces, furniture (indoor) and vehicles, foliage (outdoor) along its path. This type of propagation in all directions reduces the signal power levels and is more prominent when the nodes are located in a highly mixed and messy environment [75].

The services of wireless network spans through a variety of environments and hence a single unique model may not be suitable to describe the radio propagation among the MN. The subsequent paragraphs provide a description of the models that have been considered by the reliability research fraternity.

5.1 Path Loss Model

As mentioned earlier, some researchers recently have made an attempt to evaluate the reliability of MANET considering imperfect vertices and Euclidian distance based binary links reliability model [65, 66]. However, in practicality, the likelihood of reliable communication can deteriorate even within the MN transmission range. Obviously, the hypothesis of original reception of the transmitted information within the defined transmission range thereby leading to wrong estimate, conclusions and managerial decisions thereof with an under designed network.

Path loss is an essential parameter to be considered for attaining reliable communication and is influenced by prominent factors like operation scenario (environment and distance between the users), terrain buildings and vegetation. Hence, to attain an acceptable reliability level, it is therefore necessary to model the link existence as per realism in a closer manner. In a path loss model, the received signal is a decreasing function of the distance (*Euclidean/Manhattan*) between the transmitter–receiver pair. Further, the path loss can be represented by power law as (1) [75].

$$P_r = P_0\left({}^{d_0}/_{d_{ij}}\right)^{\eta} \tag{1}$$

where, P_r is the actual received power, P_0 is the received power at a reference distance d_0 (the reference distance d_0 is typically chosen to be 1 m in indoor environment and 100 m or 1 km in outdoor environment [73]. The η is the path loss exponent that varies between 2 (FS) to 6 (heavily built urban environment where a higher value of η indicates a faster decay of radio signals, i.e., severe randomness in received power of the radio signal. The path loss exponent for different environments is provided in Table 1.

The most commonly used radio propagation models [viz., Free Space (FS) propagation model and the Two Ray Ground (TRG)] based on the path loss phenomenon are briefly described in the ensuing section.

Table 1 Path loss exponents for different environments [73]

Building type	Path loss exponent, η
Line-of-sight in buildings	1.6–1.8
Grocery store	1.8
Vacuum, infinite/free space	2.0
Retail store	2.2
Office with soft/hard partition	2.6–3
Urban areas cellular radio	2.7–3.5
Shadowed urban cellular radio	3.0–5.0
Obstructed in factories	2.0–3.0
Obstructed in buildings	4.0–6.0

5.2 FS Propagation Model

In a FS propagation model, the radio wave can propagate without being affected by the obstacles and is also referred line-of-sight (LOS) channel propagation. This model predicts that the radio signal strength or the received power decays as a function of some power of the distance as in Eq. (3).

Depending on the signal radio frequency (*f*), its additional system losses (*L*), and transmission power (P_0), the FS received power ($P_{r,\ \mathrm{FS}}$) is defined as (by *Friis* formula) (2).

$$P_{r,\mathrm{FS}}\left(d_{ij}\right) = \frac{P_t G_t G_r \lambda^2}{L\left(4\pi\, d_{ij}\right)^2} \tag{2}$$

where, G_t and G_r are the transmitter and receiver antenna gains, respectively, and *L* is the system loss. It is generally common to select $G_t = G_r = 1$ and $L = 1$; $\lambda = c/f$, is the wavelength of the carrier; *c* is the speed of light (3 x 10^8 m/s). The Euclidian distance, d_{ij} (τ), between a pair of nodes (u_i, u_j) at time "τ," given by (3);

$$d_{ij}(\tau) = \left(\left(x_j(\tau) - x_i(\tau)\right)^2 + \left(y_j(\tau) - y_i(\tau)\right)^2\right)^{1/2} \tag{3}$$

It is clear that (2) does not hold for $d_{ij} = 0$, hence, many models use the representation of close-in or reference distance d_0. Let, P_0 be received signal power at a reference distance d_0 of 1 m as (4)

$$P_0 = \frac{P_t G_t G_r}{L}(\lambda/4\pi)^2 \tag{4}$$

and Eq. (2) can be rewritten as (5)

$$P_{r,\mathrm{FS}}\left(d_{ij}\right) = P_0 \big/ d_{ij}^2 \tag{5}$$

The FS model is best suitable for transmitter–receiver pair operating in LOS or in other words the propagation in this model is through a single direct path between the mobile users with no physical obstacles between them [72].

5.3 TRG Propagation Model

The relationship observed for FS propagation in Eq. (4) does not hold good for all types of environments as the signal not only travels using a direct but can also reach to the receiver through several other paths. The TRG model is an improvement over FS model where the signal propagates through direct path at shorter distances and takes multipath for long distance coverage. This implies that the TRG model comprises of one direct path and one reflected path between (u_i, u_j). The reflected path reflects off the earth surface, and is a good estimate for propagation along highways, rural roads, and over water [72]. The received signal power of the TRG model can be expressed as (6).

$$P_{r,\mathrm{TRG}}\left(d_{ij}\right) = \frac{P_t G_t G_r h_t^2 h_r^2}{L d_{ij}^4} \tag{6}$$

where, the h_t and h_r denote the antenna heights above the ground and is, generally, considered to be constant. G_t is the antenna gain of the transmitter and G_r is the receiver antenna gain; $G_t = G_r = 1$. L is the system loss, not related to propagation, and generally $L = 1$. Note that this model combines both direct and a multipath and is independent of the carrier frequency. Further, we know that the received signal strength (P_0) at the first unit distance is defined using (7)

$$P_0 = \frac{P_t G_t G_r \lambda^2}{(4\pi)^2} \tag{7}$$

where, $\lambda = c/f$, is the wavelength of the carrier, c is the speed of light (3×10^8 m/s) and f is the frequency in Hertz. (7) is modified as (8)

$$P_t G_t G_r = \frac{P_0 (4\pi)^2}{\lambda^2} \tag{8}$$

By substituting (8) in (6), we get (9)

$$P_{r,\mathrm{TRG}} = P_0 \left(\frac{4\pi}{\lambda}\right)^2 \frac{h_t^2 h_r^2}{d_{ij}^4} \tag{9}$$

It is also known that

$$\Delta d = \frac{\lambda}{2} = \frac{2\pi h_t h_r}{d_c}$$
$$d_c = \frac{4\pi h_t h_r}{\lambda}$$

where, d_c is the critical distance (threshold distance/crossover distance). Now,

$$d_c^2 = \left(\frac{4\pi h_t h_r}{\lambda}\right)^2 \tag{10}$$

Substituting (10) in (9), (6) can be reduced to (11) as

$$P_{r,\mathrm{TRG}} = P_0 \frac{d_c^2}{d_{ij}^4} \tag{11}$$

It may be noted that when $d_{ij} < d_c$, (5) can be used, i.e., signals propagate using FS model and when $d_{ij} < d_c$, (11) is used, i.e., signal propagation is through TRG model. The performance would be the same when $d_{ij} = d_c$.

5.4 *Model I: FS–TRG Propagation Model—A Hybrid Model*

The models discussed in Sect. 5.2 and Sect. 5.3 visualize the power of the receiver in terms of distance. Referring back in Eqs. (7) and (15), respectively, it can be noted that P_0 is inversely proportional to d_{ij}^2 for free space and d_{ij}^4 for two ray ground model, respectively. Note that in a binary link reliability model, when the $d_{ij} \leq r_j$, then the link reliability is assumed as unity in theoretical sense [65, 66]. On the other hand, the signal strength weakens even up to the MN transmission range due to channel noise, fading effect or interference, thereby, decreasing the reliability. Then, the link reliability can be modeled as in Eq. (12)

$$R_l(d) = \begin{cases} R_l(d) = 1 & d \leq \delta r \\ R_l(d) = R_{FS}(d) = \frac{A}{d^2} + B & \delta r \leq d \leq \gamma r \\ R_l(d) = R_{TRG}(d) = \frac{C}{d^4} + E & \gamma r \leq d \leq r \\ R_l(d) = 0 & r \leq d \end{cases} \tag{12}$$

where, $0 \leq R_l\ (d) \leq 1$; $0 < \delta \leq 1$; $\delta \leq \gamma \leq 1$ and A, B, C, and E are all constants. δ represents the threshold distance of FS model; γ is the distance greater than δ for TRG model, d represents the Euclidean distance and r represents the transmission range. Using this model, the link reliabilities can be derived as (13) [76]

$$R_l(d) = \begin{cases} 1 & d \leq \delta r \\ \frac{\delta^2}{(1-\delta^2)}\left(\frac{r^2}{d^2} - 1\right) & \delta r \leq d \leq \gamma r \\ \frac{\gamma^2\delta^2}{(1-\delta^2)(1+\gamma^2)}\left(\frac{r^4}{d^4} - 1\right) & \gamma r \leq d \leq r \\ 0 & r \leq d \end{cases} \tag{13}$$

The link reliability, represented as a function of distance (d_{ij}) and transmission range (r_j) can be expressed as (14)

$$R_l\big(d_{ij}(\tau)\big) = \begin{cases} 1 & d_{ij}(\tau) \le \delta r_j \\ \frac{\delta^2}{(1-\delta^2)}\left(\frac{r_j^2}{d_{ij}(\tau)^2} - 1\right) & \delta r_j \le d_{ij}(\tau) \le \gamma r_j \\ \frac{\gamma^2\delta^2}{(1-\delta^2)(1+\gamma^2)}\left(\frac{r_j^4}{d_{ij}(\tau)^4} - 1\right) & \gamma r_j \le d_{ij}(\tau) \le r_j \\ 0 & r_j \le d_{ij}(\tau) \end{cases} \tag{14}$$

Figure 11 shows the link reliability model based on FS–TRG propagation wherein FS propagation can be used when the distance between the nodes lie within δr_j and γr_j, and for distances beyond γr_j, the TRG propagation model is utilized [76].

Figure 12a–c depict the changing trend in MANET reliability on varying the transmission range, network size, and coverage area, respectively, for a fixed

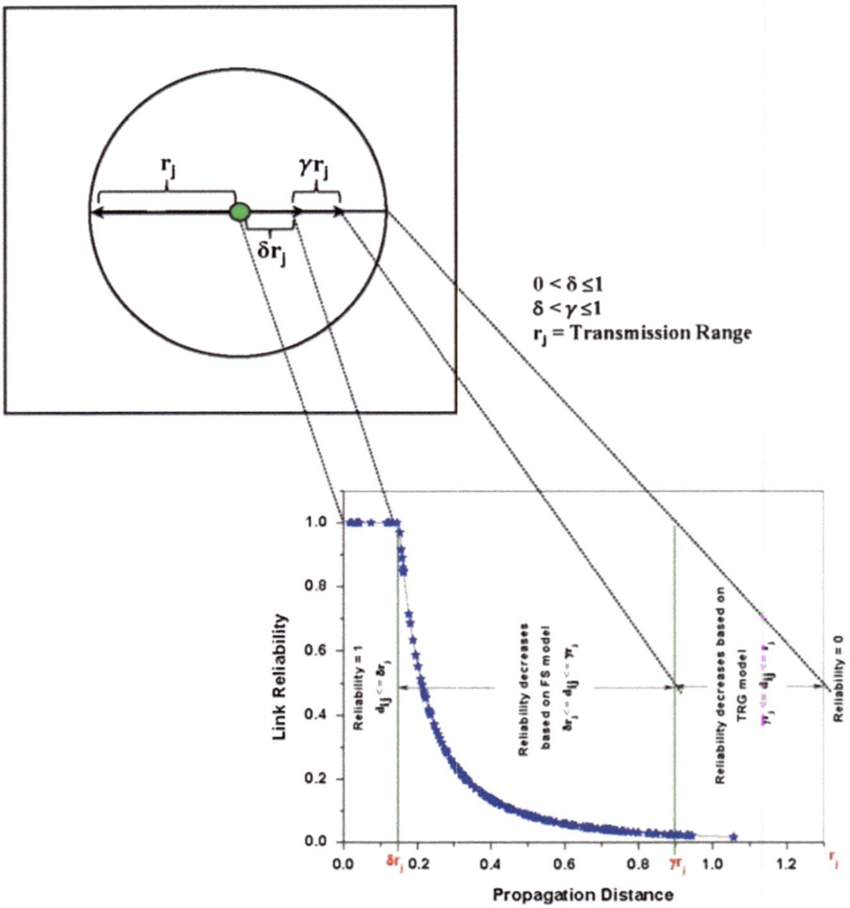

Fig. 11 FS–TRG propagation based link reliability model [73]

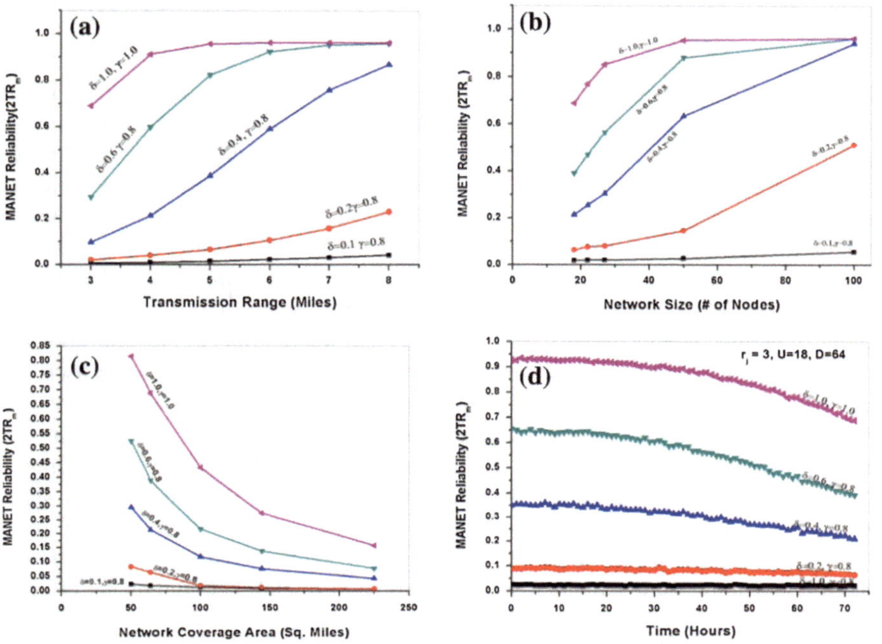

Fig. 12 Effect of varying (δ,γ) on MANET reliability $(2TR_m)$ with **a** transmission range **b** network size **c** coverage area and **d** operation time

mission duration of 72 h on the considered scenario in [65] and as stated in Sect. 4. Similarly, the simulation result in Fig. 12d portrays the effect on the MANET reliability with varying δ and γ values with respect to change in mission duration. When $(\delta, \gamma) = 1$, the reliability attained with the proposed model is the same as that of free space model. These values clearly indicate that the obtained MANET reliability is very much affected by environmental surroundings and hence the network designed solely based on the FS propagation model would yield a much inadequate and more prone-to-failure network design. For instance, radio with 3 miles radio range is an inappropriate choice, and moreover deploying the network in a pure FS environment for the given scenario provides an overemphasized MANET reliability leading to unexpected and undesired results in MN deployment.

In other words, such type of analysis and figures can encourage the designers to opt for a network with a suitable nodes transmission range, coverage area and network size to achieve the desired reliability based on the propagation model. Other reliability metrics such as $AoTR_m$ and ATR_m have also been assessed using this approach.

5.5 Model 2: Log-Normal Shadowing Propagation Model

In most of the practical channels, signal propagation takes place both in the atmosphere and near the ground. A signal transmitted via wireless media actually experience random variation in received power at a given distance due to the presence of objects in the path of the radio waves. That is, as the signal traverse from the transmitter to the receiver various signal corruptions occurs because of interactions of several factors such as obstacles (buildings, trees), environment scenarios, fluctuations in the signal amplitude, and is commonly referred as *fading*. Large-scale fading, medium-scale fading, and small-scale fading are the different types of fading. The fast fading describes the swift fluctuations in the amplitude of a radio wave over a short duration of time that is being experienced by the user. The medium-scale fading captures the randomness in the received radio wave over larger distances. The large-scale fading represents the dependency of the expected received signal mean power to the distance between the MN. Besides, depending on the environment, surroundings, and the location of the MN, the received signal strength of the MN also varies even when it lies at a same distance from the source node. This variation in signal strength is called *shadow fading*. *Shadow fading* model appears to be a statistical wave propagation model that studies the effect of several environments with respect to carrier frequency, antenna heights, polarization, and distance. Egli [77] was first to analyze the model and also provided an empirical formula.

Therefore, by considering fading, we can say that the distance between any two nodes is no longer sufficient to determine the existence of the link as is considered in the FS model/binary link reliability model. This model was purely distance-dependent model (see Fig. 13a), which does not consider the randomness present in the radio communication.

The shadowing phenomenon is generally referred to as large-scale or slow fading model. The implication of this model is that it is uncertain to have a connectivity to be established for those nodes within the defined range of the

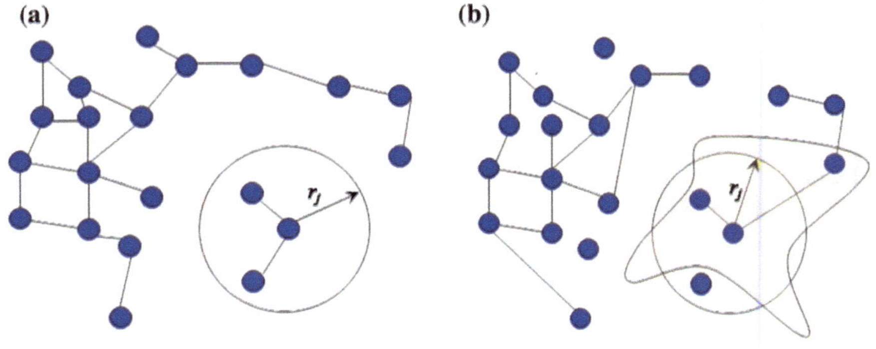

Fig. 13 **a** Purely distance dependent. **b** Shadow fading link model

transmitter whereas a node outside the defined range guarantees a communication link with the transmitter. That is, due to the random strength, possibly two events might occur, i.e., a link between two nodes can exist when $d_{ij} > r_j$; no link exist when $d_{ij} \leq r_j$ and hence the connectivity properties changes, making the connectivity analyses more complex [78]. The literature indicates wireless channels modeled based on shadow fading (see Fig. 13b) has been used in several studies, say, in connectivity analysis, protocol studies, etc. [45, 78–81].

A log-normal shadow model consists of two parts: a path loss model and a shadowing model. The path loss model predicts the mean received power as a deterministic function of distance (d_{ij}), and the shadow model reflects the variations of the received power at certain distance. The *log-normal shadowing radio propagation* model [82, 83] is represented as (15)

$$P_{r,sh} = P_0 \left(d_0 / d_{ij} \right)^{\eta} \times 10^{X/10} \tag{15}$$

Substituting Eq. (1), we get

$$P_{r,sh} = P_r 10^{X/10} \tag{16}$$

Equation (16) can be represented in decibels as (17)

$$10 \log_{10}\left(P_{r,sh}\right) = 10 \log_{10}(P_r) + X \tag{17}$$

where, X is a *Gaussian* distributed random variable (in dB) with a standard deviation σ (in dB). Table 2 provides some typical values of the shadowing deviation.

A zero standard deviation implies the absence of randomness in the received signal and hence $P_{r,sh} = P_r$, i.e., the shadow model behaves as path loss model. Reliable communication between the designated nodes is assumed to be possible when received power, $P_{r,sh}$ is greater than desired received signal threshold, γ_0, i.e., $P_{r,sh} > \gamma_0$. That is, the condition for correct reception is $P_{r,sh}/\gamma_0 > 1$ or log $(P_{r,sh}/\gamma_0) > 0$. Hence, the probability of having a link between the designated nodes at normalized distance from each other, that is, the link probability is (18) [67]:

Table 2 Shadowing deviation for different environments [73]

Environment	Shadowing deviation, σ_{dB}
Vacuum, infinite space	0
Factory, line-of-sight	3–6
Outdoor (shadowed urban cellular radio)	4–12
Grocery store	5.2
Factory, obstructed	6.8
Office, hard partition	7
Retail store	8.7
Office, soft partition	9.6

$$p(\hat{d}_{ij}) = Pr\left[10\log_{10}\left(\widehat{P}(\hat{d}_{ij})\right) > 0\right] = \frac{1}{\sqrt{2\pi}\,\sigma}\int_0^{\infty}\exp\left[-\frac{\left(t - 10\log_{10}\left(\hat{d}_{ij}^{-\eta}\right)\right)^2}{2\sigma^2}\right]dt$$
$$= \frac{1}{2}\left[1 - \operatorname{erf}\left(v\frac{\log \hat{d}_{ij}}{\xi}\right)\right], \tag{18}$$

where, $\xi \triangleq \sigma/\eta$, $v = 10/\left(\sqrt{2}\log 10\right) = 3.07, \hat{d}_{ij} = d_{ij}/r_j$.

$\hat{d}_{ij}$ is the normalized distance and ξ is the ratio between the standard deviation of power fluctuations of radio signal (σ) and the path loss exponent (η). Clearly, it is directly proportional to the signal power variations. Theoretically, the ξ can be varied between 0 and 6. When $\xi = 0$, shadowing model is equivalent to the path loss model, therefore, the log-normal shadow model is the generalization of path loss model, and can be represented as (19).

$$\lim_{\xi \to 0} p(\hat{d}_{ij}) = \begin{cases} 1, & \text{if } \hat{d}_{ij} < 1 \\ 0, & \text{if } \hat{d}_{ij} > 1 \end{cases} \tag{19}$$

Figure 14 shows the variation of the probability of the existence of a link (link reliability) with respect to the normalized distance for different values of standard deviation of shadowing model. When $\xi > 0$, there exists a nonzero probability such that nodes at distances greater than the normalized distance, ($\hat{d}_{ij} = 1$) get connected. Similarly, there exists a nonzero probability that the nodes at distances less than the

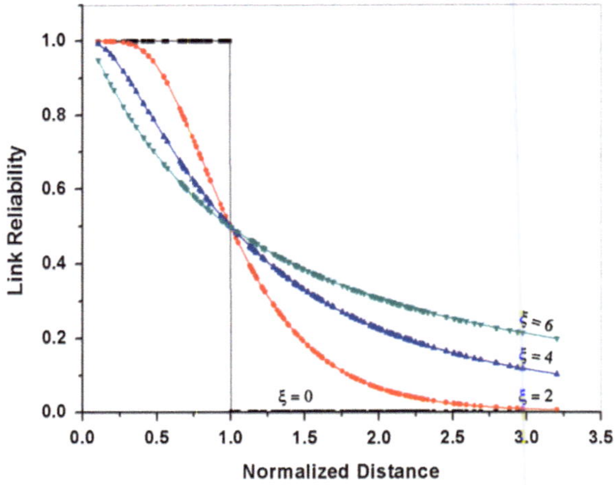

Fig. 14 Link reliability as a function of normalized distance for different values of ξ

normalized distance, ($\hat{d}_{ij} = 1$) are disconnected. It can also be seen that as ξ increases, the probability of link existence increases with distance. Further, at low value of ξ the signal fluctuations are low and hence the link probability reduces.

Therefore, the probability of link existence between a pair of nodes (u_i, u_j) at a Euclidean distance $d_{ij}(\tau)$ in terms of Euclidean distance, transmission range and ξ using (18), i.e.,

$$R_L\left(\hat{d}_{ij}(\tau)\right) = \frac{1}{2}\left[1 - \operatorname{erf}\left(\nu \frac{\log \hat{d}_{ij}}{\xi}\right)\right] \tag{20}$$

where, $\hat{d}_{ij}$ is the normalized distance and $\hat{d}_{ij} = d_{ij}/r_j$ and the Euclidean distance between the node pairs at time "τ" can be computed using (3).

Further at any instant "τ," the link status $L_{ij}(\tau)$ can be determined using (21), where, "b" is a uniform random number between (0, 1).

$$L_{ij}(\tau) = \begin{cases} 1, & \text{if } b \leq R_L\left(\hat{d}_{ij}(\tau)\right) \\ 0, & \text{if } b > R_L\left(\hat{d}_{ij}(\tau)\right) \end{cases} \tag{21}$$

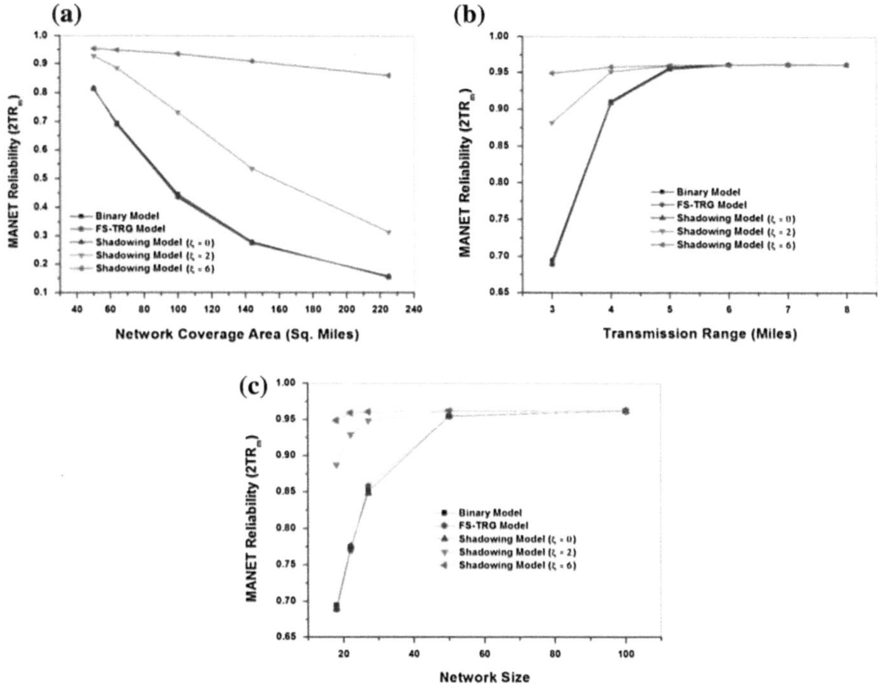

Fig. 15 Effect of **a** NCA **b** TR **c** NS on $2TR_m$ for different link probability models (binary model, FS–TRG model, shadowing Model)

The 2-terminal MANET (for the scenario taken up by [64, 65] for FS–TRG model and stated in Sect. 4), when (δ, γ) = 1, and Shadow model when $\xi = 0$, 2 and 6 are depicted in Fig. 15a–c, respectively. From the results it is clear that the achieved $2TR_m$ for all the cases, seen as single line—(binary model (black line), FS–TRG model (red line), and Shadow model ($\xi = 0$) (dark blue line)) turns out to be same with a negligible variation due to random number generator used in our MCS approach. It was also noticed that as the severity of the environment increases (say, $\xi = 6$), the network reliability increases by almost 25 % (small coverage area) and 70 % (large coverage area) *vis-a-vis* with no severity (say, $\xi = 0$).Similarly, it may be observed that 25 % of increase in reliability is observed for a small network size with each node's transmission range of 3 miles and a negligible increase for a large network size with larger transmission range. Furthermore, the results also show that when shadowing is severe, the $2TR_m$ reliability is almost maximum, which do not exceed R_i^2(source = terminal reliability = R_i) irrespective of change in simulation area, MN range and number of nodes. The estimation of reliability measures with respect to network coverage area, transmission range and network size for different shadow parameters (Results for terminal reliability are shown in Fig. 16) are reported in [67].

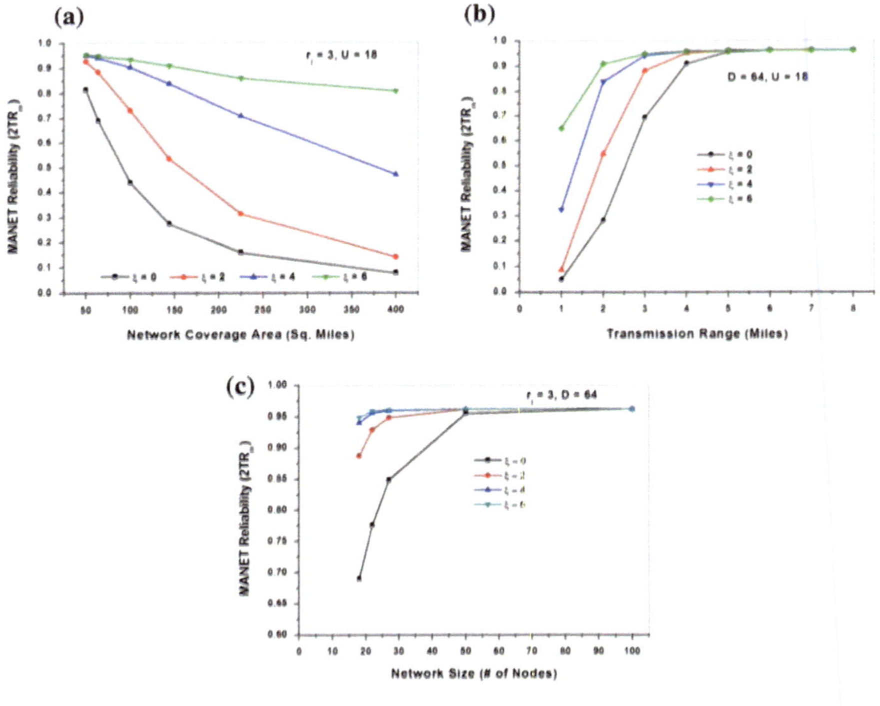

Fig. 16 Effect of **a** NCA **b** TR **c** NS on $2TR_m$ for different values of ξ

6 Conclusion

This chapter has provided some intriguing issues that are normally absent in infrastructure-based communication networks and made an attempt to highlights the current state of research through and extensive literature survey related to attributes, characteristics and models of MANET.

Although, several researchers have been setting up a trendy path related to routing protocols; multicasting; medium access control; packet forwarding protocol; soft handoff; security; third-party incentives; heterogeneous internetworking; sensor integration; mobility exploitation; motion privacy protection; mobility pattern; routing; battery life estimate, performance of MANET, etc. Nevertheless, reliability evaluation of such networks is still a demanding issue and is still at its infancy stage.

In this chapter, a Monte Carlo simulation based approach has been presented and using the same metrics that have been in vogue for infrastructure-based networks, the results for $2TR_m$ of a MANET scenario (deployed for a hypothetical military operation) have been discussed. The presented algorithm took into account the formation of network topology (in a rectangular/square area wherein the MNs move following the RWP mobility model), from one instant to another, using geometrical random graph. The probabilistic behavior of MNs has been assumed to follow a Weibull distribution whereas a hybrid model (FS–TRG model) has been used to simulate the link reliability. The methodology and analysis presented in this work can help the designers/decision makers to design/deploy an application-oriented MANET by considering several influencing factors for an assigned reliability target. We have also made an attempt to provide an insight on the impact of shadowing ξ and scenario metrics on the MANET reliability. We have shown that when $\xi = 0$, the model behaves as a path loss model and thereafter as ξ increases the MANET reliability also increases because of higher connectivity between the MN though the power fluctuations are severe. The simulated results are of practical significance for the design of MANET and further can be applied for planning of such networks by considering suitable number of nodes with desired transmission range that are needed to achieve a reliable communication within the defined simulation boundary operable in a severe environment. Further, the MANET reliability evaluation wherein the link existence (no existence) depends on the relation between the capacity demand (C_d) and the link capacity (C_{ij}) can be seen in [84].

References

1. Toh C. K., Ad Hoc Mobile Wireless Networks: Protocols and Systems, Prentice Hall Publications, 2002.
2. Subir K. Sarkar, T. G.Basavaraju, and C. Puttamadappa, "Ad Hoc Mobile Wireless Networks: Principles, Protocols, and Applications, Auerbach Publications, 2007.

3. Xiang Yang Li, "Wireless Ad Hoc and Sensor Networks: Theory and Applications", Cambridge University Press, 2008.
4. Conti M and Silvia G., Multihop Ad Hoc Networking: The Theory, IEEE Communication Magazine, pp. 78 - 86, 2007.
5. Upkar V., Snow A. P. and Malloy AD., Designing Survivable Wireless and Mobile Networks, IEEE Conference on Wireless Communications and Networking, WCNC, Vol. 1, pp. 30 - 34, 1999.
6. Frodigh M, Johansson P and Larsson P., Wireless Ad Hoc Networking—The Art of Networking Without a Network, Ericsson Review, Issue 4, pp. 248 - 263, 2000
7. Upkar V., Wireless I: Mobile and Wireless Information Systems: Applications, Networks and Research Problems, Communications of the Association for Information Systems, Vol. 12, pp. 155 - 166, 2003.
8. Imrich C, Conti M, Jennifer J and Lic N., Mobile Ad Hoc Networking: Imperatives and Challenges, Ad hoc Networks, No. 1, pp. 13 - 16, 2003.
9. Gerla M., From Battlefields to Urban Grids: New Research Challenges in Ad Hoc Wireless Networks, Pervasive and Mobile Computing, No.1, pp. 77—93, 2005.
10. Sasha D, Shim J. P., Upkar V and Knoerzer G., Evolution and Emerging Issues in Mobile Wireless Networks, Communications of the ACM, Vol. 50, No. 6, pp. 38 - 43, 2007.
11. Makki S. K., Li X-Y, Pissinou N, Makki S, Karimi M and Makki K., Sensor and Ad-Hoc Networks: Theoretical and Algorithmic Aspects, Springer Publishers, 2008.
12. Subbarao M. W., Ad Hoc Networking Critical Features and Performance Metrics, NIST Group, pp. 1 - 11, 1999.
13. Fan Bai and Ahmed Helmy, Chapter 1 A Survey of Mobility Models in Wireless Adhoc Networks, http://www.cise.ufl.edu/~helmy/papers/Survey-Mobility-Chapter-1.pdf
14. Ishibashi B and Boutaba R., Topology and Mobility Considerations in Mobile Ad Hoc Networks, Ad Hoc Networks, Vol. 3, No. 6, pp. 762 - 776, 2005.
15. Deo N., Graph Theory with Application to Engineering and Computer Science, Prentice-Hall, Englewood Cliffs, N.J., 1974.
16. Hsu L-H and Lin C-K., Graph Theory and Interconnection Networks, Taylor & Francis Group, CRC Press, 2009.
17. Erdős P and Rényi A., On the evolution of Random Graphs, Publications of the Mathematical Institute of the Hungarian Academy of Sciences, Vol. 5, pp. 17 - 60, 1960.
18. Dall J and Michael C., Random Geometric Graphs, Physical Review E, 66, 016121, pp. 1 - 9, 2002.
19. Hekmat R and Mieghem PV., Degree Distribution and Hop count in Wireless Ad Hoc Networks, In Proceedings of IEEE ICON 2003, 2003.
20. Muthukrishnan S and Gopal P., Thresholding Random Geometric Graph Properties Motivated by Ad Hoc Sensor Networks, International Journal of Computer and System Sciences, Vol. 76, pp. 686 - 696, 2010.
21. Hekmat R., Ad-hoc Networks, Fundamental Properties and Network Topologies, 2006.
22. Afonso Ferreira, "On models and algorithms for dynamic communication networks: the case for evolving graphs", http://www.hipercom.inria.fr/algotel2002/actes/59.ps.gz.
23. Matthias W., Modeling the Network Topology, Chapter 22, Modeling and Tools for Network Simulation, Springer Publisher, pp. 471 - 486, 2010.
24. Francesco De Pellegrini, Daniele Miorandi, Iacopo Carreras, and ImrichChlamtac, "A graph based model for disconnected Ad hoc networks", INFOCOM 200, 26th IEEE International Conference on Computer Communications, IEEE, pp. 373-381, 2007.
25. Dimitar T, Sonja F, Bekim C and Aksenti G., Link Reliability Analysis in Ad Hoc Networks, XII Telekomunikacioni forum TELFOR, 2004.
26. Z. Ye, Srikanth V. K., Satish K. T., A routing framework for providing robustness to node failures in mobile ad hoc networks, Ad Hoc Networks, 2(1) (2004) 87–107.
27. Shastri A, Pallavi K, Shilpi J, Performance analysis of ad hoc network under node failure, International Journal of Computer Technology and Electronics Engineering, 1(2) (2010) 168-175.

28. Tekiner F, Ashish S, Node failures on MANET routing with multimedia traffic, CSNDSP 2010, (2010) pp. 71-76.
29. Shivashankar, Sivakumar B, Varaprasad G, Identification of critical node for efficient performance in MANET, International Journal of Advanced Computer Science and Application, 3(1) (2012) 166-171.
30. Zadin A, Thomas F, Maintaining path stability with node failure in mobile ad hoc networks, Procedia Computer Science, 19 (2013) 1068-1073.
31. Fei Z, Huang N, Chen J, Impact analysis of communication network reliability based on node failure, Proceedings of the 2nd International Symposium on Computer, Communication, Control and Automation (ISCCCA-13), (2013) pp. 0239-0242.
32. S. Merkel, Sanaz M, Hartmut S, Hop count based distance estimation in mobile ad hoc networks – Challenges and consequences, Ad Hoc Networks, 15 (2014) 39–52.
33. Zhao X, Zhiyang Y, Hai W, A novel two-terminal reliability analysis for MANET, Journal of Applied Mathematics, (2013) 1-10.
34. Migov DA, Vladimir, Reliability of ad hoc networks with imperfect nodes, Multiple Access Communications, Lecture Notes in Computer Science, 8715, (2014) 49-58.
35. Rappaport T. S and Sandip S., *Radio Wave Propagation for Emerging Wireless Personal-Communication Systems*, IEEE Antennas and Propagation Magazine, Vol. 36, No. 5, pp. 14 - 24, 1994.
36. Tam W. K. and Tran V. N., *Propagation Modelling for Indoor Wireless Communication*, Journal of Electronics and Communication Engineering, pp. 221 - 228, October 1995.
37. Juki W. T, Foh C. H, Qiu D. On link reliability in wireless mobile adhoc networks. 64th IEEE Vehicular Technology Conference 2006: 1–5.
38. Dong Q and Banerjee S., *Minimum Energy Reliable Paths using Unreliable Wireless Links*, Proceedings of 6th ACM International Symposium on Mobile Ad Hoc Networking and Computing, MobiHoc'2005, pp. 449 - 459, 2005.
39. Khandani A. E., Modiano E, Abounadi J and Zheng L., *Reliability and Route Diversity in Wireless Networks*, Conference on Information Sciences and Systems, pp. 1 - 8, 2005.
40. Peiravi A and Kheibari HT., *Fast Estimation of Network Reliability Using Modified Manhattan Distance*, Journal of Applied Sciences, Vol. 8, No. 23, pp. 4303 - 4311, 2008.
41. Wei T and Guo W., *A Path Reliable Routing Protocol in Mobile Ad hoc Networks*, IEEE Computer Society, pp. 203 - 207, 2008.
42. Khandani A. E, Abounadi J, Modiano E and Zheng L., *Reliability and Route Diversity in Wireless networks*, IEEE Trans. on Wireless Communications, Vol. 7, No. 12, pp. 4772 - 4776, 2008.
43. Moses E. E. and Isabona J., *Probabilistic Link Reliability Model for Wireless Communication Networks*, International Journal of Signal System Control and Engineering Application, Vol. 2, No. 1, pp. 22 - 29, 2009.
44. Zhou X, Durrani S and Jones H. M., *Connectivity of Ad Hoc Networks: Is Fading Good or Bad?*, IEEE, pp. 1 - 5, 2004.
45. Han S. Y and Abu-Ghazaleh N. B., *On the Effect of Fading on Ad-Hoc Networks*, arXiv:cs/0504002v1 [cs.NI], pp. 1 - 12, 2005.
46. Xing L, Xu H, Amari SV and Wang W., *A New Framework for Complex System Reliability Analysis: Modeling, Verification and Evaluation*, pp. 1 - 17, http://www.cis.umassd.edu/~hxu/Papers/UMD/JATC2007-9-21.pdf, 2007.
47. Chaturvedi S. K., Network Reliability: Measures and Evaluation, Scrivener Publishing, LLC, 2016.
48. Kubat P., Estimation of Reliability for Communication / Computer Networks –Simulation / Analytic Approach, IEEE Transactions on Communications, Vol. 37, No. 9, pp. 927 - 933, 1989.
49. Xio Y-F, Chen S-Z, Li X and Li Y-H., *Reliability evaluation of wireless Sensor Networks using an Enhanced OBDD Algorithm*, The Journal of China Universities of posts and Telecommunications, Vol. 16, No. 5, pp. 62 - 70, 2009.

50. Mo Y, Han J, Zhang Z, Pan Z and Zhong F., *Approximate Reliability Evaluation of Large-scale Distributed Systems*, pp. 1 - 20, http://journal.iis.sinica.edu.tw/paper/1/120007-3.pdf?cd=874B305A3B701DC5B.
51. Wang C, Xing L, Vokkarane VM and Yan (Lindsay) Sun., *Reliability analysis of wireless sensor networks using different network topology characteristics*, International Conference on Quality, Reliability, Risk, Maintenance, and Safety Engineering (ICQR2MSE), pp. 12 - 16, 2012.
52. Shrestha A, Xing L and Liu H., *Modeling and Evaluating the Reliability of Wireless Sensor Networks*, In Proceeding of IEEE, pp. 186 - 191, 2007, http://ieeexplore.ieee.org/stamp/stamp.jsp?arnumber= 04126347.
53. Hallani H and Shahrestani SA., *Enhancing the Reliability of Ad-hoc Networks through Fuzzy Trust Evaluation*, Proceedings of the 8th WSEAS International Conference on APPLIED COMPUTER SCIENCE, pp. 93 - 98, 2008.
54. Golnoosh G, Dana A, Ghalavand A and Rezahosieni M., *Reliable Routing Algorithm based on Fuzzy logic for Mobile Ad Hoc Network*, 3rd International IEEE Conference on Advanced Computer Theory and Engineering (*ICACTE*), pp. V5-606 - V5-609, 2010.
55. Su B-L, Wang W-L and Huang Y-M., Fuzzy Logic Weighted Multi-Criteria of Dynamic Route Lifetime for Reliable Multicast Routing in Ad Hoc Networks, Expert Systems with Applications, Vol. 35, pp. 476 - 484, 2008.
56. Dana A, Ghalavand G, Ghalavand A and Farokhi F., *A Reliable Routing Algorithm for Mobile Adhoc Networks based on Fuzzy Logic*, International Journal of Compute r Science Issues, Vol. 8, Issue 3, No. 1, pp. 128 - 133, 2011.
57. Mala C, Sankaran S, Prasad S, Gopalan N and Sivaselvan B., *Routing for Wireless Mesh Networks with Multiple Constraints using Fuzzy Logic,* The International Arab Journal of Information Technology, Vol. 9, No. 1, pp. 1 - 8, 2012.
58. Mohan B. V, Kumar B. V. S and Nath J. V., *A Reliable Routing Algorithm in MANET Using Fuzzy*, International Journal of Advanced Research in Computer Science and Software Engineering, Vol. 2, Issue 10, pp. 451 - 455, 2012.
59. Lessmann, Janacik P, Lachev L and Orfanus D., *Comparative Study of Wireless Network Simulator*, IEEE The Seventh International Conference on Networking, pages 517 - 523, 2008.
60. Upkar V, Snow AP and Malloy AD., Measuring the Reliability and Survivability of Infrastructure-oriented Wireless Networks, Proceedings of 26th Annual IEEE Conference on Local Computer Networks, LCN 2001, pp. 611 - 618, 2001.
61. Suh J-J and Han C-M., *System Reliability Estimation Using Simulation Combined with Network Reductions*, Microelectronics Reliability, Vol. 36, No. 9, pp. 1263 - 1267, 1996.
62. Cancela H and Khadiri M El., *A Simulation Algorithm for Source - Terminal Communication Network Reliability*, Proceedings of Simulation' 96, pp. 155 - 161, 1996.
63. Wang H and Pham H., Survey of Reliability and Availability Evaluation of Complex Networks using Monte Carlo Technique, Microelectronic Reliability, Vol. 37, No. 2, pp. 187 - 209, 1997.
64. Konak A., *Combining Network Reductions and Simulation to Estimate Network Reliability*, Proceedings of the 2007 Winter Simulation Conference, pp. 2301- 2305, 2007.
65. Cook J. L. and Ramirez-Marquez J. E., *Mobility and Reliability Modeling for a Mobile Ad Hoc Network*, IIE Transactions, Vol. 41, No. 1, pp. 23 - 31, 2009.
66. Chaturvedi S. K, Padmavathy N. The influence of scenario metrics on network reliability of mobile ad hoc network. International Journal of Performability Engineering 2013;9(1):61–74.
67. Padmavathy, N., and S. K. Chaturvedi. Reliability evaluation of capacitated mobile ad hoc network using log-normal shadowing propagation model, International Journal of Reliability and Safety, 2015; 9 (1): 70-89.
68. ShawqiKharbash, and Wenye Wang., "Computing Two-Terminal Reliability in Mobile Ad hoc Networks, IEEE Communication Society, pp. 2833-2838, 2007.

69. Brooks R. R, Pillai B, Racunas S and Rai S., *Mobile Network Analysis Using Probabilistic Connectivity Matrices*, IEEE Transactions on Systems, Man and Cybernetics – Part C: Applications and Review, Vol. 37, No. 4, pp. 694 - 702, 2007.
70. Cook J. L. and Ramirez-Marquez J. E., *Two-Terminal Reliability Analyses for a Mobile Ad Hoc Wireless Network,* Reliability Engineering and System Safety, Vol. 92, pp. 821 - 829, 2007.
71. Padmavathy N. and S. K. Chaturvedi, Mobile Ad Hoc Networks: Reliability Measures and Evaluation, LAMBERT Academic Publishing, Germany, 2016.
72. Goldsmith A., *Path Loss and Shadowing*, Chapter 2, Wireless Communications, Cambridge University Press, pp. 24 — 57, 2005.
73. Rappaport T. S., *Mobile Radio Propagation: Large Scale Path Loss*, Chapter 3, Wireless Communications, Prentice Hall, 2nd Edition, pp. 68 - 138, 2002.
74. Lee W. C. Y., *Mobile Communications Engineering: Theory and Applications*, Second Edition, Tata McGraw Hill Publishers, 2008.
75. Pahlavan K and Krishnamurthy P., Characteristics of Wireless Medium, Chapter 2, Principles of Wireless Networks – A Unified Approach, Prentice – Hall of India Pvt. Ltd., pp. 39 - 80, 2002.
76. Padmavathy, N., and S. K. Chaturvedi. Evaluation of Mobile Ad Hoc Network Reliability using Propagation-based Link Reliability Model. International Journal of Reliability Engineering and System Safety, 2013; 115: 1-9.
77. Egli J. J., *Radio Propagation above 40 MC Over Irregular Terrain*, Proceeding of IRE, Vol. 45, No. 10, pp. 1383 - 1391, 1957.
78. Bettstetter C and Hartmann C., *Connectivity of Wireless Multihop Networks in a Shadow fading Environment, Wireless Networks*, Springer Science + Business Media, Inc., Vol. 11, pp. 571 - 579, 2005.
79. Zhou X, Durrani S and Jones HM.,*Connectivity of Ad Hoc Networks: Is Fading Good or Bad?*, IEEE, pp. 1 - 5, 2004.
80. Hekmat R and Mieghem PV., Connectivity in Wireless Ad Hoc Networks with a Log-normal Radio Model, Mobile Networks and Applications, Springer Science + Business Media, Inc., Vol. 11, pp. 351 - 360, 2006.
81. Banerjee S and Arunabha Sen., *Impact of Region-Based Faults on the Connectivity of Wireless Networks in Log-Normal Shadow Fading Model*, Proceedings of IEEE ICC 2011, IEEE Communications Society, pp. 1 - 6, 2011.
82. Stuedi P and Alonso G., *Log-Normal Shadowing meets* SINR*: A Numerical Study of Capacity in Wireless Networks*, Proceedings of IEEE SECON 2007, IEEE Computer Society, pp. 550 - 559, 2007.
83. Stuedi P., From Theory to Practice-Fundamental Properties and Services of Mobile Ad Hoc Networks, 2008.
84. Padmavathy, N., and S. K. Chaturvedi. *Reliability Evaluation of Capacitated Mobile Ad Hoc Network using Log-Normal Shadowing Propagation Model,* International Journal of Reliability and Safety, Vol. 9, No. 1, pp. 70-89, 2015.

Software Reliability Modeling with Impact of Beta Testing on Release Decision

Adarsh Anand, Navneet Bhatt, Deepti Aggrawal and Ljubisa Papic

Abstract Increased dependence of humans on technologies has made it necessary for developing the software with high reliability and quality. This has led to an increased interest of firms toward the development of software with high level of efficiency; which can be achieved by incorporating beta tests for improving and ensuring that the software is safe and completely free from errors. In a software release life cycle, beta testing is the last important step that software developers carry out before they launch new software. Beta testing is a unique testing process that helps software developers to test a software product in different environments before its final release in the market. In this chapter of the book, we develop a SRGM by inculcating the concept of beta testing in the fault removal process to account for situations that might occur when the software is used in diverse environments. This is done to evade the chances of system being failed in the field. Conducting beta tests results in enhancement of software reliability and has been widely acknowledged. Furthermore, we have developed an optimal scheduling model and showed the importance of beta test while determining the general availability time of the software and making the system more cost effective. For validating the accuracy and predictive capability of the proposed model, we analyzed it on real software data set.

Keywords Software reliability · Software release life cycle · Beta testing · Optimal release scheduling

A. Anand (✉) · N. Bhatt
Department of OR, University of Delhi, New Delhi 110007, India
e-mail: adarsh.anand86@gmail.com

N. Bhatt
e-mail: navneetbhatt@live.com

D. Aggrawal
Keshav Mahavidyalaya, University of Delhi, New Delhi 110034, India
e-mail: deepti.aggrawal@gmail.com

L. Papic
DQM Research Centre, Cacak, Serbia
e-mail: dqmcenter@open.telekom.rs

M. Ram and J.P. Davim (eds.), *Advances in Reliability and System Engineering*,
Management and Industrial Engineering, DOI 10.1007/978-3-319-48875-2_5

1 Introduction

The growth of the internet—and the lucrative opportunities it presents—is bringing with it an explosion in software application development. Software has become an inherent part of every fabric of our lives. In today's connected economy, almost every government as well as private and nonprofit enterprise rely on software as a core business function. The growth in software advancement and rapid delivery of new features led to a major shift in the way to meet customer demands, and therefore an organized environment for development and testing becomes an integral part of the value chain. In 2015, according to Gartner, the worldwide size of the security software market was US$22.1 billion, an increase of 3.7 % over 2014 [1]. In India, the IT sector has increased its impact on India's GDP from 1.2 % in 1998 to 9.5 % in 2014, further aggregating a revenue of US$143 billion in FY2016, where export revenue raised to US$108 billion and domestic to US$ billion, rising by over 8.5 % [2].

As software application becomes ingrained in our day-to-day life, its failures result in disastrous situations which are becoming even more serious. Reports of tragic effects of software failure exist in large numbers. Some well-known failures such as programming errors in the radiation therapy machine result in the death of three persons due to the massive overdose of radium [3]. An on-board software program failure caused an explosion in the Ariane 5 heavy lift launch vehicle on June 4, 1996, which cost more than US$7.0 billion to the European Space Agency [4], and a software bug present in the engine control system of Royal Air Force helicopter caused its crash, killing more than 25 persons [5]. Also in the very last year some of the famous software glitch which resulted in severe disruption were entertained that includes the Amazon 1p price glitch which caused products on sale in marketplace for just one penny. This flaw resulted in a loss of $100,000 for the vendors. In September 2014, Apple had a major embarrassment when it had been forced to pull the iOS 8 update merely after its release due to the various malfunctions in the software [6].

For an increasing demand of delivering reliable software products to the users, software quality has become more vital recently due to the increased security concerns arising from software vulnerabilities. The quality of a software encountered by an end user is an association of the faults in a software product when it is released, plus the efforts that the developer makes to patch the imperfections after release. Once a software is up for general availability it has to be free from all the flaws. A software product has usually gone through a number of stages in its development before it is available for general availability. Software testing is one of the phases which is usually performed for various purposes: first, to improve quality; second, for verification and validation, and for reliability estimation [7]. In this regard, software reliability models can give significant level of reliability for a software during the development process. Over the past four decades, research activities in field of reliability engineering have been done, and various software reliability growth models (SRGMs) have been proposed in the literature [8].

SRGMs have been effective in assessing the software reliability and the potential number of faults in the software.

Selecting a suitable software reliability growth model for attaining a desired level of reliability; software engineer makes it sure that the testing has been performed till the time sufficient number of bugs or faults have been removed and the software offering is ready to be released. A crucial decision for the firms lies at time when the software has to be released in the market or to know the exact time till when the execution of testing activities should be done. Many researchers have established various software scheduling problems considering different aspects; to analyze the period of software testing phase in the literature [8]. Selecting an appropriate software reliability model uniquely identifies a time point suggesting the time at which the software is up for release in the market. But the problem aforesaid is not just associated with its general availability, it involves many factors and attributes that a firm have to take care while determining the optimum release time. If testing stops prematurely, pending errors in the software may leave the software developers with dissatisfied customers and can invite a high cost of fixing faults during its operational stage. On the other hand, if shipping a software is too late, it surely increases its reliability but the cost due to economic factors constitute may charge a firm with high testing cost, penalty cost due to late delivery. Hence, while deciding the optimum release time both factors have to be taken care judiciously.

In this chapter, we have focused our objective on testing the releases of a software during its lifetime and further implying that the reliability can be improved when the software undergoes phase transformation from alpha testing to beta testing phase. Furthermore, an optimal release scheduling model is provided which incorporates various costs and a preferred level of software reliability in determination of optimum release time of a software. The rest of this chapter is organized as follows. First, in the following section, a brief background of our study is provided. Section 3 provides the literature review of our study. In Sect. 4, we derive a SRGM based on Non-Homogenous Poisson Process (NHPP) to represent a fault detecting/removal process during the testing and further integrate the beta testing activities. Section 5 evaluates the proposed SRGM with a numerical example of a real software failure data. In Sect. 6 an optimal release time cost model is presented, finally conclusions and acknowledgement are given in Sects. 7 and 8.

2 Background and Motivation

2.1 Software Release Life Cycle

Prior to the testing, a software product has been put through the stages of development and maturity which include the activities: requirement overview, design making, coding, and system testing. The life cycle of a software is the collection of

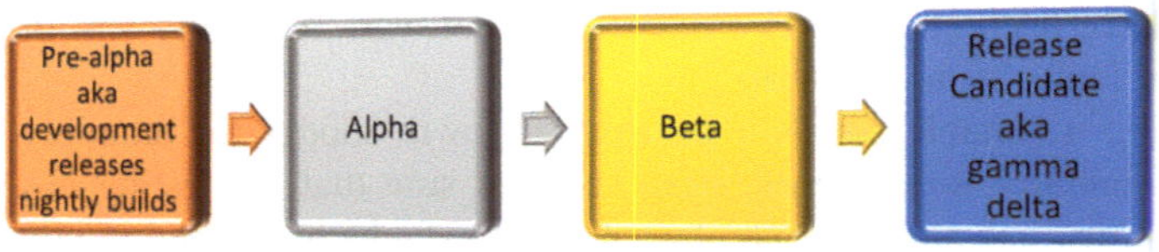

Fig. 1 Stages of development [9]

all the phases of development of an application: varying since its preliminary requirement to its ultimate release, along with the updated versions of the previous releases in order to fix loopholes and to improve the features present in the software. As shown in Fig. 1, a software release is characterized by different versions in its lifetime [9].

With the continuous improvement in the development of software in every stage; each version of the software is released either in private or public for testing. A final software release is preceded by the deployment of alpha and then beta versions. The final quality of a software mainly depends on the activities that are performed during the testing phase; that is, with the debugging of faults, reliability of the software improves. As mentioned earlier, the software testing process is very complex. Due to this reason there are many types of software testings available in the literature [10]. One of the types of software testing is alpha testing. Usually it is performed at the development site by a number of in-house testers or an independent test team. Alpha testing allows a developer to perform internal acceptance testing, which is normally employed prior to the beta testing [10, 11].

2.2 *Beta Testing*

As shown in Fig. 1, prior to release of beta version of software, after the completion of in-house testing activities for each pre-alpha and alpha versions of the software by the developer then the software ends with a feature freeze version, representing that, there will be no more design and feature changes in the software. Then, the software version is termed as feature complete. Beta phase usually commences after the transition of a software program to a feature complete application and the software is ready for the open and closed beta testing [9].

Every enterprise or organization follows its own software testing life cycle. We emphasize on a specific stage of software testing called beta testing. It is the first "user test" of a software that allows a developer to assess the usability and functionality feedback by the end users. During beta testing, a pre-release version "*Betaware*" is given to a group of users to test in "real world" environments; in order to validate the software offering. The motive of beta test is to improve the software prior to its release [12]. Contributing to the beta testing activities allows an enterprise to get benefits in several ways. First, the system environment of beta testers can vary with respect to the both hardware and software point of view as it is often difficult to match in lab testing environment—since a software application is likely to have different performance related issues and can trigger bugs while running on different system environments, beta testers can help discover faults or compatibilities issues that cannot be detected in testing labs. Second, beta testers are independent and are likely to be less biased. Third, beta testing can save the costs that may result due to the reputation damage or warrant expenses if the offering has to be recalled from the market after launch [13].

Beta testing has been traditionally meant to a small number of selected and trained beta testers that check out a product and review the product based on their experiences. As in this competitive environment, an effective beta test can save valuable time and enable a firm to launch its offering earlier in this fast moving world. With the rapid growth of the Internet, most software firms release multiple beta versions of the software application for the open beta testing. Jiang et al. [13] and Fine [14] mentioned the term public beta testing, where the betaware is made available to websites and the interested users or testers who are comfortable living on the absolute bleeding edge may download the latest release of software before it hits the public and put a hand in volunteering the release with the goal of discovering and reporting bugs. Figure 2 shows a screenshot of an application Snap beta version; a free Google Play client for BlackBerry 10 operating system, mentioning the known issues and the changelog of the releases prior to the beta version3 reported by the beta testers [15].

Fig. 2 Screenshot of snap beta version [15]

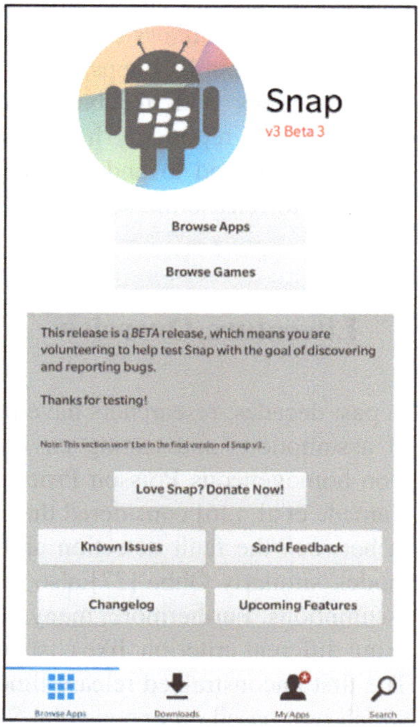

Like *snap*, most software releases various beta versions and after testing with each release publically, then arrives as a release candidate. A release candidate is generally considered as a final product and in this, all of its software features have been tested thoroughly in order to minimize the likelihood of fatal flaws while releasing a software. The final version of the software is generally termed as general availability (GA). As a brief review of some examples of well-known public betas, beta testing has been received a great response. In September 2000, a preview of Apple's next-generation operating system Mac® OS X was released for the beta testing [16]. On the same lines Microsoft released a community technology previews (CTPs), a beta testing program back in 2005 and 2006 for its operating system Windows Vista, and for its latest operating system Windows 10; Microsoft scheduled the operating system for the public beta testing through the Windows Insider program from October 2014. A total of 10 builds were released as of May 29, 2015 for the public beta testing before its general availability on July 29, 2015. Over 1.5 million users have already downloaded and installed the Windows 10 through the Microsoft insider program by the end of 2014. Other very popular betas are Google's Gmail, Google Calendar, and Google News [9, 17–21].

The public beta testing phenomenon has also influenced the mobile applications for smartphones and tablets. For example, services such as BlackBerry Beta Zone offered in BlackBerry 10 smartphones provide the beta versions of various mobile applications for the public beta testing. WhatsApp Messenger being available for the beta testing program in the BlackBerry Beta Zone has reported more than 450 bugs as of April, 2016 [22]. There are even websites being developed in bringing together developers and beta testers for Android and iPhone apps. For example, websites like TestFairy.com and iBetaTest.com allow iPhone and Android app developers to publish apps, and testers may download the apps and provide the general issues and feedbacks to the developers [13]. As of April, 2016, iBetaTest.com reportedly has around 16,600 beta testers around the world. This chapter is focused to understand the importance of beta testing before making a final call to release the software.

3 Literature Overview

In past decades, researchers have proposed a number of SRGMs under different set of assumptions and testing environment and most of them are based upon the Non-homogeneous Poisson Process (NHPP) in the literature [8, 23–25]. In 1983, Yamada et al. [26] considered the testing process as two stages by taking a time lag in between the fault detection and removal and called it as the delayed S-shaped model, similarly Ohba [27] also developed S-shaped models using different set of assumptions. Furthermore, many optimal release time models have been developed using different criterion, like cost and reliability, provided by the management team. The first unconstrained release time policy was derived by Okumoto and Goel [28] which was based on exponential SRGM. Later, Yamada and Osaki [29] developed the constrained software release time problem with the objective as cost minimization and reliability maximization. Kapur et al. [30] discussed and studied the bi-criterion optimal release time problem. Jain et al. [31] considered warranty time and warranty cost in the development of software release time problem. Recently, Singh et al. [32] have formulated an optimal release time problem using the multi-attribute utility theory. Further, Singh et al. [33] used different set of attributes in the construction of release time problem. Moreover, Arora et al. [7] studied the impact of patching a software when the software product is available for purchase. Many authors investigated the importance of patching after the release of software in the market [34, 35]. It should be noted that the SRGMs proposed earlier [8] only considered the testing activities that are performed when the software has been tested under the lab environment. With consideration of faults detected under the beta testing environment and providing the significant measures to the reliability of software, this study tries to capture the behavior of reliability growth when both alpha and beta testing phases are studied consecutively.

Among the family of studying this problem of optimal release time, the concept of beta testing has received less attention. As an aspect of software engineering, a few authors have considered the idea of beta testing in their study. Wiper and Wilson [36] described a fault detection process during the beta testing phases using the Bayesian statistical method and further developed a cost model for the optimal testing strategy. Kocbek and Heričko [11] showed the use of beta testing to identify the optimal number of testers for the mobile application. Another study by Jiang et al. [13] considered the impact of beta testing activities on the acceptance of final software product. Furthermore, Mäkinen et al. [37] examined the effect of open virtual forum on the adoption of beta products. However, no methodology has been proposed to measure the effectiveness of beta testing during the software testing.

The lack of research on beta testing activities concerning the development of software reliability modeling is a notable shortage that the critical role it plays in the software testing. This study tries to deviate the focus of software engineers to this vital area by studying and scheduling the release plan of software based on beta testing. To do that, emphasis is paid on the stages of the software release cycle with special attention to beta test phase. In the succeeding section, we have developed a mathematical model incorporating the process of beta testing. Furthermore, an optimal release time model is developed for the determination of software release time.

4 Model Formulation

In this section, we propose a new NHPP SRGM that incorporates the fault discovery process "beta testing" when the in-house alpha testing phase has been completed.

Research has been done to take some practical issues into the consideration of software reliability modeling. For example, post-release testing and software release policies can be found in [38]. In this chapter, we derive a new model incorporating the faults detected during the beta testing phases into software reliability

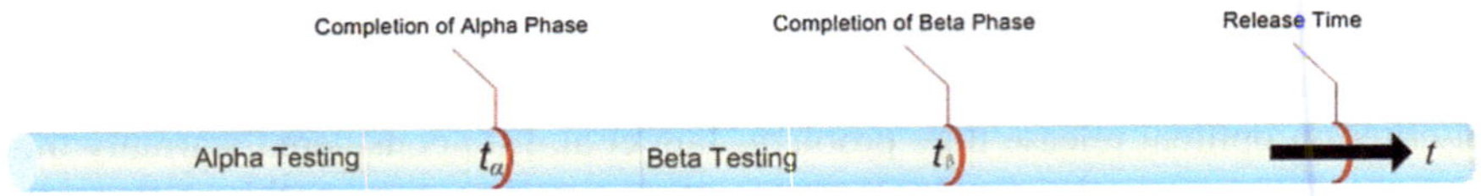

Fig. 3 Timeline of Testing

assessment. The framework has been modeled into three different phases considering the detection and correction, respectively. As illustrated in Fig. 3, a software being tested during its release life cycle undergoes testing to identify and remove defects before it is released to the customers. Following are the notations used in the proposed model.

$m_1(t)$	Expected no. of faults removed by time 't_α'
$m_2(t)$	Expected no. of faults detected by time 't_β'
$m_3(t)$	Expected no. of faults removed by time 't'
a	Initial no. of faults
b_1	Fault detection/removal rate by the testers in time period $(0, t_\alpha)$
b_2	Fault detection/removal rate by the users in time period (t_α, t_β)
b_3	Fault detection/removal rate by the testers in time period (t_β, t).

4.1 *Phase I: Alpha Testing Phase*

In Phase I, the failure pattern is observed in the time period during which the SRGM undergoes alpha testing. Most of SRGMs proposed in the literature were based with the assumption that software testing has been performed only till the in-house testing phase and the software is up for release.

In the time interval $[0, t_\alpha)$, the failure intensity is proportional to the number of software faults remaining in the software till the time software undergoes alpha testing. The failure intensity during this period can be described using the following differential equation:

$$\frac{\mathrm{d}m_1(t)}{\mathrm{d}t} = b_1(a - m_1(t)) \quad 0 \leq t < t_\alpha \tag{1}$$

where

$$m_1(0) = 0$$

and a is the potential faults present in the software initially and b_1 is the fault detection rate in the alpha testing period. Solving the above equation, the mean value function for the first phase of testing can be obtained, as follows:

$$m_1(t) = a(1 - e^{-b_1 t}), \quad \text{when} \quad 0 \le t < t_\alpha \tag{2}$$

4.2 Phase II: Beta Testing Phase

In Phase II, the testing process is considered as the fault discovering process called beta testing. After the completion of alpha testing phase at time t_α, a feature complete software version is introduced to the market which is likely to contain a number of known or unknown bugs as a betaware for beta testing. Here, the software is tested under user environment and the failure occurrences are reported back by the beta testers.

During the beta testing phase, the number of faults discovered can be modeled as follows:

$$\frac{dm_2(t)}{dt} = b_2[(a - m_1(t_\alpha)) - m_2(t)] \quad t_\alpha < t \le t_\beta \tag{3}$$

where

$$m_2(t_\alpha) = m_0 \ge 0$$

and b_2 is the fault detection rate due to the beta testers. It represents that the fault intensity is proportional to the outstanding faults reported by time t_β. Solving Eq. (3), the mean value function for the faults discovered during the beta testing phase can be obtained as:

$$m_2(t) = ae^{-b_1 t_\alpha} + e^{-b_2(t - t_\alpha)}(m_0 - ae^{-b_1 t_\alpha}), \quad \text{when} \quad t_\alpha < t \le t_\beta. \tag{4}$$

We refer to m_0 as the beta bugs, representing a finite number of bugs which are being detected instantly by the beta testers who use the software.

4.3 Phase III

Phase III corresponds to the testing process of SRGM as the fault removal phenomenon. After the completion of beta testing, the faults reported by the beta testers during the Phase II period are corrected after time t_β. In Phase III, the fault removal intensity is proportional to the residual faults that were being observed. The following system of differential equation describes the same:

$$\frac{\mathrm{d}m_3(t)}{\mathrm{d}t} = b_3\left[m_2(t_\beta) - m_3(t)\right] \quad t_\beta \le t \tag{5}$$

where

$$m_3(t_\beta) = m_1 \ge 0$$

and $m_2(t_\beta)$ is the cumulative faults at time t_β and b_3 is the fault removal rate. Using the initial condition provided, the mean value function is obtained as:

$$m_3(t) = m_2(t) + (m_1 - m_2(t))\mathrm{e}^{-b_3(t-t_\beta)}$$
$$m_3(t) = (a\mathrm{e}^{-b_1 t_\alpha} + (m_0 - a\mathrm{e}^{-b_1 t_\alpha})\mathrm{e}^{-b_2(t-t_\alpha)})(1 - \mathrm{e}^{-b_3(t-t_\beta)}) + m_1\mathrm{e}^{-b_3(t-t_\beta)}, \quad \text{when} \quad t_\beta \le t \tag{6}$$

Here we mention $m_1 \ge 0$ as those beta bugs which are being removed as in when the fault removal phenomenon is started after beta testing. The faults captured by the beta tester were reported back during the beta phase allowing the software engineer to remove a certain number of beta bugs by the time Phase III has been started. Thus the total faults removed under overall testing of the software are the bugs from Phase I and Phase III which has the mathematical form as given below:

$$m^*(t) = m_1(t) + m_3(t)$$
$$m^*(t) = a(1 - \mathrm{e}^{-b_1 t}) + (a\mathrm{e}^{-b_1 t_\alpha} + (m_0 - a\mathrm{e}^{-b_1 t_\alpha})\mathrm{e}^{-b_2(t-t_\alpha)})(1 - \mathrm{e}^{-b_3(t-t_\beta)}) + m_1\mathrm{e}^{-b_3(t-t_\beta)} \tag{7}$$

Above Eq. (7) denotes the total faults being removed during the total testing time which is divided into three phases as mentioned above.

5 Numerical Analysis and Model Validation

5.1 Data Description

For the validation of the proposed model, we have used data of a Brazilian switching software project. The data size of the software was about 300 KB and it was written in assembly language [39]. In the total execution period of 81 weeks, 461 faults have been removed. Also the entries of the data set comprises of different phases of testing, i.e., first 30 entries are from validation phase and next entries are from field trials and system operations. The data used in the estimation procedure gives a clear indication that the software project has undergone through its release life cycle with alpha testing and beta testing phases.

5.2 Performance Analysis

In this subsection, we examine the model given by Goel and Okumoto [23] with our proposed model. We have estimated the parameters of SRGM by using the methods of LSE. Nonlinear regression module of SPSS software package has been used for the estimation of parameters and the evaluation of goodness of fit criteria.

The parameter estimates and the goodness of fit criteria for the data set under consideration are given in Tables 1 and 2, respectively. Figure 4 represents the goodness of fit curve for the data set and the estimated and predicted values appear to be closely related.

From Table 2, we can see that the MSE, RMSPE, and Variation of the proposed model incorporating beta testing are less than the G-O model. We also see that the Biasness of our model attains smaller values. Moreover, the R^2 also conclude that the proposed model fit the data excellently. Altogether, it is sensible to conclude that the proposed model incorporating beta testing attains better goodness of fit criteria. The improvement achieved by our model is obvious due to the release of software for beta testing; which reflects a significant variation of the fault detection rates and instead of considering a constant fault detection rate throughout its release life cycle. On the basis of our results, we can conclude that software developing firms should practice pre-release testing activities for enhancing the product for general availability.

Table 1 Parameter estimates

Parameters	GO model	Proposed model
a	546.082	602.779
b_1	0.02421	0.023
b_2	–	0.115
b_3	–	0.026
m_0	–	16.410
m_1	–	47.850

Table 2 Goodness of fit

Parameters	GO model	Proposed model
MSE	99.86442	87.4553
Bias	−0.921008	0.000000101
Variation	10.01268	9.410026
RMSPE	10.05495	9.410026
R^2	0.994	0.995

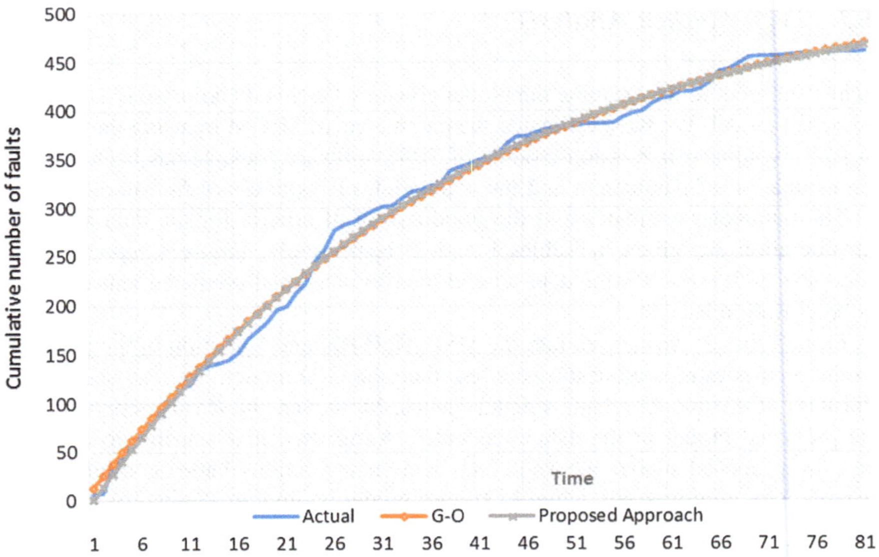

Fig. 4 Goodness of fit curve

6 A Cost Model for Optimal Release Time

In this section, we apply the NHPP model to establish the optimal scheduling policy for a software. Since, in practice every software developer needs to determine for how long software should be tested, such that the expected cost is optimally minimal and the reliability of the software product must satisfy customers' requirements as well. In the literature of software reliability engineering, these kinds of problems are termed as software release time decision problems. There may be a solution that no more testing should be done but generally speaking, the main goal is to achieve a balance between under-testing and over-testing. If a premature software is released in the market, the developer may suffer from loss of user confidence, and if there is a delay in release or over-testing it will impose the burden of penalty cost and overly time consuming. Therefore, optimal release policies are important and realistic issue. Research on software development cost estimation has been conducted in literature [8]. In this study, we have compared the well-known Okumoto and Goel [28] model with our proposed optimal release time problem which has taken the consideration of software cost when the software testing has been performed consecutively as alpha testing and beta testing.

6.1 Review of Cost Model (Okumoto and Goel [28])

In 1983, Okumoto and Goel [28] developed a mathematical model comprising various costs that occurred in the testing and maintenance of a software and is given as

$$K(t) = K_1 \times m(t) + K_2 \times (m(t_L) - m(t)) + K_3 \times t \tag{8}$$

Okumoto and Goel [28] used the mean value function given by Goel and Okumoto [23] in order to formulate the cost model. The cost function comprised of different cost components, i.e., the cost incurred due to the removal of a fault observed during testing and operational phases; as K_1 and K_2, respectively, and K_3 as per unit testing cost. The optimal release time of a software is determined by minimizing the following unconstrained function.

$$\text{Min}\, K(t) = K_1 \times m(t) + K_2 \times (m(t_L) - m(t)) + K_3 \times t \tag{9}$$

6.2 Cost Model Formulation

In contrast to policies given earlier, we here provide a scheduling policy. The proposition is based on certain basic assumptions which are as follows:

(a) Fault removal process is described by NHPP.
(b) During the alpha testing phase fault is removed perfectly.
(c) Faults detected during the field trial are removed after the completion of beta testing phase.
(d) Total fault content in the software is fixed.
(e) The cost incurred in testing is proportional to per unit testing time.
(f) The cost of fixing faults during the alpha and beta phase is proportional to the time required to isolate all the faults found by the end of testing.

In our case, the advantage of beta testing that we are claiming in the aforesaid sections provides a better fault count prediction for the software reliability if a developer releases its software for the field trials. As previously mentioned, software firms are keen to know the release time for their software product. We try to capture the optimal testing time, such that the firms expect a minimum cost with an achievable reliability.

The idea is to optimize the value t^* with respect to the cost involved. We have assumed the following costs in our study:

i. a cost C_1 for isolating a fault during the alpha testing. This reflects the cost incurred to the firm to remove a fault during the in-house testing activities. Typically, in the lab environment situation, this cost is likely to be less as fixing a detected fault incurs a small amount of CPU hours.

ii. a cost C_2 for isolating a fault after the beta testing. This reflects the cost of fixing the discovered bugs at the end of beta testing. Generally, in the beta testing situation, the cost of collecting the feedbacks and responding to the comments and queries of beta testers must be considered and this cost contributes a higher value to the firm.
iii. a cost C_3 per failure per unit time after the software is released. We would normally set this cost to be much higher than the previous costs as the damage caused by leaving high complex faults in the software would require many resources for the correction.
iv. a cost C_4 per unit time. This reflects the cost of testing incurred to the firm per unit time.

This implies that the overall expected cost function of testing is

$$C(t) = C_1 \times m_1(t) + C_2 \times m_3(t) + C_3 \times (a - m^*(t)) + C_4 \times t \tag{10}$$

In the above Eq. (10), the first component corresponds to the cost involved in testing the software during the alpha phase [where $m_1(t)$ is as given in Eq. (2)]. Second component represents the cost incurred while removing the faults detected in the beta phase and $m_3(t)$ is taken as given in Eq. (6). Third component talks about the cost required to correct the faults during the operational phase and $m*(t)$ incorporates the fault correction phenomenon during the overall testing phase; as is given in Eq. (7). The last component of the cost function denotes the per unit testing cost. Hence, we find the optimal release time of the software that minimizes the total testing cost subject to achieving a reliability level, R_0. Then the optimization problem can be expressed as

$$\begin{aligned} &\text{Min. } C(t) = C_1 \times m_1(t) + C_2 \times m_3(t) + C_3 \times (a - m^*(t)) + C_4 \times t \\ &\text{s.t} \\ &R(x/t) \geq R_0 \end{aligned} \tag{11}$$

Solving the aforesaid problem under the minimization criteria provides the optimal time to release the software in the market.

6.3 Numerical Illustration

The estimates of G-O model and the proposed model are used in the cost optimization problem (Table 1). Making use of OPTMODEL procedure of SAS software [41] to determine the optimal release time and the associated testing cost, and for checking the appropriateness of the proposed approach, we consider two cases corresponding to the optimization problem as below.

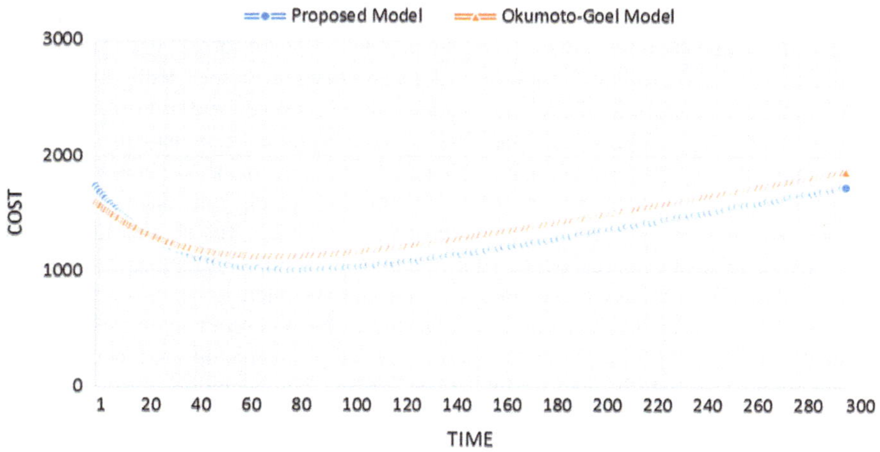

Fig. 5 Expected cost function

Case 1: Okumoto and Goel optimal release time model

We have assumed the cost coefficients to be $K_1 = 1.25$, $K_2 = 3$ and $K_3 = 4$ (the cost in thousands). After solving Eq. (9), we get the optimal release time $t^* = 73.438$ and expected cost $K(t^*) = 1137.84$.

Case 2: Anand optimal release time model (Proposed)

The cost coefficients in the cost model have been assumed to be $C_1 = 0.5$, $C_2 = 0.75$, $C_3 = 3$, $C_4 = 4$ and the desired reliability is assumed to be 0.8. On solving Eq. (11), we get the optimal release time $t^* = 79.77$ and expected cost $C(t^*) = 1011.75$.

After solving both the cases, it is obvious that the optimal release time of a software incorporating beta testing coincide with the actual release time than the optimal time obtained on solving traditional releases policy. Also, when considering best testing activity the optimal time is much larger than the optimal time without beta testing. Moreover, while following the proposed strategy in determination of optimal release time, we observe that the total cost incurred is minimum when compared with the traditional cost model. The detailed view of the cost function versus testing time is illustrated in Fig. 5 (both traditional and proposed methodology).

7 Conclusion

With the availability and easy accessibility of web, beta testing has attained enormous growth in the software market. However, the advantages of beta testing have not been examined deeply. For a software developer, testing a software before its launch is not only vital but also a critical task. The acceptability of a software by the end users can be only judged by exposing the software to real life usage. Beta testing can be considered an important task as any bugs, errors, or problems noticed during the beta tests will make the product viable, stable, and robust before its general availability. The developed SRGM is unique as it considers the faults discovered during the beta testing phase. Moreover, in this study we have emphasized on providing beta releases for software testing and determination of optimal release time. The modeling framework presented in this chapter aims to fill the void by releasing the software after examining the faults detected by the beta testers rather than releasing the product after the completion of in-house testing activities as considered in the previous literature. The outcomes from this study to develop model incorporating beta testing are very promising and showing an improvement in reliability and optimal release time. The accuracy of the proposed model has been verified by the numerical example.

Acknowledgments The research work presented in this chapter is supported by grants to the first author from University of Delhi, R&D Grant No-RC/2015/9677, and Delhi, India.

References

1. Press Release (2016) Gartner Says Worldwide Security Software Market Grew 3.7 Percent in 2015. http://www.gartner.com/newsroom/id/3377618. Accessed 12 August 2016
2. India IT-BPM Overview (2016) http://www.nasscom.in/indian-itbpo-industry. Accessed 15 June 2016
3. Levenson NG, Turner CS (1993) An investigation of the Therac-25 accidents. Computer 26 (7): 18-41
4. Dowson M (1997) The Ariane 5 software failure. Software Engineering Notes 22 (2): 84.
5. Rogerson S (2002) The Chinook helicopter disaster. IMIS Journal 12(2).
6. Charlotte Jee (2015) Top 10 software failures of 2014. http://www.computerworlduk.com/galleries/infrastructure/top-10-software-failures-2014-3599618/. Accessed 15 June 2016
7. Arora A, Caulkins JP, Telang R (2006) Research Note: Sell First, Fix Later: Impact of Patching on Software Quality. Management Science 52(3): 465-471
8. Kapur PK, Pham H, Gupta A, Jha PC (2011) Software Reliability assessment with OR application. Springer, Berlin
9. Software release life cycle (2016) https://en.wikipedia.org/wiki/Software_release_life_cycle. Accessed 15 June 2016
10. Pradhan T (2012) All Types of Software Testing. http://www.softwaretestingsoftware.com/all-types-of-software-testing/. Accessed 15 June 2016
11. Kocbek M, Hericko M (2013) Beta Testing of a Mobile Application: A Case Study. SQAMIA 29-34
12. Buskey CD (2005) A Software Metrics Based Approach to Enterprise Software Beta Testing Design. Dissertation, Pace University
13. Jiang Z, Scheibe KP, Nilakanta S (2011) The Economic Impact of Public Beta Testing: The Power of Word-of-Mouth. Supply Chain and Information Management Conference Papers: Posters and Proceedings Paper 10
14. Fine MR (2002) Beta testing for better software. John Wiley & Sons, New York
15. Bla1ze (2016) Snap v3 Beta sees over 11,000 downloads in the first 24 hours. http://crackberry.com/snap-v3-sees-over-11000-downloads-first-24-hours. Accessed 15 June 2016

16. Apple Press Info (2000) Apple Releases Mac OS X Public Beta. http://www.apple.com/pr/library/2000/09/13Apple-Releases-Mac-OS-X-Public-Beta.html. Accessed 15 June 2016
17. Bogdan Popa (2014) Windows 10 Preview Was Installed by 1.5 Million Users. http://news.softpedia.com/news/Windows-10-Preview-Was-Installed-by-1-5-Million-Users-467767.shtml. Accessed 15 April 2016
18. Posts by Gabe Aul (2016) http://blogs.windows.com/bloggingwindows/author/gabeaul/. Accessed 15 June 2016
19. Press Release (2005) Microsoft Windows Vista October Community Technology Preview Fact Sheet. http://www.microsoft.com/presspass/newsroom/winxp/WinVistaCTPFS.mspx. Accessed 15 June 2016
20. Terry Myerson (2015) Hello World: Windows 10 Available on July 29. http://blogs.windows.com/bloggingwindows/2015/06/01/hello-world-windows-10-available-on-july-29/. Accessed 15 June 2016
21. Fester P (2005) A long winding road out of beta. ZDNet. http://www.zdnet.com/article/a-long-winding-road-out-of-beta/. Accessed 15 June 2016
22. BlackBerry Beta Zone (2015) WhatsApp Messenger Bug Reports. https://appworld.blackberry.com/webstore/content/35142896/?lang=en. Accessed 15 June 2016
23. Goel AL, Okumoto K (1979) Time dependent error detection rate model for software reliability and other performance measures. IEEE Trans Reliability 28(3): 206–211
24. Kapur PK, Garg RB, Kumar S (1999) Contributions to hardware and software reliability. World Scientific Publishing Co. Ltd, Singapore
25. Musa JD, Iannino A, Okumoto K (1987) Software reliability: measurement, prediction, applications. McGraw Hill, New York
26. Yamada S, Ohba M, Osaki S (1984) S-shaped software reliability growth models and their applications. IEEE Trans Reliability 33(4): 289–292
27. Ohba M (1984) Software reliability analysis models. IBM J Res Dev 28: 428–443
28. Okumoto K, Goel AL (1983) Optimal release time for computer software. IEEE Trans Softw Eng SE 9(3): 323–327
29. Yamada S, Osaki S (1987) Optimal Software Release Policies with simultaneous Cost and Reliability Requirements. European Journal of Operational Research 31: 46-51
30. Kapur PK, Agarwal S, Garg RB (1994) Bicriterion release policy for exponential software reliability growth model. Proceedings of the 3rd International Symposium on Software Reliability Engineering 28:165–180
31. Jain M, Handa BR (2001) Cost analysis for repairable units under hybrid warranty. In: M.L. Agarwal, K. Sen, eds. Recent Developments in Operational Research. Narosa Publishing House: New Delhi 149–165
32. Singh O, Kapur PK, Anand A (2012) A Multi Attribute Approach for Release Time and Reliability Trend Analysis of a Software. International Journal of System Assurance and Engineering Management (IJSAEM) 3 (3): 246-254
33. Singh O, Aggrawal D, Kapur PK (2012) Reliability Analysis and Optimal Release Time for a Software using Multi-Attribute Utility Theory. Communications in Dependability and Quality Management -An International Journal 5(1): 50-64
34. Anand A, Agarwal M, Tamura Y, Yamada S (2016) Economic Impact of Software Patching and Optimal Release Scheduling. Quality and Reliability Engineering International. doi:10.1002/qre.1997
35. Das S, Anand A, Singh O, Singh J (2015) Influence of Patching on Optimal Planning for Software Release & Testing Time. Communication in Dependability and Quality Management: An International Journal 18(4): 82-93
36. Wiper MP, Wilson SP (2006) A Bayesian analysis of beta testing. TEST 15(1): 227-255
37. Mäkinen SJ, Kanniainen J, Peltola I (2014) Investigating Adoption of Free Beta Applications in a Platform-Based Business Ecosystem. Journal of Product Innovation Management 31(3): 451-465
38. Jiang Z, Sarkar S, Jacob VS (2012) Postrelease testing and software release policy for enterprise-level systems. Information Systems Research 23(3-part-1): 635-657

39. Kanoun K, Martini M, Souza J (1991) A method for software reliability analysis and prediction application to the TROPICO-R switching system. IEEE Trans. on Software Engineering 17 (4): 334–344
40. Huang CY, Lyu MR (2011) Estimation and analysis of some generalized multiple change-point software reliability models. IEEE Transactions on Reliability 60(2): 498-514
41. SAS Institute Inc (2010) SAS/ETS® 9.22 User's Guide. Cary, NC

Switching-Algebraic Analysis of System Reliability

Ali Muhammad Rushdi and Mahmoud Ali Rushdi

Abstract This chapter deals with the paradigm of handling system reliability analysis in the Boolean domain as a supplement to (rather than a replacement to) analysis in the probability domain. This paradigm is well-established within the academic circles of reliability theory and engineering, albeit virtually unknown outside these circles. The chapter lists and explains arguments in favor of this paradigm for systems described by verbal statements, fault trees, block diagrams, and network graphs. This is followed by a detailed exposition of the pertinent concept of the Real or Probability Transform of a switching (two-valued Boolean) function, and that of a Probability-Ready Expression (*PRE*). Some of the important rules used in generating a *PRE* are presented, occasionally along with succinct proofs. These include rules to *achieve* disjointness (orthogonality) of ORed formulas, and to *preserve* statistical independence, as much as possible, among ANDed formulas. Recursive relations in the Boolean domain are also discussed, with an application to the four versions of the AR algorithm for evaluating the reliability and unreliability of the k-out-of-n:G and the k-out-of-n:F systems. These four versions of the algorithm are explained in terms of signal flow graphs that are compact, regular, and acyclic, in addition to being isomorphic to the Reduced Ordered Binary Decision Diagram (ROBDD). An appendix explains some important properties of the concept of Boolean quotient, whose expectation in Boolean-based probability is the counterpart of conditional probability in event-based probability.

A.M. Rushdi (✉)
Department of Electrical and Computer Engineering,
Faculty of Engineering, King Abdulaziz University,
P.O. Box 80204, Jeddah 21589, Saudi Arabia
e-mail: arushdi@kau.edu.sa

M.A. Rushdi
Department of Electronics and Communications Engineering,
Faculty of Engineering, Cairo University, Giza 12613, Egypt
e-mail: mahmoud.rushdi@gmail.com

M. Ram and J.P. Davim (eds.), *Advances in Reliability and System Engineering*,
Management and Industrial Engineering, DOI 10.1007/978-3-319-48875-2_6

Keywords System reliability · Boolean domain · Probability transform · Probability-ready expression · Recursive relation · Signal flow graph · k-out-of-n system

1 Introduction

This chapter deals with the paradigm of handling system reliability analysis in the Boolean domain as a supplement to (rather than a replacement to) analysis in the probability domain. This paradigm is called *switching-algebraic analysis of reliability* or (two-valued) Boolean analysis of reliability. It is also termed *Logical Probabilistic Analysis (LPA)* by Ryabinin [1], who traced it back to great pioneers such as Boole, Poretskii, and Bernstein. Ryabinin also rebutted dubious criticism cast against *LPA*, and lamented that *LPA* is missing in prominent publications on probability theory, including those by celebrated scholars such as Markov and Kolmogorov. A modern revival of *LPA* was started in 1963 by Merekin [2] and Premo [3] and later in 1973 by Fratta and Montanari [4]. After that, applications of LPA in system reliability analysis were covered by literally hundreds of publications in prestigious journals (see, e.g., [5–62]). It is therefore astonishing that *LPA* has so far not found its way to popular texts on probability (see, e.g., [63–72]), and is still virtually unknown outside the circles of reliability theorists and engineers. This state of affairs is possibly an obstacle hindering fruitful interplay between system reliability and many of its potential fields of application. This chapter is, therefore, an attempt to review reliability applications of *LPA* and enhance awareness of the scientific community about it. It is also an invitation to explore its utility in other parts of probability theory that deal with generalized Bernoulli trials.

System reliability analysis deals with expressing the reliability of a system in terms of the reliabilities of its constituent components. Hence, this analysis entails advanced applications of probability theory, and is consequently "*based on the algebra of events (a version of set algebra), which is isomorphic to the bivalent or 2-valued Boolean algebra (switching algebra)*" [56]. Besides using the algebra of events in what is termed probabilistic or arithmetic analysis [73, 74], modern system reliability analysis also utilizes Boolean algebra by employing the indicator variables for probabilistic events instead of the events themselves. Subsequently, expectations of these indicator variables are used to represent probabilities of the corresponding events. In particular, expectations of system success and failure replace system reliability and unreliability, respectively.

An approach bearing certain similarities to the paradigm exposed herein (without going to the Boolean domain) studies probability via expectations [65]. It is based on an axiomatization of probability theory using expectations instead of probabilities (i.e., using linear operators instead of measures). A related approach is suggested by Feller [63, p. 219] who states that "*The reduction of probability theory to random variables is a short-cut to the use of analysis and simplifies the theory in many ways. However, it also has the drawback of obscuring the probability background.*" The paradigm exposed herein handles probability in terms of indicator random Bernoulli variables, but it does not obscure the probability background, since it stresses that the random variables are indicators for events, and it keeps two parallel schemes of event-based probability and Boolean-based probability. These schemes are compared in Table 1 which cites well-known results from elementary probability theory. The first scheme utilizes the concepts of probabilities of events, union and intersection operators, and conditional probability, while the second scheme replaces these by expectations of event indicators, logical "OR" and "AND" operators, and Boolean quotient (see Appendix A). Table 1 shows how probability formulas involving the union and intersection operators become simpler when the pertinent events are mutually exclusive or statistically independent, respectively. As indicated in Table 2, the indicator variable I_A is a Bernoulli variable of two values 0 and 1, such that

$$I_A = 0 \quad \text{iff the event } A \text{ does not occur,} \tag{1a}$$

$$I_A = 1 \quad \text{iff the event } A \text{ occurs.} \tag{1b}$$

Table 2 also indicates definitions of the Bernoulli variables $\overline{I_A}$, $I_A \vee I_B$, $I_A \wedge I_B$, so that the following expectations are obtained

$$E\{I_A\} = Pr\{A\}, \tag{2a}$$

$$E\{\overline{I_A}\} = Pr\{\bar{A}\}, \tag{2b}$$

$$E\{I_A \vee I_B\} = Pr\{\bar{A} \cap B\} + Pr\{A \cap \bar{B}\} + Pr\{A \cap B\} = Pr\{A \cup B\}, \tag{2c}$$

$$E\{I_A \wedge I_B\} = Pr\{A \cap B\}. \tag{2d}$$

In the sequel, we will use the symbols S and $\bar{S}$ to denote system success and failure, respectively, with their expectations $R = E\{S\}$ and $U = E\{\bar{S}\}$ representing system reliability and unreliability, respectively. Subscripts of obvious meanings will be added to distinguish systems of particular types. The corresponding symbols X_i, $\bar{X}_i$, $p_i = E\{X_i\}$, and $q_i = E\{\bar{X}_i\}$ will denote the success, failure, reliability, and unreliability, respectively, for component of number $i, 1 \leq i \leq n$. Unless otherwise stated, component successes are assumed statistically independent all throughout this chapter. A Boolean expression for the indicator of system success will be presented as a function of the indicators of component successes. Transition from the Boolean domain to the probability domain is achieved via the *Real Transform* [75–82] which can be viewed as an expression of system reliability (unreliability) as a function of component reliabilities (unreliabilities).

The organization of the rest of this chapter is as follows. Section 2 lists arguments for handling system reliability in the Boolean domain before going to the probability domain. Section 3 reviews the basic concepts of the Real or Probability Transform of a switching function. Section 4 presents the definition of a

Table 1 Comparison of event-based and boolean-based probability relations

	Event-based probability	Boolean-based probability
Union (disjunction/ORing)	$Pr(A \cup B) = Pr(A) + Pr(B) - Pr(A \cap B)$	$E(I_A \vee I_B) = E(I_A) + E(I_B) - E(I_A \wedge I_B)$
	$Pr(A \cup B) = Pr(A) + Pr(B)$ when A and B are mutually-exclusive events	$E(I_A \vee I_B) = E(I_A) + E(I_B)$ when I_A and I_B are disjoint/orthogonal indicators
Intersection (conjunction/ANDing)	$Pr(A \cap B) = Pr(A)Pr(B\|A)$ where $Pr(B\|A)$ denotes the conditional probability of B given A	$E(I_A \wedge I_B) = E(I_A)E(I_B/I_A)$ where (I_B/I_A) denotes the Boolean quotient of I_B given I_A
	$Pr(A \cap B) = Pr(A)Pr(B)$ *when A and B are statistically – independent events*	$E(I_A \wedge I_B) = E(I_A)E(I_B)$ *when I_A and I_B are statistically – independent indicators*
Total probability	$Pr\{A\} = \sum_{i=1}^{m} Pr\{A\|B_i\}\, Pr\{B_i\}$ where the events B_i are mutually exclusive and exhaustive	$E\{I_A\} = \sum_{i=1}^{m} E\{I_A/I_{B_i}\}E\{I_{B_i}\}$ where the indicators I_{B_i} constitute an orthonormal set
Bayes' theorem	$Pr(B_i\|A) = Pr(A \cap B_i)/Pr(A)$, $Pr(A) \neq 0$.	$E\{I_{B_i}/I_A\} = E(I_A \wedge I_{B_i})/E(I_A)$, $E(I_A) \neq 0$.

Table 2 Indicator variable for set complementation

A	$\bar{A}$	I_A	$\overline{I_A}$
Does not occur	Occurs	0	1
Occurs	Does not occur	1	0

Table 3 Indicator variables for set union and intersection

A	B	$A \cup B$	$A \cap B$	I_A	I_B	$I_A \vee I_B$	$I_A \wedge I_B$
Does not occur	Does not occur	Does not occur	Does not occur	0	0	0	0
Does not occur	Occurs	Occurs	Does not occur	0	1	1	0
Occurs	Does not occur	Occurs	Does not occur	1	0	1	0
Occurs	Occurs	Occurs	Occurs	1	1	1	1

Probability-Ready Expression (*PRE*) and lists some of the rules used in generating a *PRE*. Section 5 discusses recursive relations in the Boolean domain. Section 6 concludes the chapter (Table 3).

2 Advantages of Working in the Switching (Boolean) Domain

There is a large potpourri of ways for describing a reliability system, including verbal statements, fault trees, block diagrams, and network graphs. Generally, reliability analysis of a system conducted in the probabilistic (arithmetic) domain is lengthy and error-prone [73, 74]. We will now discuss other arguments for working in the switching (Boolean) domain for each specific type of system description:

- For systems described by *verbal statements*, work in the Boolean domain allows a separation and distinction between the mathematical formalization of a reliability problem, and the algorithmic derivation of its formal solution. Table 4 shows the verbal description versus the Boolean specification of some common reliability systems, mostly of the k-out-of-n or related types. For lack of space, we omitted more sophisticated systems including some dual, multidimensional, multi-valued extensions of the reported systems. It is clear that the word statement and the mathematical description of each reported system are immediately and obviously equivalent. There is no mental strain or conceptual difficulty in going from the somewhat vague realm of language to the exactly precise arena of mathematics. The primary Boolean description is in terms of

Table 4 Boolean specification of some common reliability systems

System name	Verbal description	Primary boolean specification	Secondary (complementary) boolean specification
Series (nonredundant) system	"A system that is good iff all its n components are good" [56]	$S_S = \bigwedge_{i=1}^{n} X_i$	$\bar{S}_S = \vee_{i=1}^{n} \bar{X}_i$
Parallel (totally-redundant) system	"A system that is failed iff all its n components are failed" [56]	$\overline{S_P} = \bigwedge_{i=1}^{n} \bar{X}_i$	$S_P = \vee_{i=1}^{n} X_i$
k-out-of-n:G (partially-redundant) system $(1 \leq k \leq n)$	"A system that is good iff at least k out of its n components are good" [27, 39, 61]	$S_G(k,n) = \bigvee_{\boldsymbol{K}} (\bigwedge_{a \in \boldsymbol{K}} X_a)$	$\bar{S}_G(k,n) = \bigwedge_{\boldsymbol{K}} (\bigvee_{a \in \boldsymbol{K}} \bar{X}_a)$
		$\boldsymbol{K}$ is a subset of $\{1, 2, \ldots, n\}$ with cardinality $\lvert\boldsymbol{K}\rvert = k$	
k-out-of-n:F (Partially-redundant) system $(1 \leq k \leq n)$	"A system that is failed iff at least k out of its n components are failed" [27, 39, 61]	$\bar{S}_F(k,n) = \bigvee_{\boldsymbol{K}} (\bigwedge_{a \in \boldsymbol{K}} \bar{X}_a)$	$S_F(k,n) = \bigwedge_{\boldsymbol{K}} (\bigvee_{a \in \boldsymbol{K}} X_a)$
		$\boldsymbol{K}$ is a subset of $\{1, 2, \ldots, n\}$ with cardinality $\lvert\boldsymbol{K}\rvert = k$	
Linear consecutive-k-out-of-n:F system	"A sequence of n components arranged in a straight line such that the system fails iff at least k consecutive components fail" [32, 52, 83–85]	$\bar{S}_L(k,n) = \bigvee_{i=1}^{n-k+1} (\bigwedge_{j=1}^{k} \bar{X}_{i+j-1})$	$S_L(k,n) = \bigwedge_{i=1}^{n-k+1} (\bigvee_{j=1}^{k} X_{i+j-1})$
Circular consecutive-k-out-of-n:F system	"A sequence of n components arranged on a circle such that the system fails iff at least k consecutive components fail" [33, 52, 83–85]	$\bar{S}_C(k,n) = \bigvee_{i=1}^{n} (\bigwedge_{j=1}^{k} \bar{X}_{\{i+j-1\}})$	$S_C(k,n) = \bigwedge_{i=1}^{n} (\bigvee_{j=1}^{k} X_{\{i+j-1\}})$
		$\{m\}_n$ is the integer m reduced modulo n to the interval [1, n]	
(n, f, k) system	"A sequence of n components arranged in a straight line or a circle such that the system fails iff at least f components fail OR at least k consecutive components fail" [52, 88–91]	$\overline{S_{OL}}(n,f,k) = \overline{S_F}(f,n) \vee \overline{S_L}(k,n)$ $\overline{S_{OC}}(n,f,k) = \overline{S_F}(f,n) \vee \overline{S_C}(k,n)$	$S_{OL}(n,f,k) = S_F(f,n) \wedge S_L(k,n)$ $S_{OC}(n,f,k) = S_F(f,n) \wedge S_C(k,n)$
$< n,f,k >$ system	"A sequence of n components arranged in a straight line or a circle such that the system fails iff at least f components fail AND at least k consecutive components fail" [89, 91]	$\overline{S_{AL}}(n,f,k) = \overline{S_F}(f,n) \wedge \overline{S_L}(k,n)$ $\overline{S_{AC}}(n,f,k) = \overline{S_F}(f,n) \wedge \overline{S_C}(k,n)$	$S_{AL}(n,f,k) = S_F(f,n) \vee S_L(k,n)$ $S_{AC}(n,f,k) = S_F(f,n) \vee S_C(k,n)$
Threshold system (weighted k-out-of-n: G system)	"A system that is good iff a certain weighted sum of its component successes equals or exceeds a certain threshold" [36, 52, 53, 55, 60, 61, 86, 87]	Both S_T and $\overline{S_T}$ are threshold switching functions $S_T(\boldsymbol{X}) = 1$ iff $\sum_{i=1}^{n} W_i X_i \geq T$ where T is the system threshold and W_i is the weight of component i	

system success or failure, depending on the nature of the system, with the complementary description delegated to a secondary role. Once in the Boolean domain, one can solve the problem of reliability evaluation by utilizing an extensive set of algorithmic tools and visual aids. For most of the systems in Table 4, both probabilistic and Boolean solutions exist. For some of the systems, such as the k-out-of-n systems, Boolean solutions [27] were earlier to develop. For other systems, such as the consecutive-k-out-of-n ones, Boolean solutions [32, 33] arrived later, as an afterthought, to provide a more insightful and easy-to-comprehend alternative.

- *Formulation* of a system reliability problem in the Boolean domain is the natural strategy to use for systems described by *fault trees* [9, 10, 44, 49, 50, 58], which are composed of logical AND–OR gates, and can similarly be utilized with systems described by *block diagrams* [15]. Fault trees are used to relate probabilistic events, with its inputs being basic events and its output being the top event. However, it could be more convenient to apply Boolean-based probability to fault trees by using indicators for events rather the events themselves [57, 62, 78, 92].
- *Formulation* in the Boolean domain is the method of choice with systems described by network graphs when these systems are handled through the enumeration of minimal pathsets and minimal cutsets [12, 23, 26, 48], rather than other graph techniques.

Work in the Boolean domain might allow the treatment of certain types of statistical dependencies among component successes. For example, in the case of *total* positive dependency $(X_i = X_j)$ idempotency leads to $(X_i \vee X_j = X_i)$ and $(X_i \wedge X_j = X_i)$, while in the case of *total* negative dependency $(X_i = \bar{X}_j)$ orthogonality leads to $(X_i \vee X_j = 1)$ and $(X_i \wedge X_j = 0)$. Cases of *partial* dependencies might be handled via expectations of Boolean quotients.

Work in the Boolean domain also provides the insight necessary to avoid traps and pitfalls, as well as to detect anomalies, inconsistencies, and errors of the third kind (errors of solving the wrong problem). A case in point is that of the so-called *strict consecutive-k-out-of-n:G(F) system* [93–96], which lacks a definition that avoids ambiguity, inconsistency, or self-contradiction. Unfortunately, many capable and great mathematicians wasted their precious time producing elegant solutions for what they mistakenly assumed to be this system. Insight of the Boolean domain allowed a critical review of literature that partially prevented the publication of more such *irrelevant* work.

Without resort to the Boolean domain, one might be obliged to use the notorious *Inclusion–Exclusion (IE) Principle* to compute the probability of the union of n events [63, 67], where n might be large. Use of the IE Principle in system reliability analysis is undesirable, since "*(a) it produces an exponential number of terms that have to be reduced subsequently* via *addition and cancellation, and (b) it involves too many subtractions, making it highly sensitive to round-off errors, and possibly leading to catastrophic cancellations*" [52, 54, 56].

3 The Real Transform of a Switching Function

Various forms of the Real Transform (also called the Probability or Arithmetic transform) are discussed in [75–82]. The definition given in [79, 82] states that the real transform $R(\boldsymbol{p}) = R(p_1, p_2, \ldots, p_n)$ of a switching function $f(\boldsymbol{X})$ possesses the following two properties:

(a) $R(\boldsymbol{p})$ is a multiaffine continuous real function of continuous real variables $\boldsymbol{p} = [p_1\, p_2 \ldots p_n]^{\mathrm{T}}$, i.e., $R(\boldsymbol{p})$ is a first-degree polynomial in each of its arguments p_i.

(b) $R(\boldsymbol{p})$ has the same "truth table" as $f(\boldsymbol{X})$, i.e.

$$R(\boldsymbol{p} = \boldsymbol{t}_j) = f(\boldsymbol{X} = \boldsymbol{t}_j), \quad \text{for}\, j = 0, 1, \ldots, (2^n - 1), \tag{3}$$

where $\mathbf{t}_j$ is the jth input line of the truth table; $\mathbf{t}_j$ is an n-vector of binary components such that

$$\sum_{i=1}^{n} 2^{n-i} t_{ji} = j, \quad \text{for}\, j = 0, 1, \ldots, (2^n - 1). \tag{4}$$

We stress that property (b) above does not suffice to produce a unique $R(\boldsymbol{p})$ and it must be supplemented by the requirement that $R(\boldsymbol{p})$ be multiaffine to define $R(\boldsymbol{p})$ uniquely [77, 79]. We also note that if the Real Transform R and its arguments $\mathbf{p}$ are restricted to discrete binary values (i.e., if $R : \{0, 1\}^n \rightarrow \{0, 1\}$) then R becomes the *multilinear form* of a switching function [11, 97]. This form is typically referred to as the *structure function* [98, 99] in system reliability, and is a way to mimic Boolean algebra in terms of arithmetic operators rather than Boolean ones.

The definition above for $R(\boldsymbol{p})$ implies that it is a function from the n-dimensional real space to the real line $(R(\boldsymbol{p}) : \boldsymbol{R}^n \rightarrow \boldsymbol{R})$. Though both R and $\boldsymbol{p}$ could be free real values, they have very interesting interpretations as probabilities, i.e., when restricted to the [0.0, 1.0] and $[0.0, 1.0]^n$ real intervals. An important property of the Real Transform $R(\boldsymbol{p})$ is that if its vector argument or input $\boldsymbol{p}$ is restricted to the domain within the n-dimensional interval $[0.0, 1.0]^n$, i.e., if $0.0 \leq p_i \leq 1.0$ for $1 \leq i \leq n$, then the image of $R(\boldsymbol{p})$ will be restricted to the unit real interval [0.0, 1.0].

The Real Transform is a bijective (one-to-one and onto) mapping from the set of switching functions to the subset of multiaffine functions such that if the function's domain is the power binary set $\{0, 1\}^n$ then its image belongs to the binary set $\{0, 1\}$. Evidently, an $R(\boldsymbol{p})$ restricted to binary values whenever its arguments are restricted to binary values can produce the "truth table" that completely specifies its inverse image $f(\boldsymbol{X})$ via (3). On the other hand, a multiaffine function of n variables is completely specified by 2^n independent conditions [77, 79], e.g., the ones in (3).

The transform $R(\boldsymbol{p})$ is related to the original function $f(\boldsymbol{X})$ via the truth table in (3), which consists of 2^n lines. Therefore, the complexity of implementing the Real

Transform, i.e., of going from $f(\boldsymbol{X})$ to $R(\boldsymbol{p})$ is exponential, and the problem of implementing the Real Transform is generally intractable. However, if $f(\boldsymbol{X})$ is cast in *PRE* form, then it can be converted trivially (at no cost) to $R(\boldsymbol{p})$. In this case, the burden of achieving the Real Transform is shifted to the Boolean domain where the initial formula for $f(\boldsymbol{X})$ is to be converted (at a potentially reduced cost) to a *PRE*.

4 Probability-Ready Expressions

A Reliability-Ready Expression (*RRE*) is an expression in the switching (Boolean) domain that can be directly transformed, on a one-to-one basis, to its Real or Probability Transform by replacing switching (Boolean) indicators by their statistical expectations, and also replacing logical multiplication and addition (ANDing and ORing) by their arithmetic counterparts [25, 42, 49, 50, 54, 56]. We now present some useful rules that might be required in the conversion of an arbitrary switching (Boolean) expression to an equivalent compact *PRE*:

1. A switching expression is a *PRE* expression if
   (a) all *ORed* terms are *orthogonal* (*disjoint*), and
   (b) all *ANDed* sums are *statistically independent.*

   The conversion is achieved by replacing Boolean variables by their expectations, AND operations by arithmetic multiplications, and OR operations by arithmetic additions. While there are literally hundreds of methods to introduce characteristic (a) of orthogonality (disjointness) into a Boolean expression (see, e.g., Bennetts [9, 15], Abrahams [13], Dotson and Gobien [14], and Rushdi [17, 20]) there is no way to induce characteristic (b) of statistical independence. The best that one can do is to observe statistical independence when it exists, and then take care to preserve it and take advantage of it [25, 42, 62].

2. As an analog to series-parallel reduction, any set of ANDed or ORed variables that do not appear elsewhere in a formula should be combined into a single variable. This might considerably reduce the number of variables involved, and helps in producing a more compact expression.
3. Two terms are disjoint if they have at least a single *opposition*, i.e., if there is at least one variable that appears complemented in one term and appears un-complemented in the other, e.g., the two terms $A\bar{B}C$ and BC are disjoint since the complemented literal $\bar{B}$ appears in $A\bar{B}C$ while the un-complemented literal B appears in BC.
4. If neither of the two terms A and B in the sum $(A \vee B)$ subsumes the other ($A \vee B \neq A$ and $A \vee B \neq B$) and the two terms are not disjoint ($A \wedge B \neq 0$), then B can be disjointed with A by the relation

$$A \vee B = A \vee B(\overline{y_1 y_2 \ldots y_e})$$
$$= A \vee B(\overline{y_1} \vee y_1\overline{y_2} \vee \ldots \vee y_1 y_2 \ldots y_{e-1}\overline{y_e}), \quad (5)$$

where $\{y_1, y_2, \ldots, y_e\}$ is the set of literals that appear in the term A and do not appear in the term B. Note that the term B is replaced by e ($\geq$1) terms that are disjoint with one another besides being disjoint with the term A [9, 15, 54, 56]. Formula (5) might be seen as an immediate extension of the familiar *Reflection Law* [100–105].

$$a \vee b = a \vee (b \wedge \bar{a}). \quad (5a)$$

A formal proof of (5) is possible via *perfect induction* over a *much reduced truth table* that consists of the $(e + 1)$ orthonormal cases $\{y_1 = 0\}$, $\{y_1 = 1, y_2 = 0\}$, $\{y_1 = y_2 = 1, y_3 = 0\}, \ldots,$ $\{y_1 = y_2 = \cdots = y_{e-1} = 1, y_e = 0\}$, and $\{y_1 = y_2 = \cdots = y_{e-1} = y_e = 1\}$. For the first e cases, the L.H.S. = the R.H.S = B. For the last case, the L.H.S. = $A/y_1y_2 \cdots y_e \vee B$, while the R.H.S. = $A/y_1y_2 \cdots y_e$. The two sides are equal since B subsumes $A/y_1y_2 \cdots y_e$, and hence is absorbed in it. The term $(A/y_1y_2 \cdots y_e)$ is the *Boolean quotient* obtained by restricting A through the assignment $\{y_1 = y_2 = \cdots y_e = 1\}$ (See Appendix A). In passing, we note that the roles of A and B can be reversed in (5), and that (5) does not guarantee minimality of the resulting disjointed expression [51]. Disjointness typically results in an (often dramatic) increase in the number of terms in sum-of-products (sop) switching expressions. However, there are more sophisticated disjointness techniques that achieve "*shellability*", i.e., they obtain a disjoint sop expression that retains the same number of terms as the original sop expression [34, 46, 55, 105]. Notable examples of shellable expressions include those of the success or failure of a k-out-of-n:G or a k-out-of-n:F system [39], and that of the success of a coherent threshold system [55].

5. Given a term P and two sums of products A and B where A has no variables in common with P, while B shares some variables with P, then the logical product $(A \wedge B)$ is disjointed with P by disjointing B and P and keeping A unchanged [42].
6. The complements of a sum and a product are given by a product and a disjoint sum, namely, *PRE* versions of De Morgan's Laws [102]

$$\overline{\vee_{i=1}^{n} A_i} = \wedge_{i=1}^{n} \bar{A}_i, \quad (6)$$

$$\overline{\wedge_{i=1}^{n} A_i} = \overline{A_1} \vee A_1\overline{A_2} \vee \cdots \vee A_1 A_2 \cdots A_{n-1}\overline{A_n}. \quad (7)$$

A formal proof of (6) is achieved by considering just two exhaustive cases, namely: (a) the case of all A_i's being 0, for which the L.H.S. = the R.H.S = 1, and (b) the case when at least one A_i is 1, for which the L.H.S. = the R.H.S. = 0.

Likewise, a formal proof of (7) is obtained via *perfect induction* over the $(n+1)$ orthonormal cases $\{A_1 = 0\}$, $\{A_1 = 1, A_2 = 0\}$, $\{A_1 = A_2 = 1, A_3 = 0\}, \ldots$, $\{A_1 = A_2 = \ldots = A_{n-1} = 1, A_n = 0\}$, and $\{A_1 = A_2 = \ldots = A_{n-1} = A_n = 1\}$. For the first n cases, the L.H.S. = the R.H.S = 1. For the last case, the L.H.S. = the R.H.S. = 0.

7. The most effective way for converting a Boolean formula into a *PRE* form is the Boole–Shannon Expansion

$$f(\boldsymbol{X}) = (\bar{X}_i \wedge f(\boldsymbol{X}|0_i)) \vee (X_i \wedge f(\mathbf{X}|1_i)), \tag{8}$$

which expresses a (two-valued) Boolean function $f(\boldsymbol{X})$ in terms of its two subfunctions $f(\boldsymbol{X}|0_i)$ and $f(\boldsymbol{X}|1_i)$. These subfunctions are equal to the Boolean quotients $f(\boldsymbol{X})/\bar{X}_i$ and $f(\boldsymbol{X})/X_i$, and hence are obtained by restricting X_i in the expression $f(\boldsymbol{X})$ to 0 and 1, respectively. If $f(\boldsymbol{X})$ is a sop expression of n-variables, the two subfunctions $f(\boldsymbol{X}|0_i)$ and $f(\boldsymbol{X}|1_i)$ are functions of at most $(n - 1)$ variables. A formal proof of (8) is achieved by considering just two exhaustive cases, namely: (a) the case $\{X_i = 0\}$, for which the L.H.S. = the R.H.S $= f(\boldsymbol{X}|0_i)$, and (b) the case $\{X_i = 1\}$, for which the L.H.S. = the R.H.S. $= f(\boldsymbol{X}|1_i)$. The expansion (8) serves our purpose very well. Once the subfunctions in (8) are expressed by *PRE* expressions, $f(\boldsymbol{X})$ will be also in *PRE* form, thanks to the facts that (a) The R.H.S. of (8) has two disjoint parts, with the first part containing the complemented literal $\bar{X}_i$ and the second part containing the un-complemented literal X_i, and (b) Each of these two parts is a product of two statistically-independent entities. The Boole–Shannon Expansion in the Boolean domain is equivalent to the *Total Probability Theorem* [67] in the probability domain and to the *Factoring Theorem* [56] in the "Graph Domain". A visual aid or tool to implement this expansion is provided by the Variable-Entered Karnaugh Map (VEKM) [17].

8. Let a success formula be written in the form

$$S = N_1M_1 \vee N_2M_2, \tag{9}$$

where the set of formulas $\{N_1, N_2\}$ is statistically independent of the set of formulas $\{M_1, M_2\}$. Expression (9) might result due to path enumeration via network decomposition [22, 106–109]. A *PRE* formula of S (that preserves the original statistical independence as much as possible) is [22]

$$S_{PRE} = N_{1(PRE)}M_{1(PRE)} \vee (\overline{N_1}N_2)_{(PRE)}M_{2(PRE)} \vee (N_1N_2)_{(PRE)}(\overline{M_1}M_2)_{(PRE)}. \tag{10}$$

A formal proof of (10) is possible via *perfect induction* over a *much reduced truth table* that consists of only the three orthonormal cases $\{N_1 = 0\}$, $\{N_1 = 1, M_1 = 0\}$, and $\{N_1 = M_1 = 1\}$. For the first two cases, the L.H.S. = the

R.H.S. = $N_{2(PRE)}M_{2(PRE)}$. For the third case, the L.H.S. = the R.H.S = 1. Note that this proof entails a partitioning of the higher-level space of N_1 and M_1 rather than a partitioning of the space of the underlying arguments $\boldsymbol{X}$.

9. Products might be kept compact by utilizing the following rules of *disjoint intelligent multiplication*

$$(a \vee x)(a \vee y) = a \vee \bar{a}xy, \tag{11}$$

$$(a \vee x)(\bar{a} \vee y) = \bar{a}x \vee ay. \tag{12}$$

Formula (11) results due to the absorption of the terms ax and ay in their subsumed term a, and then using the Reflection Law (5a). The term xy that would have appeared in formula (12) is deleted since it is the consensus of $\bar{a}x$ and ay, and hence it is covered by their disjunction $\bar{a}x \vee ay$. A corollary of (11) is

$$\overline{ab} \wedge \overline{ac} \wedge \overline{ad} = \bar{a} \vee a\bar{b}\bar{c}\bar{d}. \tag{13}$$

The complemented form of (13) is a ANDed with the complement of $\bar{b}\bar{c}\bar{d}$. This complemented form allows separating the common factor a in the formula without introducing the OR ($\vee$) operator in it, and hence leads to full utilization of statistical independence [25].

10. When one has a *partially-factored* expression of system success or failure such as the following one [56]

$$S = X_3 \vee X_7 \vee (X_5 \vee X_8)(X_4 \vee X_2X_9) \vee (X_1 \vee X_6)(X_2 \vee X_4X_9), \tag{14}$$

then it should be converted into *PRE* form via (10) without spoiling or expanding the factored form, so as to *preserve statistical independence* [22, 25, 42, 56]

$$\begin{aligned} S_{PRE} = {} & X_3 \vee \overline{X_3}\big(X_7 \vee \overline{X_7}\big(\big(X_5 \vee \overline{X_5}X_8\big)\big(X_4 \vee \overline{X_4}X_2X_9\big) \\ & \vee\big(X_1 \vee \overline{X_1}X_6\big)\big(\overline{X_5}\overline{X_8}\big(X_2 \vee \overline{X_2}X_4X_9\big) \vee \big(X_5 \vee \overline{X_5}X_8\big)X_2\overline{X_4X_9}\big)\big)\big). \end{aligned} \tag{15}$$

This transforms on a one-to-one basis to the reliability expression [56]

$$\begin{aligned} R = {} & p_3 + q_3(p_7 + q_7((p_5 + q_5p_8)(p_4 + q_4p_2p_9) \\ & + (p_1 + q_1p_6)(q_5q_8(p_2 + q_2p_4p_9) + (p_5 + q_5p_8)p_2q_4q_9))). \end{aligned} \tag{16}$$

5 Recursive Relations in the Boolean Domain

This Section provides a novel discussion about the utility of the Boolean domain in deriving recursive relations for the two versions of the k-out-of-n system. The reader is referred to [32, 33] for Boolean-domain derivations of recursive relations for consecutive systems. The success $S_G(k,j,\boldsymbol{X}_j)$ of a k-out-of-j:G system is a symmetric monotonically nondecreasing switching function, and hence satisfies recursive relations obeyed by such a function [27]. Equivalently, it is given by its Boole–Shannon expansion about its last argument X_j (namely, Eq. (5.40a) of Rushdi [39]):

$$S_G(k,j,\boldsymbol{X}_j) = \overline{X}_j S_G(k,j-1,\boldsymbol{X}_{j-1}) \vee X_j S_G(k-1,j-1,\boldsymbol{X}_{j-1}), 1 \le k \le j \le n, \quad (17a)$$

where $\boldsymbol{X}_j = [X_1, X_2, \ldots, X_j]^{\mathrm{T}}$ is the vector of the first j component successes. The R.H.S. of Eq. (17a) involves two subfunctions $S_G(k,j-1,\boldsymbol{X}_{j-1})$ and $S_G(k-1, j-1,\boldsymbol{X}_{j-1})$ of the original success function. The region of validity of (17a) in the kj-plane is bordered by two straight lines on which the following two boundary conditions hold

$$S_G(k,j,\boldsymbol{X}_j) = 1, \quad k = 0, j \ge 1, \quad (17b)$$

$$S_G(k,j,\boldsymbol{X}_j) = 0, \quad k = j+1, j \ge 1, \quad (17c)$$

Equations (17) are in *PRE* form, with Probability Transforms

$$R_G(k,j,\boldsymbol{p}_j) = q_j R_G(k,j-1,\boldsymbol{p}_{j-1}) + p_j R_G(k-1,j-1,\boldsymbol{p}_{j-1}), \quad 1 \le k \le j \le n, \quad (18a)$$

$$R_G(k,j,\boldsymbol{p}_j) = 1, \quad k = 0, j \ge 1, \quad (18b)$$

$$R_G(k,j,\boldsymbol{p}_j) = 0, \quad k = j+1, j \ge 1, \quad (18c)$$

which govern the reliability of the k-out-of-j:G system. Based on the recursive relation (18a) together with the boundary conditions (18b and 18c), a quadratic-time *iterative* algorithm was developed by Rushdi [27], and later named the AR algorithm [37, 39]. The AR algorithm has the beautiful characteristic of having the same form (and hence same complexity) for computing both the reliability and unreliability of either the k-out-of-n:G system or its dual the k-out-of-n:F system. Table 5 illustrates this point by listing various recursive relations together with boundary conditions for the two dual types of k-out-of-n systems in the Boolean and probabilistic domains. Entries in the lower two rows of Table 5 are Probability Transforms of the corresponding entries of the upper two rows, which happen to be in *PRE* form. Entries in the left column of Table 5 are inversions of the

Table 5 Various recursive relations together with boundary conditions for the two dual types of k-out-of-n systems in the Boolean and probabilistic domains

k-out-of-n:G system	$\bar{S}_G(k,j,\boldsymbol{X}_j) = \overline{X_j}\ \bar{S}_G(k,j-1,\boldsymbol{X}_{j-1}) \vee X_j\ \bar{S}_G(k-1,j-1,\boldsymbol{X}_{j-1}),\ 1 \le k \le j \le n,$ $\bar{S}_G(k,j,\boldsymbol{X}_j) = 0,$ $k = 0, j \ge 1,$ $\bar{S}_G(k,j,\boldsymbol{X}_j) = 1,$ $k = j+1, j \ge 1,$	$S_G(k,j,\boldsymbol{X}_j) = \overline{X_j}\ S_G(k,j-1,\boldsymbol{X}_{j-1}) \vee X_j\ S_G(k-1,j-1,\boldsymbol{X}_{j-1}),\ 1 \le k \le j \le n,$ $S_G(k,j,\boldsymbol{X}_j) = 1,$ $k = 0, j \ge 1,$ $S_G(k,j,\boldsymbol{X}_j) = 0,$ $k = j+1, j \ge 1,$
k-out-of-n:F system	$S_F(k,j,\boldsymbol{X}_j) = X_j\ S_F(k,j-1,\boldsymbol{X}_{j-1}) \vee \overline{X_j}\ S_F(k-1,j-1,\boldsymbol{X}_{j-1}),\ 1 \le k \le j \le n,$ $S_F(k,j,\boldsymbol{X}_j) = 0,$ $k = 0, j \ge 1,$ $S_F(k,j,\boldsymbol{X}_j) = 1,$ $k = j+1, j \ge 1,$	$\bar{S}_F(k,j,\boldsymbol{X}_j) = X_j\ \bar{S}_F(k,j-1,\boldsymbol{X}_{j-1}) \vee \overline{X_j}\ \bar{S}_F(k-1,j-1,\boldsymbol{X}_{j-1}),\ 1 \le k \le j \le n,$ $\bar{S}_F(k,j,\boldsymbol{X}_j) = 1,$ $k = 0, j \ge 1,$ $\bar{S}_F(k,j,\boldsymbol{X}_j) = 0,$ $k = j+1, j \ge 1,$
k-out-of-n:G system	$U_G(k,j,\boldsymbol{p}_j) = q_j\ U_G(k,j-1,\boldsymbol{p}_{j-1}) + p_j\ U_G(k-1,j-1,\boldsymbol{p}_{j-1}),\ 1 \le k \le j \le n,$ $U_G(k,j,\boldsymbol{p}_j) = 0,$ $k = 0, j \ge 1,$ $U_G(k,j,\boldsymbol{p}_j) = 1,$ $k = j+1, j \ge 1,$	$R_G(k,j,\boldsymbol{p}_j) = q_j\ R_G(k,j-1,\boldsymbol{p}_{j-1}) + p_j\ R_G(k-1,j-1,\boldsymbol{p}_{j-1}),\ 1 \le k \le j \le n,$ $R_G(k,j,\boldsymbol{p}_j) = 1,$ $k = 0, j \ge 1,$ $R_G(k,j,\boldsymbol{p}_j) = 0,$ $k = j+1, j \ge 1,$
k-out-of-n:F system	$R_F(k,j,\boldsymbol{p}_j) = p_j\ R_F(k,j-1,\boldsymbol{p}_{j-1}) + q_j\ R_F(k-1,j-1,\boldsymbol{p}_{j-1}),\ 1 \le k \le j \le n,$ $R_F(k,j,\boldsymbol{p}_j) = 0,$ $k = 0, j \ge 1,$ $R_F(k,j,\boldsymbol{p}_j) = 1,$ $k = j+1, j \ge 1,$	$U_F(k,j,\boldsymbol{p}_j) = p_j\ U_F(k,j-1,\boldsymbol{p}_{j-1}) + q_j\ U_F(k-1,j-1,\boldsymbol{p}_{j-1}),\ 1 \le k \le j \le n,$ $U_F(k,j,\boldsymbol{p}_j) = 1,$ $k = 0, j \ge 1,$ $U_F(k,j,\boldsymbol{p}_j) = 0,$ $k = j+1, j \ge 1,$

corresponding entries of the right column. Power of working in the Boolean domain is manifested in the *complementation* and *dualization* operations that transform $S_G(k,j,\boldsymbol{X}_j)$ in (17) to $\bar{S}_G(k,j,\boldsymbol{X}_j)$ and $S_F(k,j,\boldsymbol{X}_j)$, respectively. First, we note that both complementation and dualization change a boundary condition of 1 to 0 and a boundary condition of 0 to 1. We next consider the recursive domain $1 \le k \le j \le n$. Inverting $S_G(k,j,\boldsymbol{X}_j)$ to produce $\bar{S}_G(k,j,\boldsymbol{X}_j)$ is straightforward since $S_G(k,j,\boldsymbol{X}_j)$ When one has a *partially-factored* expression of system success or failure such as the following one [56] basis $\{\bar{X}_j, X_j\}$ with coefficients $S_G(k,j-1,\boldsymbol{X}_{j-1})$ and $S_G(k-1,j-1,\boldsymbol{X}_{j-1})$, respectively, and hence $\bar{S}_G(k,j,\boldsymbol{X}_j)$ is expressed in terms of the *same* orthonormal basis, but with inverted coefficients $\bar{S}_G(k,j-1,\boldsymbol{X}_{j-1})$ and

$\bar{S}_G(k-1, j-1, \boldsymbol{X}_{j-1})$. Since $S_F(k, j, \boldsymbol{X}_j)$ is the *dual* function of $S_G(k, j, \boldsymbol{X}_j)$, it is obtained from it by inverting both its output S_G and inputs (arguments) $\boldsymbol{X}_j$, namely $S_F(k, j, \boldsymbol{X}_j) = \bar{S}_G(k, j, \bar{\boldsymbol{X}}_j)$, and hence it is given by the complement of (17a) with inputs inverted, i.e., it is given by the complement of

$$X_j S_G(k, j-1, \bar{\boldsymbol{X}}_{j-1}) \vee \overline{X_j} S_G(k-1, j-1, \bar{\boldsymbol{X}}_{j-1}).$$

This complement is given by

$$\begin{aligned} & X_j \bar{S}_G(k, j-1, \bar{\boldsymbol{X}}_{j-1}) \vee \overline{X_j} \bar{S}_G(k-1, j-1, \bar{\boldsymbol{X}}_{j-1}) \\ & \quad = X_j S_F(k, j-1, \boldsymbol{X}_{j-1}) \vee \overline{X_j} S_F(k-1, j-1, \boldsymbol{X}_{j-1}). \end{aligned}$$

Rushdi [52] noted that the AR algorithm is, in fact, an implementation of the Reduced Ordered Binary Decision Diagram (ROBDD) strategy when this strategy is adapted for computing the k-out-n-reliability. The ROBDD strategy was proposed by Bryant [110] as an extension of the BDD methodology of Akers [111]. The ROBDD deals with *general* switching (two-valued Boolean) functions, and is now considered the state-of-the-art data structure for handling such functions, with extensive applications in reliability [112–122]. As stated earlier, the AR algorithm has a domain of applicability that is narrower than that of the ROBDD algorithm, as it is restricted to switching functions that are both *monotonically nondecreasing* and *totally symmetric*. Apart from this, the AR algorithm has exactly the same features as the ROBDD algorithm, namely:

1. Both the AR and ROBDD algorithms are based on the Boole–Shannon expansion in the Boolean domain.
2. Both algorithms visit the variables in a certain order, typically monotonically ascending or monotonically descending.
3. Both algorithms reduce the resulting expansion tree (which is exponential in size) to a rooted acyclic graph that is both canonical and hopefully compact or sub-exponential. The reduction rules [122] require 3(a) merging isomorphic subtrees, and 3(b) deletion of useless nodes whose outgoing edges point to the same child node.

Figure 1 translates the lower two rows of Table 5 into a quad of Mason Signal Flow Graph (SFGs) for computing the reliability and unreliability of the k_1-out-of-$(k_1 + k_2)$:G system and the k_1-out-of-$(k_1 + k_2)$:F system for $k_1 \leq 0$ and $k_2 \leq -1$. Figure 1 is drawn over a rectangular grid of coordinates $k_1 = k$, $k_2 = n - k_1$. The graphs in Fig. 1 are the essence of the four versions of the iterative AR algorithm. Striking similarities (and trivial, albeit subtle differences) of these graphs with ROBDDs are discussed in Rushdi and Alturki [61].

Though the recursive relation (17a) was discovered initially in the Boolean domain [27], it can be derived, as an afterthought, in the probability domain. For this purpose, we employ the Total Probability Theorem with the two mutually

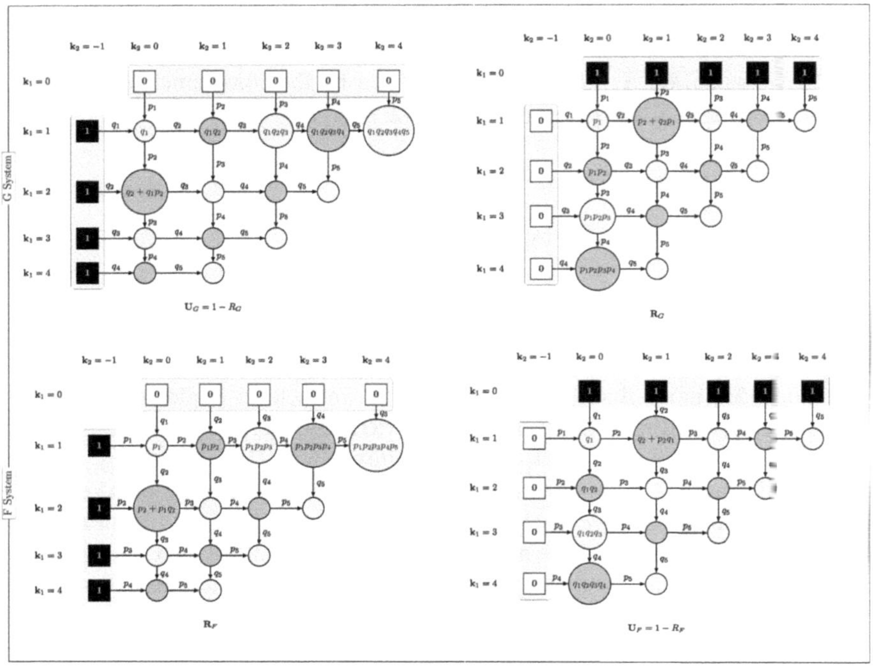

Fig. 1 Translation of the lower two rows of Table 5 into a quad of Mason signal flow graph (SFGs)

exclusive and exhaustive events {component j is failed} and {component j is good}, for $1 \leq k \leq j \leq n$, as follows

$$\begin{aligned}
R_G(k,j,\boldsymbol{p}_j) &= Pr\{\text{at least } k \text{ components out of } j \text{ are good}\} \\
&= Pr\{\text{at least } k \text{ components out of } j \text{ are good} | \text{component } j \text{ is failed}\} \\
&\quad Pr\{component\ j\ is\ failed\} \\
&\quad + Pr\{\text{at least } k \text{ components out of } j \text{ are good} | \text{component } j \text{ is good}\} \\
&\quad Pr\{\text{component } j \text{ is good}\} \\
&= Pr\{\text{at least } k \text{ components out of } (j-1) \text{ are good}\} q_j \\
&\quad + Pr\{\text{at least } (k-1) \text{ components out of } (j-1) \text{ are good}\} p_j \\
&= q_j R_G(k, j-1, \boldsymbol{p}_{j-1}) + p_j R_G(k-1, j-1, \boldsymbol{p}_{j-1}).
\end{aligned}$$

The derivation above should not be viewed as a defeat of purpose of the paradigm presented herein. Simple as it may be, this derivation was arrived at only after (18a) became known, as a transform of the Boolean result (17a).

6 Conclusions

This chapter gives a detailed overview of the current practice of solving system reliability problems by first formulating and manipulating them in the Boolean domain, and subsequently going to the probability domain. This practice is a supplementary alternative of the more dominant practice of working solely in the probability domain all throughout the solution process. The equivalence between event-based probability (which uses probabilities of events) and Boolean-based probability (which employs expectations of indicator variables of such events) is demonstrated. In particular, the expectation of a Boolean quotient is identified as a conditional probability, and the Boole–Shannon expansion is recognized as a Boolean-domain counterpart of the Total Probability Theorem.

Formulation of a system reliability problem in the Boolean domain is conceptually very simple for systems described by verbal statements. It is the natural strategy to use for systems described by fault trees, which are composed of logical AND–OR gates, and can similarly be utilized with systems described by block diagrams. It is the method of choice with systems described by network graphs, when these systems are handled through the enumeration of minimal pathsets and minimal cutsets, rather than via other graph techniques. *Manipulations* in the Boolean domain are facilitated for small problems by the existence of insightful manual tools such as the Karnaugh map or the variable-entered Karnaugh map, and for large problems due to the availability of many scalable efficient algorithms such as those utilizing the ROBDD. *Transfer* from the Boolean domain to the probability domain is achieved via the Real or Probability Transform of a switching (two-valued Boolean) function, which is generally an intractable process of exponential complexity. However, computational burden can be shifted to the Boolean domain, with a potential decrease in complexity, if the pertinent Boolean function is first cast into the form of a *PRE*. Some of the important rules used in generating a *PRE* are presented, occasionally along with succinct proofs. These include rules to *achieve* disjointness (orthogonality) of ORed formulas, and to *preserve* statistical independence as much as possible among ANDed formulas.

The chapter reports a special case of Boolean-domain success by discussing recursive relations governing the reliability and unreliability of the k-out-of-n:G and the k-out-of-n:F systems. The complementation (inversion) and dualization operations in the Boolean domain are utilized to interrelate these four entities, and to evaluate them via four versions of the same algorithm. These four versions of the algorithm are explained in terms of signal flow graphs that are compact, regular, and acyclic, in addition to being isomorphic to the ROBDD, which is reputed to be the most efficient technique for manipulating Boolean functions.

From a pedagogical point of view, the paradigm and concepts discussed herein might be too involved to be of utility in elementary probability education. However, a somewhat bold suggestion is to employ the present paradigm and concepts in areas constituting advanced applications of probability (beside that of system reliability) such as telecommunications, genetics, and finance.

Appendix A: Boolean Quotient

Let us define a literal to be a letter or its complement, where a letter is a constant or a variable. A Boolean term or product is a conjunction or ANDing of m literals in which no letter appears more than once. For $m = 1$, a term is a single literal and for $m = 0$, a term is the constant 1. Note that, according to this definition the constant 0 is not a term. Given a Boolean function f and a term t, the Boolean quotient of f with respect to t, denoted by (f), is defined to be the function formed from f by imposing the constraint $\{t = 1\}$ explicitly [103], i.e.,

$$f/t = [f]_{t=1}, \tag{19}$$

The Boolean quotient is also known as a ratio, a subfunction, or a restriction. Brown [103] lists and proves several useful properties of Boolean quotients, of which we reproduce the following ones:

$$f/1 = f, \tag{20}$$

$$f/st = (f/s)/t = (f/t)/s\,, \quad \text{for } st \neq 0, \tag{21}$$

{for n-variable functions f and g and an m-variable term t with $m \leq n$},

$$f \leq g \;\Rightarrow\; f/t \leq g/t \tag{22}$$

$$t \wedge f = t \wedge (f/t) \tag{23}$$

$$\bar{t} \vee f = \bar{t} \vee (f/t) \tag{24}$$

$$t \wedge f \leq f/t \leq \bar{t} \vee f \tag{25}$$

In this Appendix, we followed Brown [103] in denoting a Boolean quotient by an inclined slash f . However, it is possible to denote it by a vertical bar f to stress the equivalent meaning (borrowed from conditional probability) of f conditioned by t or f given t.

References

1. **Ryabinin, I. A.,** Logical probabilistic analysis and its history. *International Journal of Risk Assessment and Management*, **18**(3-4): 256-265, (2015).
2. **Merekin, Yu. V.,** Solution of problems of probabilistic calculation of single-cycle schemes by an orthogonalization method, *Computing Systems*, Issue 4, (1963) (In Russian).
3. **Premo, A. F.**, The use of Boolean algebra and a truth table in the formulation of a mathematical model of success, *IEEE Transactions on Reliability*, **R-12**(3): 45-49, (1963).
4. **Fratta, L.,** and **Montanari, U.,** A Boolean algebra method for computing the terminal reliability in a communication network, *IEEE Transactions on Circuit Theory*, **20**(3): 203-211, (1973).

5. **Hurley, R. B.,** Probability maps, *IEEE Transactions on Reliability*, **R-12**(3): 39-44, (1963).
6. **Krauss, P. H.**, Representation of conditional probability measures on Boolean algebras, *Acta Mathematica Hungarica*, **19**(3-4), 229-241, (1968).
7. **Parker, K. P.**, and **E. J. McCluskey**, Probabilistic treatment of general combinational networks, *IEEE Transactions on Computers*, **24**(6): 668-670, (1975)
8. **Ogus, R. C.,** The probability of a correct output from a combinational circuit, *IEEE Transactions on Computers*, **24**(5): 534-544, (1975).
9. **Bennetts, R. G.,** On the analysis of fault trees, *IEEE Transactions on Reliability*, **R-24** (3): 175-185, (1975).
10. **Aggarwal, K. K.**, Comments on "On the analysis of fault trees," *IEEE Transactions on Reliability*, **R-25**(2): 126-127, (1976).
11. **Schneeweiss, W. G.,** Calculating the probability of Boolean expression being 1, *IEEE Transactions on Reliability*, **R-26**(1): 16-22, (1977).
12. **Locks, M. O.**, Inverting and minimalizing path sets and cut sets, *IEEE Transactions on Reliability*, **R-27**(2): 107-109, (1978).
13. **Abraham, J. A.,** An improved algorithm for network reliability, *IEEE Transactions on Reliability,* **R-28** (1): 58-61, (1979).
14. **Dotson, W.,** and **Gobien J.,** A new analysis technique for probabilistic graphs, *IEEE Transactions on Circuits and Systems,* **CAS-26**(10): 855-865, (1979).
15. **Bennetts, R. G.,** Analysis of reliability block diagrams by Boolean techniques, *IEEE Transactions on Reliability,* **R-31**(2): 159-166, (1982).
16. **Locks, M. O.**, Recursive disjoint products: a review of three algorithms, *IEEE Transactions on Reliability*, **R- 31**(1): 33-35, (1982).
17. **Rushdi, A. M.,** Symbolic reliability analysis with the aid of variable-entered Karnaugh maps, *IEEE Transactions on Reliability*, **R-32**(2): 134-139, (1983).
18. **Rushdi, A. M.,** and **Al-Khateeb, D. L.,** A review of methods for system reliability analysis: A Karnaugh-map perspective, *Proceedings of the First Saudi Engineering Conference, Jeddah, Saudi Arabia*, vol. 1, pp. 57-95, (1983).
19. **Gupta, P. P.,** and **Agarwal, S. C.,** A Boolean algebra method for reliability calculations, *Microelectronics and Reliability*, **23**(5): 863-865, (1983).
20. **Rushdi, A. M.**, Overall reliability analysis for computer-communication networks, *Proceedings of the Seventh National Computer Conference*, Riyadh, Saudi Arabia, pp. 23-38, (1984).
21. **Schneeweiss, W. G.,** Disjoint Boolean products via Shannon's expansion, *IEEE Transactions on Reliability,* **R-34** (4): 329-332, (1984).
22. **Rushdi, A. M.**, On reliability evaluation by network decomposition, *IEEE Transactions on Reliability*, **R-33**(5): 379-384, (1984), Corrections: *ibid*, **R-34**(4): 319 (1985).
23. **Shier, D. R.,** and **Whited, D. E.**, Algorithms for generating minimal cutsets by inversion, *IEEE Transactions on Reliability*, **R-34**(4): 314-319, (1985).
24. **Locks, M. O.,** Recent developments in computing of system reliability, *IEEE Transactions on Reliability*, **R-34**(5): 425-436, (1985).
25. **Rushdi, A. M.,** and **Goda, A. S.,** Symbolic reliability analysis via Shannon's expansion and statistical independence, *Microelectronics and Reliability*, **25**(6): 1041-1053, (1985).
26. **Rushdi, A. M.,** Map derivation of the minimal sum of a switching function from that of its complement, *Microelectronics and Reliability*, **25**: 1055-1065, (1985).
27. **Rushdi, A. M.,** Utilization of symmetric switching functions in the computation of k-out-of-n system reliability, *Microelectronics and Reliability*, **26**(5): 973-987, (1986).
28. **Rushdi, A. M.,** Efficient computation of k-to-l-out-of-n system reliability, *Reliability Engineering*, **17**(3): 157-163, (1987), Erratum: *ibid*, **19**(4): 321, (1987).
29. **Rushdi, A. M.**, and **Dehlawi F.**, Optimal computation of k-to-l-out-of-n system reliability, *Microelectronics and Reliability*, **27**(5): 875-896, (1987), Erratum : *ibid*, **28**(4): 671, (1988).
30. **Locks, M. O.,** A minimizing algorithm for sum of disjoint products. *IEEE Transactions on Reliability*, **R-36**(4): 445-453, (1987).

31. **Rushdi, A. M.**, On computing the syndrome of a switching function, *Microelectronics and Reliability*, **27**(4): 703-716, (1987).
32. **Rushdi, A. M.**, A switching-algebraic analysis of consecutive-k-out-of-n: F systems, *Microelectronics and Reliability*, **27**(1):171-174, (1987).
33. **Rushdi, A. M.**, A switching-algebraic analysis of circular consecutive-k-out-of-n: F systems, *Reliability Engineering & System Safety*, **21**(2): 119-127, (1988).
34. **Ball M.**, and **Provan J.**, Disjoint products and efficient computation of reliability, *Operations Research*, **36**(5): 703-715, (1988).
35. **Heidtmann, K. D.**, Smaller sums of disjoint products by subproduct inversion, IEEE Transactions on Reliability, **38**(3): 305-311, (1989).
36. **Rushdi, A. M.,** Threshold systems and their reliability, *Microelectronics and Reliability*, **30** (2): 299-312, (1990).
37. **Rushdi, A. M.,** Comment on: An efficient non-recursive algorithm for computing the reliability of k-out-of-n systems, *IEEE Transactions on Reliability*, **40**(1): 60-61, (1991).
38. **Veeraraghavan, M.,** and **Trivedi, K. S.**, An improved algorithm for symbolic reliability analysis, *IEEE Transactions on Reliability*, **40**(3): 347-358, (1991).
39. **Rushdi, A. M.**, *Reliability of k-out-of-n Systems*, Chapter 5 in Misra, K. B. (Editor), *New Trends in System Reliability Evaluation*, Vol. **16**, Fundamental Studies in Engineering, Elsevier Science Publishers, Amsterdam, The Netherlands, 185-227, (1993).
40. **Soh, S.,** and **Rai, S.**, Experimental results on preprocessing of path/cut terms in sum of disjoint products technique, *IEEE Transactions on Reliability*, **42**(1): 24-33, (1993).
41. **Liu, H. H., Yang, W. T.,** and **Liu, C. C.**, An improved minimizing algorithm for the summation of disjoint products by Shannon's expansion, *Microelectronics and Reliability*, **33**(4), 599-613, (1993).
42. **Rushdi, A. M.,** and **AbdulGhani A. A.,** A comparison between reliability analyses based primarily on disjointness or statistical independence, *Microelectronics and Reliability*, **33**: 965-978, (1993).
43. **Rai, S., Veeraraghavan, M.,** and **Trivedi, K. S.**, A survey of efficient reliability computation using disjoint products approach, *Networks*, **25**(3): 147-163, (1995).
44. **Schneeweiss, W. G.**, Advanced hand calculations for fault tree analysis and synthesis, *Microelectronics and Reliability*, **37**(3): 403-415, (1997).
45. **Tsuchiya, T., Kajikawa, T.**, and **Kikuno, T.**, Parallelizing SDP (Sum of disjoint products) algorithms for fast reliability analysis, *IEICE Transactions on Information and Systems*, **83** (5): 1183-1186, (2000).
46. **Boros, E., Crama, Y., Ekin, O., Hammer, P. L., Ibaraki, T.,** and **Kogan, A.**, Boolean normal forms, shell ability, and reliability computations, *SIAM Journal on Discrete Mathematics*, **13**(2): 212-226, (2000).
47. **Balan, A. O.,** and **Traldi, L.**, Preprocessing minpaths for sum of disjoint products, *IEEE Transactions on Reliability*, **52**(3): 289-295, (2003).
48. **Miltersen, P. B., Radhakrishnan, J.**, and **Wegener, I.,** On converting CNF to DNF. *Theoretical Computer Science*, **347**(1): 325-335, (2005).
49. **Rushdi, A. M.,** and **Ba-Rukab O. M.,** A doubly-stochastic fault-tree assessment of the probabilities of security breaches in computer systems, *Proceedings of the Second Saudi Science Conference, Part Four: Computer, Mathematics, and Statistics, Jeddah, Saudi Arabia*, pp. 1-17, (2005).
50. **Rushdi, A. M.,** and **Ba-Rukab O. M.,** Fault-tree modelling of computer system security, *International Journal of Computer Mathematics*, **82** (7): 805-819, (2005).
51. **Traldi, L.,** Non-minimal sums of disjoint products, *Reliability Engineering & System Safety*, **91**(5): 533-538, (2006).
52. **Rushdi, A. M.,** Partially-redundant systems: Examples, reliability, and life expectancy, *International Magazine on Advances in Computer Science and Telecommunications*, **1**(1): 1-13, (2010).

53. **Higashiyama Y**, and **Rumchev V.**, New version of SDP method for weighted–k–out–of–n system, *Proceedings of the 15th World Multi-Conference on Systemics, Cybernetics, and Informatics 2012*; **1**: 120-125, (2012).
54. **Rushdi, A. M. A.**, and **Hassan A. K.**, Reliability of migration between habitat patches with heterogeneous ecological corridors, *Ecological Modelling*, **304**: 1-10, (2015).
55. **Rushdi, A. M. A.** and **Alturki, A. M.**, Reliability of coherent threshold systems, *Journal of Applied Sciences*, **15**(3): 431-443, (2015).
56. **Rushdi, A. M. A.**, and **Hassan, A. K.**, An exposition of system reliability analysis with an ecological perspective, *Ecological Indicators*, **63**, 282-295, (2016).
57. **Rushdi, A. M. A.**, and, **Al-Qwasmi, M. A.**, Exposition and comparison of two kinds of a posteriori analysis of fault trees, *Journal of King Abdulaziz University: Computing and Information Technology*, **5**(1): 55-74, (2016).
58. **Bamasak, S. M**, and **Rushdi, A. M. A.**, Uncertainty analysis of fault-tree models for power system protection. *Journal of Qassim University: Engineering and Computer Sciences*, **8**(1): 65-80, (2016).
59. **Rushdi, A. M. A.**, and **Hassan, A. K.**, Quantification of uncertainty in the reliability of migration between habitat patches. *Computational Ecology and Software*, **6**(3): 66-82, (2016).
60. **Alturki, A. M.**, and **Rushdi A. M. A.**, Weighted voting systems: A threshold-Boolean perspective, *Journal of Engineering Research*, **4**(1): 125-143, (2016).
61. **Rushdi A. M. A.**, and **Alturki, A. M.**, Unification of mathematical concepts and algorithms of k-out-of-n system reliability: A perspective of improved disjoint products, *Journal of Engineering Research*, **5**(1), (2017).
62. **Rushdi, A. M. A.**, and, **Al-Qwasmi, M. A.**, Utilization of basic-event statistical independence in simplifying the uncertainty analysis of fault trees, *Journal of King Abdulaziz University: Computing and Information Technology*, **6**(2), (2017.(
63. **Feller, W.**, *An Introduction to Probability Theory and Its Applications, Vol. 1*, 3rd Edition, Wiley, (1968).
64. **Mosteller, F.**, *Fifty Challenging Problems in Probability with Solutions*, Revised Edition, Dover Publications, (1987).
65. **Whittle, P.**, *Probability via Expectation*, Fourth Edition, Springer, New York, (2000).
66. **Chung, K. L.**, *A Course in Probability Theory*, Third Edition, Academic Press, (2001).
67. **Trivedi, K. S.**, *Probability & Statistics with Reliability, Queuing, and Computer Science Applications*, 2nd Edition, Prentice-Hall, Englewood Cliffs, NJ, USA, (2002).
68. **Jaynes, E. T.**, *Probability Theory: The Logic of Science*, Cambridge University Press, (2003).
69. **Bertsekas, D. P.**, and **Tsitsiklis, J. N**, *Introduction to Probability*, 2nd Edition, Athena Scientific, (2008).
70. **Ross, S.**, *A First Course in Probability*, 8th Edition, Pearson Prentice Hall, (2009).
71. **Ross, S. M.**, *Introduction to Probability Models*, 10th Edition, Academic Press, (2014).
72. **Morin, D. J.**, *Probability: For the Enthusiastic Beginner*, Create Space Independent Publishing Platform, (2016).
73. **Falkowski, B. J.**, **Lozano, C. C.**, and **Łuba, T**. Reliability analysis based on arithmetic and Boolean representations, *Proceedings of IET 15th International Conference on Mixed Design of Integrated Circuits and Systems (MIXDES 2008)*, pp. 505-509, (2008).
74. **Falkowski, B. J.**, **Lozano, C. C.**, and **Łuba, T.**, Arithmetic and Boolean techniques for derivation of system reliability expressions, *Electronics and Telecommunications Quarterly*, **54**(2): 167-177, (2008).
75. **Papaioannou, S. G.**, and **Barrett, W. A.**, The real transform of a Boolean function and its applications, *Computers & Electrical Engineering*, **2**(2): 215-224, (1975).
76. **Kumar, S. K.**, and **Breuer, M. A.**, Probabilistic aspects of Boolean switching functions via a new transform, *Journal of the ACM (JACM)*, **28**(3): 502-520, (1981).
77. **Rushdi, A. M.**, How to hand-check a symbolic reliability expression, *IEEE Transactions on Reliability*, **R-32**(5): 402-408, (1983).

78. **Rushdi, A. M.**, Uncertainty analysis of fault-tree outputs, *IEEE Transactions on Reliability*, **R-34**(5): 458-462, (1985).
79. **Rushdi, A. M.**, On computing the spectral coefficients of a switching function, *Microelectronics and Reliability*, **27**(6): 965-979, (1987).
80. **Heidtmann, K. D.,** Arithmetic spectrum applied to fault detection for combinational networks, *IEEE Transactions on Computers*, **40**(3): 320-324, (1991).
81. **Jain, J.,** Arithmetic transform of Boolean functions, in *Representations of Discrete Functions*, in **Sasao, T., and M. Fujita** (Editors), Kluwer Academic Publishers, Dordrecht, the Netherlands, pp. 133-161, (1996).
82. **Rushdi, A. M. A.** and **Ghaleb, F. A. M.**, The Walsh spectrum and the real transform of a switching function: A review with a Karnaugh-map perspective, *Journal of Qassim University: Engineering and Computer Sciences,* **7**(2): 73-112, (2015).
83. **Kossow, A.,** and **Preuß, W.** Zuverlässigkeitsanalyse konsekutiver Systeme—eine Übersicht (Reliability analysis of consecutive systems-an overview), *Operations-Research-Spektrum*, **11**(3): 121-130, (1989).
84. **Preuss, W. W.,** and **Boehme, T. K.,** On reliability analysis of consecutive-k out-of-n: F systems and their generalizations—A survey. In *Approximation, Probability, and Related Fields*, pp. 401-411, Springer US, (1994).
85. **Malinowski, J.,** and **Preuss, W**. On the reliability of generalized consecutive systems—A survey, *International Journal of Reliability, Quality and Safety Engineering*, 2(02): 187-201, (1995).
86. **Eryilmaz, S.,** Capacity loss and residual capacity in weighted k-out-of-n: G systems, *Reliability Engineering & System Safety*, **136**: 140-144, (2015).
87. **Li, X., You, Y.,** and **Fang, R.,** On weighted k-out-of-n systems with statistically dependent component lifetimes, *Probability in the Engineering and Informational Sciences*, 1-14, published online (2016).
88. **Chang G. J., Cui, L.** and **Hwang, F. K.,** Reliabilities for (n, f, k) systems, *Statistics & Probability Letters*, **43**: 237-242, (1999).
89. **Guo, Y., Cui, L., Li, J.,** and **Gao, S**., Reliabilities for (n, f, k (i, j)) and ⟨n, f, k (i, j)⟩ systems, *Communications in Statistics-Theory and Methods*, **35**(10): 1779-1789, (2006).
90. **Eryilmaz, S**., and **Aksoy, T.**, Reliability of Linear (n, f, k) Systems with weighted components, *Journal of Systems Science and Systems Engineering*, **19**(3): 277-284, (2010).
91. **Kamalja, K. K.,** and **Shinde, R. L.**, On the reliability of (n, f, k) and ⟨n, f, k⟩ systems, *Communications in Statistics-Theory and Methods*, **43**(8): 1649-1665, (2014).
92. **Rushdi, A. M.,** and **Ba-Rukab, O. M**. The modern syllogistic method as a tool for engineering problem solving. *Journal of Engineering and Computer Sciences: Qassim University*, **1**(1): 57-70, (2008).
93. **Rushdi, A. M.**, A conditional probability treatment of strict consecutive-k-out-of-n: F systems, *Microelectronics and Reliability*, **29**(4): 581-586, (1989).
94. **Rushdi, A. M.,** Effect of statistical dependencies in strict consecutive-k-out-of-n: F systems. Microelectronics Reliability, **28**(2): 309-318, (1988).
95. **Rushdi, A. M., S. G. Papastavridis**., and **R. A. Evans**, Some open questions (and replies) on: Strict consecutive-k-out-of-n:F systems, *IEEE Transactions on Reliability*, **39**(3): 380-381, (1990).
96. **Hwang, F. K.,** Comment on strict consecutive-k-out-of-n:F systems, *IEEE Transactions on Reliability*, **40**(3): 264, 270, (1991).
97. **Schneeweiss, W. G.,** *Boolean Functions with Engineering Applications and Computer Programs*, Springer-Verlag, New York, NY, USA, (1989).
98. **Barlow, R. E.**, and **F. Prochan,** *Mathematical Theory of Reliability*, Wiley, New York, NY, USA, (1996).
99. **Aven, T.**, and **U. Jensen,** *Stochastic Models in Reliability*, Springer-Verlag, New York, NY, USA, Vol. 41, (1999).
100. **Hammer, P. L.** and **S. Rudeanu**, *Boolean Methods in Operations Research and Related Areas*, Springer Verlag, Berlin, Germany, (1968).

101. **Rudeanu, S.**, *Boolean Functions and Equations*, North-Holland Publishing Company & American Elsevier, Amsterdam, the Netherlands (1974).
102. **Muroga, S.,** *Logic Design and Switching Theory,* Wiley, New York, NY, USA, (1979).
103. **Brown, F. M.**, *Boolean Reasoning: The Logic of Boolean Equations*, Kluwer Academic Publishers, Boston, MA, USA (1990).
104. **Brown, F. M.**, *Boolean Reasoning: The Logic of Boolean Equations, 2nd Ed.*, Dover Publications, Mineola, NY, USA, (2003).
105. **Crama, Y.,** and **Hammer, P. L.,** *Boolean functions: Theory, Algorithms, and Applications,* Cambridge, United Kingdom, Cambridge University Press, (2011).
106. **Pecht, M. G., Jieyu, S. H. E.**, and **Barbe, D. F.**, Evaluating terminal pair system reliability, *IEICE Transactions on Fundamentals of Electronics, Communications and Computer Sciences*, **76**(4): 555-564, (1993).
107. **Butler, A. C., Rao, S. S.,** and **LeClair, S. R.**, Reliability analysis of complex systems using symbolic logic, *Reliability Engineering & System Safety*, **40**(1): 49-60, (1993).
108. **Nahman, J. M.**, Exact enumeration of minimal paths of partitioned networks, *Microelectronics and Reliability*, **34**(7): 1167-1176, (1994).
109. **Shen, Y.**, Computerization of the R-ABC algorithm, *Microelectronics and Reliability,* **36**(9): 1219-1221, (1996).
110. **Bryant, R. E.,** Graph-based algorithms for Boolean function manipulation, *IEEE Transactions on Computers,* **100**(8): 677-691, (1986).
111. **Akers S.**, Binary decision diagrams, *IEEE Transaction on Computers*, **C-27**(6): 509-516, (1960).
112. **Singh, H., Vaithilingam, S., Anne, R. K.,** and **Anneberg, L.,** Terminal reliability using binary decision diagrams, *Microelectronics and Reliability*, **36**(3): 363-365, (1996).
113. **Rauzy, A.,** A brief introduction to binary decision diagrams, *Journal Européen des Systèmes Automatisés*, **30**(8): 1033-1050, (1996).
114. **Jinglun, Z.,** and **Quan, S.,** Reliability analysis based on binary decision diagrams. *Journal of Quality in Maintenance Engineering*, **4**(2): 150-161, (1998).
115. **Zang, X., Sun, N.,** and **Trivedi, K. S.**, A BDD-based algorithm for reliability analysis of phased-mission systems, *IEEE Transactions on Reliability*, **48**(1): 50-60, (1999).
116. **Andrews, J. D.,** and **Dunnett, S. J.**, Event-tree analysis using binary decision diagrams, *IEEE Transactions on Reliability*, **49**(2): 230-238, (2000).
117. **Reay, K. A.,** and **Andrews, J. D.,** A fault tree analysis strategy using binary decision diagrams, *Reliability Engineering & System Safety*, **78**(1): 45-56, (2002).
118. **Dutuit, Y.,** and **Rauzy, A.**, Approximate estimation of system reliability via fault trees, *Reliability Engineering & System Safety*, **87**(2): 163-172, (2005).
119. **Xing, L.**, An efficient binary-decision-diagram-based approach for network reliability and sensitivity analysis, *IEEE Transactions on Systems, Man, and Cybernetics-Part A: Systems and Humans*, **38**(1): 105-115, (2008).
120. **Rauzy A.**, Binary Decision Diagrams for Reliability Studies, Chapter 25 in **Misra, K. B.** (Editor), *Handbook of Performability Engineering*, Springer, London, UK, pp. 381-396, (2008).
121. **Mo, Y.,** A multiple-valued decision-based approach to solve dynamic fault tree, *IEEE Transactions on Reliability*, **63**(1): 81-93, (2014).
122. **Mo, Y., Xing , L., Amari, S. V.,** and **Dugan, J. B.**, Efficient analysis of multi-state k-out-of-n system, *Reliability Engineering & System Safety*, **133**: 95-105, (2015).

Reliability Optimization: A Particle Swarm Approach

Sangeeta Pant, Anuj Kumar and Mangey Ram

Abstract In recent years, substantial efforts related to the applications of Particle Swarm Optimization (PSO) to various areas in engineering problems have been carried out. This chapter briefly gives the details of PSO development and its applications to reliability optimization.

Keywords Reliability · Optimization · Particle swarm optimization

1 Recent Works and Advances of PSO

Presently, we have many variants of Particle Swarm Optimization (PSO) and are expected to grow further rapidly. Figure 1 describes the basic variants and modifications in PSO over the years. Various modifications to the original PSO has been proposed so far [43]. Also, novel ideas from other disciplines such as evolutionary algorithms have been imported to the framework of PSO. PSO algorithms can be divided into the global version (*gbest* model) and the local version (*lbest* model) types, with the ability of the *lbest* model to prevent a solution being trapped in local minima. The *gbest* model, on the other hand, has more chance to get trapped into a local optimum. However, the global version is superior to the local version in terms of the speed of convergence to the optimum solution and the computation time.

To reduce the possibility of particles flying out of the problem space, Eberhart et al. [42] put forward a clamping scheme that limited the speed of each particle to a range $[-V_{\max}, V_{\max}]$. To assist with the balance between exploration and exploitation a modified PSO, incorporating an inertia weight, w was introduced [128]. The initial experiments suggested that a value between 0.8 and 1.2 provided

S. Pant · A. Kumar (✉)
Department of Mathematics, University of Petroleum and Energy Studies, Dehradun 248007, India
e-mail: anuj4march@gmail.com

M. Ram
Department of Mathematics, Graphic Era University, Dehradun 248002, India

M. Ram and J.P. Davim (eds.), *Advances in Reliability and System Engineering*, Management and Industrial Engineering, DOI 10.1007/978-3-319-48875-2_7

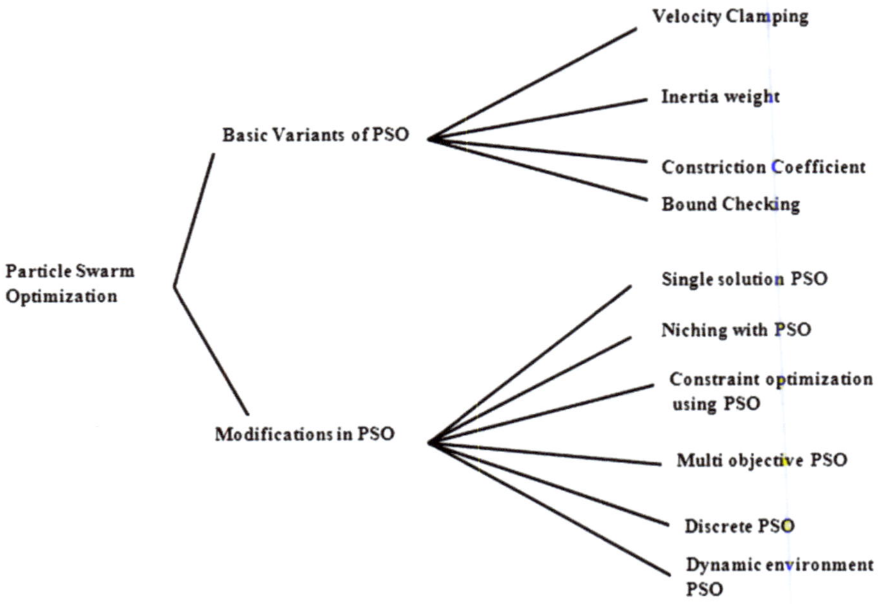

Fig. 1 Basic variants of PSO

good results, although in later work Eberhart and Shi [40] indicated that the value is typically set to 0.9 (reducing the stepwise movement of each particle, allowing greater initial exploration) reducing linearly to 0.4 (speeding convergence to the global optimum) during an optimization run. A summary of various existing inertia weight strategies is given in Table 1 [94]. Constriction is an alternative method for controlling the behaviour of particles in the swarm. Rather than applying inertia to the velocity memory, Clerc and Kennedy (developed 1999, published [20]) applied a constriction factor to the new velocity. Eberhart and Shi [40] showed that with judicious parameter settings, the two approaches were algebraically equivalent and improved performance could be achieved across a wide range of problems.

PSO has also been successfully applied to solve the constrained optimization problems. A variety of approaches [4, 15, 56, 103, 134, 137] have been developed to work with constrained optimization methods.

Although the basic PSO was developed to find single solutions to optimization problems, but later it is observed that PSO has an inherent ability to find multiple solutions. Niching is an important technique for multimodal optimization. PSO can be used as an effective niching method for maintaining stable subpopulations (or niches) [44, 76, 95, 135, 144].

PSO was originally developed and applied to static problems where the objective function does not change. Later, It is realized that PSO be adapted to solve dynamic problems as well. Several simple modifications [39, 41, 54, 57, 80, 105, 157] have been applied to improve its performance in dynamic environment. One of the first

Table 1 Description of different inertia weight strategies

Label	Inertia weight strategy	Adaption mechanism	Feedback parameter	References
W1	$w = c$	Constant w	–	Shi and Eberhart [128, 132]
W2	$w = 0.5 + \frac{\text{rand}()}{2}$	Random w	–	Eberhart and Shi [41]
W3	$w(\text{iter}) = w_{\min} + \frac{\text{maxiter}-\text{iter}}{\text{maxiter}}(w_{\max} - w_{\min})$	Linear time varying	–	Eberhart and Shi [41], Shi and Eberhart [129, 130]
W4	$w(\text{iter}) = w_{\min} + \left(\frac{\text{maxiter}-\text{iter}}{\text{maxiter}}\right)^n (w_{\max} - w_{\min})$	Nonlinear time varying	–	Chatterjee and Siarry [17]
W5	$w(\text{iter}) = w_{\text{initial}} \times U^{\text{iter}}$	Nonlinear time varying	–	Jiao et al. [60]
W6	$w(\text{iter}) = w_{\min} \times Z + \frac{\text{maxiter}-\text{iter}}{\text{maxiter}}(w_{\max} - w_{\min})$	Linear time varying with random changes	–	Feng et al. [46, 47]
W7	$w(\text{iter}) = w_{\min} + \frac{1-\left(\frac{\text{iter}}{\text{maxiter}}\right)}{1-s\left(\frac{\text{iter}}{\text{maxiter}}\right)}(w_{\max} - w_{\min})$	Nonlinear time varying	–	Lei et al. [72]
W8	$w(\text{iter}) = \left(\frac{2}{\text{iter}}\right)^{0.3}$	Nonlinear time varying	–	Fan and Chiu [45]
W9	$w(\text{iter}) = w_{\max} + \frac{\text{maxiter}-\text{iter}}{\text{maxiter}}(w_{\min} - w_{\max})$	Linear time varying	–	Zheng et al. [158, 159]
W10	Fuzzy rules	Adaptive	Best fitness	Saber et al. [122], Shi and Eberhart [131]
W11	$w_i^t = w_{\text{initial}} - \alpha(1 - h_i^t) + \beta s$	Adaptive	Fitnees of the current and previous iterations	Yang et al. [151]
W12	$w = 1.1 + \frac{gbest}{(pbest_i)_{\text{average}}}$	Adaptive	Global best and average local best fitness	Arumugam and Rao [6]

(continued)

Table 1 (continued)

Label	Inertia weight strategy	Adaption mechanism	Feedback parameter	References
W13	$w_i = w_{\min} + (w_{\max} - w_{\min}) \frac{\text{Rank}_i}{\text{total population}}$	Adaptive	Particle rank	Panigrahi et al. [99]
W14	$w = 1 - \alpha\left(\frac{1}{1+e^{ISA_{ij}}}\right),\ ISA_{ij} = \frac{\lvert x_{ij} - P_{ij}\rvert}{\lvert p_{ij} - p_{gi}\rvert + \varepsilon}$	Adaptive	Distance to particle and global best positions	Qin et al. [110]
W15	$w = w_{\text{initial}} + \left(1 - \frac{\text{dist}_i}{\text{max\_dist}}\right),$ $\text{dist}_i = \left(\sum_{d=1}^{D} (gbest_d - x_{i,d})^2\right)^{1/2}$	Adaptive	Distance to global best position	Suresh et al. [138]

studies in the application of PSO to dynamic environment came from Carlisle and Dozier [16], where the efficiency of different velocity models has been evaluated. Coelho et al. [22] successfully applied the split adaptive PSO design proportional integral (PI) controllers for nonlinear time-varying process in discrete time domain.

As far as the main body of PSO development work is concerned it concentrated on optimization in continuous search spaces, although some research has also been conducted into the application of the algorithm to discrete problems. Kennedy and Eberhart [61] developed the first discrete version of PSO for binary problems. Later, Al-kazemi and Mohan [3] compared this with an alternative velocity update technique called Multiphase Discrete PSO (M-DiPSO). Laskari et al. [71] presented an implementation of PSO that initially operated in continuous space but truncated the real number values to integers. Afshinmanesh et al. [1] have developed a hybrid approach, combining PSO and Artificial Immune Systems (AIS), for the optimization of binary problems.

For more developments in PSO one can refer the review papers by Parsopoulos and Vrahatis [104], Banks et al. [10, 11], Dian et al. [38].

A number of approaches have been proposed to extend the PSO for multiple objective problems. One of the earliest proposals for progression of PSO strategy for solving MOOP was made by Moore and Chapman [91] in an unpublished manuscript from 1999. Since then there have been several recent attempts to use PSO for multi-objective optimization problems only. Some of these concepts have been surveyed briefly in this section.

Dynamic Neighbourhood PSO Hu and Eberhart [55] proposed this method in which, in each generation, after calculating distances to every other particles, each particle finds its new neighbourhood. Among the new neighbours each particle finds the local best particle.

The Multi-objective Particle Swarm Optimizer (MOPSO) Coello and Lechuga [24] found PSO particularly suitable for MOOP mainly because of the high speed of convergence that the PSO presents for single objective optimization problem and proposed multi-objective particle swarm optimization (MOPSO). MOPSO used two archives, one for storing globally non-dominated solutions found so far by search process, while the other for storing the individual best solutions attained by each particle. They used method inspired by Knowles and Corne [65], for maintaining diversity. An adaptive grid feature used in this method, based upon objective functions values of archive members is applied to the archive, with the goal of producing well distributed Pareto optimal front. In this method first hyperspace is divided into small hypercube and each cube is assigned weight which is inversely proportional to the number of non-dominated solutions inside the cube. Then roulette wheel selection is used to select one of the hypercubes from which the *gbest* will be picked.

The Multi-objective Particle Swarm Optimizer (MOPSO) Coello et al. [25] improved the aforementioned MOPSO by incorporating a genetic operator as mutation operator. The mutation operator boosts the exploration capability of MOPSO presented in Coello and Lechuga [24].

The Swarm Metaphor Ray and Liew [117] proposed the Pareto dominance and combining concepts of evolutionary techniques with the PSO. All non-dominated individuals which are performing better and having less constraint violations are highly ranked based on Pareto ranking and saved as a set of leaders (SOL). The selection of a group leader as *gbest* from the SOL is based on roulette wheel selection which ensures SOL members with a large crowding radius have a higher probability of being selected as a leader.

The Approach of Mostaghim and Teich [92] proposed the sigma method for finding the suitable *gbest* for each particle, they also used turbulence factor to enhance the exploration capability.

The Algorithm of Fieldsend and Singh [48] suggested an approach in which they used an unconstrained elite archive (in which a special data structure called dominated tree is adopted) to store the non-dominated individuals found along the search process. The concept of the turbulence was also incorporated by them.

Bartz-Beielstein et al. [12] proposed an idea of using elitism (through the use of external archive) into PSO. They analysed different methods of selecting and deleting particles from the archive to generate a satisfactory approximation of the Pareto optimal front.

The Non-dominated Sorting PSO Li [77] developed Non-dominated sorting particle swarm optimization (NSPSO) which incorporated the main mechanism of Non-dominated sorting genetic algorithm (NSGA-II) [34]. In his algorithm, the population of particles was combined with the personal best position and the best was selected from the new population to compose the next population.

Another Multi-objective Particle Swarm Optimization (AMOPSO) Pulido and Coello [109] further improved the performance of PSO, and proposed an MOPSO algorithm, which is called AMOPSO. Their algorithm implements the subdivision of the decision space into multiple sub-swarms via clustering techniques. Their goal was to improve the diversity of solutions on the Pareto optimal front. At some point during the search process, different sub-swarms exchange information, as each sub-swarm chooses a different leader other than its own to preserve diversity.

The Algorithm of Parsopoulos et al. [106] developed parallel vector evaluated particle swarm optimization (VEPSO), is a multi-swarm variant of PSO, which is inspired by the vector evaluated genetic algorithm (VEGA). In VEPSO, each swarm is evaluated using only one of the objective functions of the problem under consideration and the information it possesses for this objective function is communicated to other swarms through the exchange of their best experience.

The Algorithm of Sierra and Coello [133] suggested a new MOPSO, which is also known as OMOPSO. In their design, the population is divided into three sub-swarms of equal size. Each sub-swarm adapted to a different mutation operator. In doing so, the ability of exploration and exploitation was enhanced during the search process.

Multi-Objective Particle Swarm Optimization with Crowding Distance (MOPSO-CD) Raquel and Naval [112] developed another PSO based approach called MOPSO-CD, which incorporated the crowding distance into PSO and the distribution of non-dominated solutions was improved on the Pareto optimal front. The crowding distance mechanism together with a mutation operator maintains the diversity of non-dominated solutions in the external archive. Raquel and Naval [112] also showed that MOPSO-CD is highly competitive in converging towards the Pareto optimal front and generated a well-distributed set of non-dominated solutions. We discuss this approach in detail in the next chapter.

A Hybrid PSO Liu et al. [79] proposed a hybrid PSO, which combined the global search ability of PSO with a synchronous local fine-tuning and used fuzzy global best to handle the premature convergence.

Time Variant MOPSO (TV-MOPSO) Tripathi et al. [141] adapted the vital parameters in PSO, namely the inertia weight and the acceleration coefficients during the iterations.

Elitist Mutated (EM)-MOPSO Reddy and Kumar [118] proposed (EM)-MOPSO, which incorporates an efficient mutation strategy called elitist mutation to enhance exploration and exploitation in the search space.

Dynamic Population Multiple Swarm MOPSO (DMOPSO) Most recently Leong and Yen [73] proposed an algorithm DMOPSO, inspired by Pulido and Coello [109] and incorporates the following four proposed strategies: (1) cell-based rank density estimation scheme to keep track of the rank and density values of the particles; (2) population growing strategy to increase the population size to promote exploration capability; (3) population declining strategy to prevent the population size from growing excessively; and (4) adaptive local archives designed to improve the distributed solutions along the sections of the Pareto optimal front that associate with each sub-swarm.

Other than aforementioned methods several other methods [2, 13, 14, 33, 50] for MOPSO have been proposed till date. For more survey on various PSO proposals reader can refer to Reyes-Sierra and Coello [121], Fieldsend [49], Padhye et al. [97].

2 Reliability Optimization (An Overview)

Almost every one of us is acquainted with the term reliability in day-to-day life. When we assign attribute 'reliable' to a component or a system (a system may be consist of collection of several components) we precisely mean to say that the same will render service for a good or at least reasonable period of time. In the modern age of sciences, a high degree of reliability is being demanded from all kinds of users whether it is in general, public sectors, industries, defense and space research programmes. There is too much at stake in terms of cost, human life, and national security to take any risks with equipments which might not function properly when required. Moreover, the present day weapons used for military purposes consist of thousands of small parts, each interwoven into a complex web which constitutes the weapons. The failure of any one of these could adversely affect the operation of the weapon. Therefore, it becomes more important that each part of the complex equipment must be highly reliable so the equipment as a whole must be reliable. The concept of high reliability is equally important outside the Military field. Computers which are complex as well as expensive play a major role in industrial and scientific activities. If a Computer does not operate even for a single day even due to any software failure, it not only spells inconvenience but also cause financial loss, i.e. the software reliability testing is very important.

So the needs of obtaining highly reliable systems and components have acquired special importance with the development of the present day technology.

The theory of reliability is not very old; usually world war second in 1939 is regarded as the starting point of reliability discipline. Before world war second, in the first quarter of twentieth century a team of workers in 'Bell Telephone Laboratories' developed statistical methods for solving there quality control problems which is strongly linked with quality control. They provided the basis for development of statistical quality control. The American Society for Testing and Materials, The American Standard Association and The American Society for Mechanical Engineers also worked for the quality control techniques. But these technique were widely used till world war second 1939. Complexity and automation of equipments used in the war resulted in severe problems of maintenance and repairs. The equipment/component failed beyond the expectation. During this war army and navy in USA set up a joint committee known as Vacuum Tube Development Committee for the study of failure in vacuum tube which is considered to be one of the root causes of the trouble. The major committee on reliability was set up by U.S Defense Department in 1950. This was latter called the Advisory Group on Reliability of Electronic Equipment (AGREE). During 1950s Germany, Japan, and Britain also took interest in such type of study. The last 20 years have seen remarkable progress in the application of reliability in industries and in other departments of all the developed and developing countries.

The theory of reliability is the new scientific discipline that studies the general regularity that must be maintained under design, experimentation, manufacture, acceptance, and use of units/components in order to maximal effectiveness from their use. The need of obtaining highly reliable systems and components has acquired Special importance with the development in the present day technology.

The problem of increasing reliability of equipments/components becomes more important and urgent in connection with the complex mechanization, modern sophistication, and automation of industrial process in many fields of industry, transportation, communication, space technology, etc. The complex system, equipments, machines, etc., are not of much use if they cannot perform the work adequately for which they are intended.

A system is a combination of elements forming a planetary whole, i.e. there is a functional relationship between its components. The properties and behavior of each component ultimately affects the properties of the system. Any system has a hierarchy of components that pass through the different stages of operations which can be operational, failure, degraded or in repair. Failure does not mean that it will always be complete; it can be partial as well. But both these types affect the performance of system and hence the reliability. Majority of the systems in the industries are repairable. The performance of these systems can influence the quality of product, the cost of business, the service to the customers, and thereby the profit of enterprises directly. Modern repairable systems tend to be highly complex due to increase in convolution and automation of systems. During the last 45 years reliability concepts have been applied in various manufacturing and technological fields. Earlier researcher discussed reliability and steady state analysis of some realistic engineering systems by using different approaches. Reliability techniques have also been applied to a number of industrial and transportation problems including automobile industry. Here the study is focused on the engine assembly process of automobiles.

A high degree of reliability is also desirable from the economic point of view so as to reduce the overall costs. Sometimes the annual maintaining cost of some system in operable state is much higher than its original cost. Insufficient reliability of units engenders great loss in servicing, partial stoppage of equipment, and there may be accidents with considerable damage to the equipment and even the cost may be more serious in term of human life, national prestige and security. All these factors and many more, demanded high reliability in the design and operations of components/systems/equipments in various reliability models of practical utility, we often across with the situations of maximizing the profit. The profit earned out by an operable system besides other parameters depends upon the cost incurred against the repairmen needed to repair the failure stages of the system.

The present day theory of reliability has been developed during the last two decades by engineers and mathematicians of various countries.

Every science is based on some fundamental concepts and definitions, same is true for theory of reliability also. Some of the basic concepts on which the theory of reliability is formulated are given below.

System A system is an arbitrary device consisting of different parts components or units.

Time It is the period during which one can expect the satisfactory performance of a system.

Adequate It indicates the criteria for operations of the device to satisfactory.

Failure Mode It is the effect by which a failure is observed.

Failure Rate It is the incremental change in the number of failures per associated incremental change in time. Or the expected rate of occurrence of failure or the number of failures in a specified time period. Failure rate is typically expressed in failures per million or billion hours. For example, if your television has a failure rate of five failures per million hours, you can watch one million hour-long television shows and likely experience a failure during only five shows.

Uptime It is total time during which the system is in the acceptable operating conditions.

Downtime The total time during which the system is not in the acceptable operating conditions.

The definition of reliability of a system is usually stated in straightforward terms as "the probability that the system will not fail during delivery of service [119], or alternatively, that the overall system performance figure of merit will not enter failure mode between the time a service is requested and when that service is delivered [136]. A system can be designed for optimal reliability either by adding redundant components or by increasing the reliability of components [68].

There are several ways for improving the system's reliability. One way of improving reliability is either to duplex some of the unit or the whole system. Other way is to provide repair and maintenance to the system at the time of need. Some important technique of reliability improvements are as under.

Redundancy In a redundant system, some additional paths are created for the proper functioning of the system. Even though one component is sufficient for successful operation of the system, we deliberately use some more components to increase probability of success, thus causing the system to become redundant. There are three types of redundancies.

Active Redundancy An active redundant system with n-units is one which operates with every one unit. Here failure of system occurs only when all the units are fails.

Standby Redundancy A standby redundant system is one in which one unit operates on line followed by a number of spare unit called standbys. On failure of the operating unit, a standby unit, if operable, is switched on to the line by perfect or imperfect switching device. Standby can be classifies as hot, warm and cold depending on how they are loaded in the standby state. Hot standbys are those which are loaded in exactly the same way as the operating unit. Warm standbys are those which are diminished load. And cold standbys are completely unloaded and never lose their operational ability and can not fail in standby state.

Partial Redundancy The redundancy where in two or more redundant items are required to perform function k-out-of: m system. The system which is good iff at least k of it m items are good.

Maintenance All recoverable systems which are used for continuous or intermittent service for some period of time are subjected to maintenance. Maintenance action can be classified in several categories, e.g. preventive, corrective, and priority maintenance.

Preventive Maintenance Preventive maintenance is such type of check which keeps the system in a condition consistent with its built in level of performance, reliability and safety.

Corrective Maintenance It deals with the system performance when the system gives wrong results. Repair/maintenance is concerned with increasing with system availability. In order to increase the system availability, failed unit are repaired to put them into operation.

Priority Maintenance A redundant system which consist of $n \geq 2$ units in which one of the units is called the priority unit (P-unit) and others are termed as non-priority units (O-units). The P-unit is the "preferred unit" for operating on line and is never used in the status of a standby. The O-units are allowed to operate on the line only when the P-unit is under failure.

Pre-emptive Priority The repair of the O-unit is interrupted and its repair is continued as soon as the repair of the P-unit is completed. The resumed repair of the O-unit can follow any one of the following rules.

Pre-emptive Resume The repair of the O-unit is continued from the point where it was left earlier.

Pre-emptive Repeat The repair of the O-unit is started as a fresh; this implies that the time for the O-unit in the repair facility before it was left from service has no influence on its service time now.

Non Pre-emptive Priority The repair of the O-unit is continued and the repair of the P-unit is entertained only when the repair of the O-unit is completed. It is also called the Head-of-line repair police.

Inspection A system requires its inspection at random epochs in order to trace out the fault in redundant, particularly in deteriorating standby system.

The reliability optimization problems can be categorized in two ways: Single objective reliability optimization problems and Multi-objective reliability optimization problems.

- **Single Objective Reliability Optimization Problems**

  Let us consider a reliability optimization problem as follows:
  Max $f_0(r_1, r_2, \ldots, r_n, x_1, x_2, x_n)$
  subject to

$$\begin{array}{ll} f_i^c(r_1, r_2, \ldots, r_n, x_1, x_2, x_n) \leq b_i, & \text{for} \quad i = 1, 2, \ldots, m \\ l_j \leq x_j \leq u_j, x_j \in Z^+, & \text{for} \quad j = 1, 2, \ldots, n \\ r_j \in (0, 1) \subset R, & \text{for} \quad j = 1, 2, \ldots, n \end{array}$$

where n is the number of components with m constraints. Component reliability of jth component is denoted by r_j. x_j is the number of identical redundant components, i.e. the number of redundancies, at the jth component; f_i is the ith constraint function; b_i is the maximum allowable amount of the ith resource; f_0 is an objective function of the problem; Z^+ is the set of non-negative integers while R denote the set of real numbers. The objective of the reliability optimization problem is to find the components reliability in such a way that it maximize the overall system reliability under the given resources constraints, or minimizes the total cost under minimum system reliability and other resource limitations.

- **Muti-Objective Reliability Optimization Problems**

$$\begin{aligned} \text{Max } F = (&f_1(r_1, r_2, \ldots, r_n, x_1, x_2, \ldots, x_n),\ f_2(r_1, r_2, \ldots, r_n, x_1, x_2, \ldots, x_n), \ldots, \\ &f_K(r_1, r_2, \ldots, r_n, x_1, x_2, \ldots, x_n)) \end{aligned}$$

subject to

$$\begin{array}{ll} f_i^c(r_1, r_2, \ldots, r_n, x_1, x_2, x_n) \leq b_i, & \text{for} \quad i = 1, 2, \ldots, m \\ l_j \leq x_j \leq u_j, x_j \in Z^+, & \text{for} \quad j = 1, 2, \ldots, n \\ r_j \in (0, 1) \subset R, & \text{for} \quad j = 1, 2, \ldots, n \end{array}$$

$f_k, \forall k = 1, 2, \ldots, K$ is one of the objective functions of the problem, K is total number of objective functions. In most practical situations involving reliability optimization, there are several mutually conflicting goals such as maximizing system reliability and minimizing cost, weight, volume and constraints required to be addressed simultaneously. Some main objectives can be expressed as.

***Objective* 1:** The most important objective is the maximization of system reliability (R_s). It enables the system to function satisfactorily throughout its intended service period

$$\text{Max } R_s$$

As in our approach we are considering all minimization problems. Hence, the above objective is equivalent to minimization of system unreliability $(Q_s = 1 - R_s)$, can be expressed as follows:

$$\text{Min } Q_S$$

***Objective* 2:** The addition of the redundant components increases not only the system reliability but also its overall cost (C_S). A manufacturer has to balance these conflicting objectives, keeping in view the importance of reducing the overall cost

$$\text{Min } C_S$$

***Objective* 3:** As with cost, every added redundant component increases the weight of the system. Usually, the overall weight of a system needs to be minimized along with its cost even as reliability is maximized (or unreliability is minimized)

$$\text{Min } W_S$$

Reliability optimization problems can be categorized as redundancy allocation, reliability allocation and reliability–redundancy allocation problems in accordance to the type of their decision variables. If the number of redundancies, x_j's for all j, are the only variables, the problem is called redundancy allocation problem. If component reliabilities, r_j's for all j, are the only variables, the problem is termed as reliability allocation and if the decision variables of the problem include both the component reliabilities and redundancies, the problem is called a reliability–redundancy allocation problem. From the mathematical programming point of view, redundancy allocation is a pure integer nonlinear programming problem (INLP) while reliability allocation problems can be viewed as a continuous nonlinear programming problem (NLP) and reliability–redundancy allocation can be termed as a mixed integer nonlinear programming problem (MINLP).

The suitability of a metaheuristics varies problem to problem. In other words, a metaheuristic which is giving promising results on a particular set of problem may show poor performance on different problems. Optimization of reliability of complex systems is an extremely important issue in the field of reliability engineering. Over the past three decades, reliability optimization problems have been formulated as nonlinear programming problems within either single objective or multi-objective environment.

As discussed above, reliability optimization problems are categorized into three typical problems according to the types of their decision variables: reliability allocation, redundancy allocation and reliability–redundancy allocation. A number of algorithms—also categorized as approximate, exact, or heuristic/metaheuristic have been used to find optimal solutions to these problems. Algorithms such as the surrogate worth trade-off, the Lagrange multiplier, and geometric programming methods and their variants, which are efficient for the exact solution of continuous problems of the type posed by reliability allocation optimization, can only approximate the solution in the case of redundancy or redundancy–reliability allocation optimization [93, 153]. The approximation techniques involve the use of trial and error approaches to obtain integer solutions [148, 153]. The approximation techniques were popular when exact solution algorithms were not well developed. The advent of the exact algorithms, such as integer programming (IP), branch and bound, and dynamic programming (DP) [78] have made the approximation techniques less popular for solving redundancy allocation problems. The approximation and exact algorithms, though efficient with small-to-moderate sized problems having desirable properties such as convexity or monotonicity, are deficient with complex large scale ones, such as real-life network reliability and redundancy allocation optimization problems [7, 8]. Although the heuristic/metaheuristic approaches (example GA, SA, ACO, PSO and TS) yield solutions which are not exact, they do have the ability to efficiently handle complexity [5] and thus become increasingly popular in the reliability optimization field. The redundancy and the redundancy–reliability allocation optimization problems are generally more difficult to solve than the reliability allocation ones. This is because the former belongs to the class of NP-hard problems (this phenomenon was demonstrated by Chern [19], Coit et al. [32], Coit and Konak [27]) which involve non-convex and combinatorial search spaces and require a considerable amount of computational effort to find exact optimal solutions [62]. The reliability allocation problems on the other hand involve continuous optimization with a number of classical solution algorithms based on gradient and direct search methods at their disposal. They are thus relatively easier to solve. Examples of the solution algorithms which were applied in the context of the three optimization problem types are presented in Tables 2 and 3 [142].

Tillman et al. [140] has extensively reviewed the several optimization techniques for system reliability design. However, they reviewed the application of only derivative-based optimization techniques, as metaheuristics were not applied to the reliability optimization problems by that time. Mohan and Shanker [90] applied random search technique to optimize complex system. Luus [81] optimized such problems by nonlinear integer programming procedure. Over the last decade, metaheuristics have also been applied to solve the reliability optimization problems. To list a few of them Coit and Smith [30, 31], were the first to employ a GA to solve reliability optimization problems. Ravi et al. [114] developed an improved version of nonequilibrium simulated annealing called INESA and applied it to solve a variety of reliability optimization problems. Further, Ravi et al. [115, 116] first formulated various complex system reliability optimization problems with single

Table 2 Different optimization techniques used in reliability optimization of SOOP category

Model type	Solution techniques	Algorithm description	Sources
Redundancy allocation	Approximate	Interval arithmetic optimization	Munoz and Pierre [93]
	Exact	Lagrange relaxation algorithm in conjunction with dynamic programming (DP)	Ashrafi and Berman [7]
		Integer programming (IP) algorithm	Coit and Liu [28]
		Lexicographic order (P&K-Ag)	Prasad and Kuo [108]
		Improved surrogate constraint (ISC) algorithm	Onishi et al. [96]
		IP (due to Misra)	Misra and Sharma [87]
	Heuristic–metaheuristic	Simulated annealing (SA)	Atiqullah and Rao [8]
		DETMAX algorithm	Kim and Yum [62]
		Genetic algorithm (GA)	Deeter and Smith [36]
		Heuristic algorithm	Bala and Aggarwal [9]
		GA	Coit and Smith [29]
		SA	Wattanapongsakorn and Levitan [146]
		Heuristic algorithm	You and Chen [153]
		Approximate linear programming heuristic	Prasad and Raghavachari [107]
		Tabu search (TS)	Kulturel-Konak et al. [66]
		Variable neighbourhood search algorithm	Liang and Chen [78]
		SA	Wattanapongsakorn and Levitan [145]
		GA	Coit and Smith [31]
		GA	Coit and Smith [30]
Reliability allocation	Exact	Cutting plane algorithm	Majety et al. [84]
	Heuristic–metaheuristic	Random search algorithm	Mohan and Shanker [90]
		PSO	Pant et al. [101]
		CSA	Kumar et al. [67]
Redundancy–reliability allocation	Exact	Surrogate dual problem under DP algorithm	Hikita et al. [52]
		Surrogate constraint algorithm	Hikita et al. [51]
		DP	Yalaoui et al. [149]
		Mixed integer programming (MIP) algorithm	Misra and Sharma [87]

Table 3 Different optimization techniques used in reliability optimization of MOOP category

Model type	Solution techniques	Algorithm description	MOA type	Sources
Redundancy allocation	Approximate	Surrogate worth trade-off (SWT) method under dual decomposition algorithm	Scaler	Sakawa [124]
		Direct search by Min–Max algorithm		Misra and Sharma [88]
	Exact	IP due to Misra		Misra and Sharma [87]
		The weighting method in conjunction with a heuristic and an IP algorithm		Coit and Konak [27]
		Weighting method under an IP software package		Coit et al. [32]
	Heuristic–metaheuristic	GA and Monte Carlo simulation		Marseguerra et al. [85]
		Multi-objective GA		Coit and Baheranwala [26]
		Elitist non-dominated sorting GA 2 (NSGA 2)	Pareto	Wattanapongskorn and Coit [147]
		GA		Taboada and Coit [139]
		NSGA		Zhao et al. [156]
		Multi-objective ant colony		Zafiropoulos and Dialynas [154]
		Simulated annealing (SA)		Yamachi et al. [150]
		Multi-objective GA		Li and Haimes [75]
Reliability allocation	Exact	Three levels decomposition approach and the Khun Tucker multiplier method	Scaler	Salazar et al. [126]
	Heuristic–metaheuristic	NSGA 2	Pareto	Salazar et al. [126]

(continued)

Table 3 (continued)

Model type	Solution techniques	Algorithm description	MOA type	Sources
		NSGA2		Kishor et al. [63, 64]
		Ant colony (AC)		Shelokar et al. [127]
		PSO		Pant et al. [100, 102]
Redundancy–reliability allocation	Approximate	SWT	Scaler	Sakawa [123]
		Direct search technique combined with the Min–Max method		Misra and Sharma [89]
		Goal programming (GP) and goal attainment methods (GAT)		Dhingra [37]
	Heuristic/metaheuristic	Evolutionary algorithm (EA)	Pareto	Ramírez-Rosado and Bernal-Agustín [111]
		GA		Huang et al. [59]

and multi-objectives as fuzzy global optimization problems. They also developed and applied the non-combinatorial version of another metaheuristic, viz., threshold accepting to solve these problems. Recently, Shelokar et al. [127] applied the ant colony optimization (ACO) algorithm to these problems and obtained comparable results to those reported by Ravi et al. [114]. Vinod et al. [143] applied GAs to Risk Informed In-Service Inspection (RI-ISI) which aims at prioritizing the components for inspection within the permissible risk level thereby avoiding unnecessary inspections. A new fuzzy MOO method is introduced and it is used for the optimization decision-making of the series and complex system reliability with two objectives is presented by Mahapatra and Roy [82]. Mahapatra [83] considered a series-parallel system to find out optimum system reliability with an additional entropy objective function. Marseguerra et al. [86] applied GA to solve the reliability problem. Salazar et al. [125, 126] solved the system reliability optimization problem by using several EAs and MOEAs. Ravi [113] developed an extended version of the great deluge algorithm and demonstrated its effectiveness in solving the reliability optimization problems. Deep and Deepti [35] applied self-organizing migrating genetic algorithm (C-SOMGA) to optimize such type of problems. Furthermore Kuo and Prasad [70], Kuo and Wan [69] reviewed different reliability optimization and allocation techniques. More recently, Pant et al. [100, 102], Kumar et al. [67] applied PSO and cuckoos search algorithm (CSA) to solve reliability optimization problems.

3 Why Particle Swarm Approach to Reliability Optimization?

Reliability optimization problem are NP-hard in nature so it is quite difficult to achieve optimal reliability design [19]. The solution of such NP-hard optimization problems, however, is more difficult using heuristics or exact algorithms. This is because these optimization problems generate a very large search space, and searching for optimal solutions using exact methods or heuristics will necessarily be extremely time consuming. Such methods are particularly advantageous when the problem is not large. Therefore, metaheuristic algorithms, particularly cuckoos search algorithm (CSA), grey wolf optimization algorithm (GWO), ant colony optimization (ACO), genetic algorithm (GA), differential evolution (DE), particle swarm optimization (PSO), etc., are suitable for solving reliability optimization problems. The main concept of PSO is based on the food searching behavior of birds flocking or fish schooling. When PSO is adopted to solve problems, each particle has its own location and velocity, which determine the flying direction and distance, respectively. Comparing with other evolutionary approaches PSO has the following advantages [21, 53, 58, 120]:

(i) It has less parameters.
(ii) It is easy in implementation.
(iii) It has fast convergence.

These advantages are good for solving the reliability optimization problems because a population of particles in PSO can operate simultaneously so that the possibility of paralysis in the whole process can be reduced. Different PSO methods have been already successfully applied by Zavala et al. [155], Chen [18], Pandey et al. [98], Levitin et al. [74], Yeh [152], Coelho [23], Zou et al. [160], Pant and Singh [101] Pant et al. [100,102], etc., in reliability optimization problems.

References

1. **Afshinmanesh, F., Marandi, A., and Rahimi-Kian, A.,** A novel binary particle swarm optimization method using artificial immune system, in *IEEE International Conference on Computer as a Tool,* 2005, 217-220.
2. **Alatas, B. and Akin, E.,** Multi-objective rule mining using a chaotic particle swarm optimization algorithm, *Knowledge-Based Systems*, 22, 2009, 455-460.
3. **Al-kazemi, B. and Mohan, C. K.,** Multi-phase discrete particle swarm optimization, in *Fourth International Workshop on Frontiers in Evolutionary Algorithms*, 2002.
4. **AlRashidi, M. R. and El-Hawary, M. E.,** Emission-economic dispatch using a novel constraint handling particle swarm optimization strategy, in *Canadian Conference on Electrical and Computer Engineering,* 2006, 664-669.
5. **Altiparmak, F., Dengiz, B., and Smith, A. E.,** Reliability optimization of computer communication networks using genetic algorithms, in *IEEE International Conference on Systems, Man, and Cybernetics,*1998, 4676-4681.

6. **Arumugam, M. S and Rao, M. V. C.,** On the improved performances of the particle swarm optimization algorithms with adaptive parameters, cross-over operators and root mean square (RMS) variants for computing optimal control of a class of hybrid systems, *Applied Soft Computing*, 8, 2008, 324-336.
7. **Ashrafi, N. and Berman, O.,** Optimization models for selection of programs, considering cost and reliability, *IEEE Transactions on Reliability*, 41, 1992, 281-287.
8. **Atiqullah, M. M. and Rao, S. S.,** Reliability optimization of communication networks using simulated annealing, *Microelectronics Reliability*, 33,1993, 1303-1319.
9. **Bala, R. and Aggarwal, K. K.,** A simple method for optimal redundancy allocation for complex networks, *Microelectronics Reliability*, 27, 1987, 835-837.
10. **Banks, A., Vincent, J., and Anyakoha, C.,** A review of particle swarm optimization. Part I: Background and Development, *Natural Computing*, 6, 2007, 467-484.
11. **Banks, A., Vincent, J., and Anyakoha, C.,** A review of particle swarm optimization. Part II: Hybridisation, combinatorial, multicriteria and constrained optimization, and indicative applications, *Natural Computing*, 7, 2008, 109-124.
12. **Bartz-Beielstein, T., Limbourg, P., Mehnen, J., Schmitt, K., Parsopoulos, K. E., and Vrahatis, M. N.,** Particle swarm optimizers for Pareto optimization with enhanced archiving techniques, in *Congress on Evolutionary Computation*, 2003, 1780-1787.
13. **Briza, A. C. and Naval Jr, P. C,** Stock trading system based on the multi-objective particle swarm optimization of technical indicators on end-of-day market data, *Applied Soft Computing*, 11, 2011, 1191-1201.
14. **Cai, J., Ma, X., Li, Q., Li, L., and Peng, H.,** A multi-objective chaotic particle swarm optimization for environmental/economic dispatch, *Energy Conversion and Management*, 50, 2009, 1318-1325.
15. **Cao, C. H., Li, W. H., Zhang, Y. J., and Yi, R. Q.,** The geometric constraint solving based on memory particle swarm algorithm, in *International Conference on Machine Learning and Cybernetics,* 2004, 2134-2139.
16. **Carlisle, A. and Dozier, G.,** Adapting particle swarm optimization to dynamic environments, in *International Conference on Artificial Intelligence*, 2000, 429-434.
17. **Chatterjee, A. and Siarry, P.,** Nonlinear inertia weight variation for dynamic adaptation in particle swarm optimization, *Computers & Operations Research*, 33, 2006, 859-871.
18. **Chen, T. C.,** Penalty guided PSO for reliability design problems, in *PRICAI 2006: Trends in Artificial Intelligence*, 2006, 777-786.
19. **Chern, M. S.,** On the computational complexity of reliability redundancy allocation in a series system, *Operations Research Letters*, 11, 1992, 309-315.
20. **Clerc, M. and Kennedy, J.,** The particle swarm-explosion, stability, and convergence in a multidimensional complex space, *IEEE Transactions on Evolutionary Computation*, 6, 2002, 58-73.
21. **Clow, B. and White, T.** An evolutionary race: A comparison of genetic algorithms and particle swarm optimization used for training neural networks, in *International Conference on Artificial Intelligence*, 2004, 582-588.
22. **Coelho, J. P., Oliviera, P. M., and Cunha, J. B.,** Non-linear concentration control system design using a new adaptive PSO, in *5th Portugese Conference on Automatic Control*, 2002.
23. **Coelho, L. S.,** An efficient particle swarm approach for mixed-integer programming in reliability-redundancy optimization applications, *Reliability Engineering & System Safety*, 94, 2009, 830-837.
24. **Coello, C. A.C. and Lechuga, M. S,** MOPSO: A proposal for multiple objective particle swarm optimization, in *Congress on Evolutionary Computation,* 2002, 1051-1056.
25. **Coello, C. A.C., Pulido, G. T., and Lechuga, M. S.,** Handling multiple objectives with particle swarm optimization, *IEEE Transactions on Evolutionary Computation,* 8, 2004, 256-279.
26. **Coit, D. W. and Baheranwala, F.,** Solution of stochastic multi-objective system reliability design problems using genetic algorithms, in *European Safety and Reliability Conference,* 2005, 391-398.

27. **Coit, D. W. and Konak, A.,** Multiple weighted objectives heuristic for the redundancy allocation problem, *IEEE Transactions on Reliability*, 55, 2006, 551-558.
28. **Coit, D. W. and Liu, J. C.,** System reliability optimization with k-out-of-n subsystems, *International Journal of Reliability Quality and Safety Engineering*, 7, 2000, 129-142.
29. **Coit, D. W. and Smith, A. E.,** Considering risk profiles in design optimization for series-parallel systems, in *Annual Reliability and Maintainability Symposium,* 1997, 271-277.
30. **Coit, D. W. and Smith, A. E.,** Reliability optimization of series-parallel systems using a genetic algorithm, *IEEE Transactions on Reliability*, 45, 1996a, 254-260.
31. **Coit, D. W. and Smith, A. E.,** Penalty guided genetic search for reliability design optimization, *Computers & Industrial Engineering*, 30, 1996b, 895-904.
32. **Coit, D. W., T. Jin, T., and Wattanapongsakorn, N.,** System optimization with component reliability estimation uncertainty: A multi-criteria approach, *IEEE Transactions on Reliability*, 53, 2004, 369-380.
33. **De Carvalho, A. B., Pozo, A., and Vergilio, S. R.,** A symbolic fault-prediction model based on multiobjective particle swarm optimization, *Journal of Systems and Software*, 83, 2010, 868-882.
34. **Deb, K., Pratap, A., Agarwal, S., and Meyarivan, T.,** A fast and elitist multiobjective genetic algorithm: NSGA-II, *IEEE Transactions on Evolutionary Computation*, 6, 2002, 182-197.
35. **Deep K. and Deepti,** Reliability Optimization of Complex Systems through C-SOMGA, Journal of Information and Computing Science, 4, 2009, 163-172.
36. **Deeter, D. L. and Smith, A. E.,** Heuristic optimization of network design considering all-terminal reliability, in *Annual Reliability and Maintainability Symposium,* 1997, 194-199.
37. **Dhingra, A. K.,** Optimal apportionment of reliability and redundancy in series systems under multiple objectives, *IEEE Transactions on Reliability*, 41, 1992, 576-582.
38. **Dian, P. R, Siti, M. S., and Siti, S. Y.,** Particle Swarm Optimization: Technique, System and Challenges, *International Journal of Computer Applications*, 14, 2011, 19-27.
39. **Du, W. and Li, B.,** Multi-strategy ensemble particle swarm optimization for dynamic optimization, *Information sciences*, 178, 2008, 3096-3109.
40. **Eberhart, R. and Shi, Y.,** Comparing inertia weights and constriction factors in particle swarm optimization, in *IEEE Congress on Evolutionary Computation,* 2000, 84-88.
41. **Eberhart, R. and Shi, Y.,** Tracking and optimizing dynamic systems with particle swarms, in *IEEE Congress on Evolutionary Computation,* 2001, 94-100.
42. **Eberhart, R., Simpson, P., and Dobbins, R.,** Computational intelligence PC tools. Academic Press Professional, Inc., USA, 1996.
43. **Engelbrecht, A. P.** Fundamentals of computational swarm intelligence, Jhon Wiley & Sons Ltd., 2005.
44. **Engelbrecht, A. P. and van Loggerenberg,** Enhancing the NichePSO, in *IEEE Congress on Evolutionary Computation,* 2007, 2297-2302.
45. **Fan, S. and Chiu, Y.,** A decreasing inertia weight particle swarm optimizer, *Engineering Optimization*, 39, 2007, 203-228.
46. **Feng, Y., Teng, G. F., Wang, A. X., and Yao, Y. M.,** Chaotic inertia weight in particle swarm optimization, in *International Conference on Innovative Computing, Information and Control*, 2007, 475-475.
47. **Feng, Y., Yao, Y. M., and Wang, A. X**., Comparing with chaotic inertia weights in particle swarm optimization, in *Conference on Machine Learning and Cybernetics, International*, 2007, 329-333.
48. **Fieldsend, J. E. and Singh, S., A** Multi-objective algorithm based upon particle swarm optimisation, an efficient data structure and turbulence., *Workshop on Computational Intelligence*, Birmingham, UK, 2002, 37–44,
49. **Fieldsend, J. E.,** Multi-objective particle swarm optimization methods, Department of Computer Science, University of Exeter, 2004.

50. **Goh, C. K., Tan, K. C., Liu, D. S., and Chiam, S. C.,** A competitive and cooperative co-evolutionary approach to multi-objective particle swarm optimization algorithm design, *European Journal of Operational Research*, 202, 2010, 42-54.
51. **Hikita, M., Nakagawa, Y., Nakashima, K., and Narihisa, H.,** Reliability optimization of systems by a surrogate-constraints algorithm, *IEEE Transactions on Reliability*, 41, 1992, 473-480.
52. **Hikita, M., Nakagawa, Y., Nakashima, K., and Yamato, K.,** Application of the surrogate constraints algorithm to optimal reliability design of systems, *Microelectronics and reliability*, 26, 1986, 35-38.
53. **Hodgson, R. J. W.** Particle swarm optimization applied to the atomic cluster optimization problem, in *Genetic and evolutionary computation conference,* 2002, 68–73.
54. **Hu, X. and Eberhart, R.,** Adaptive particle swarm optimization: Detection and response to dynamic systems, in *Congress on Evolutionary Computation,* 2002a, 1666-1670.
55. **Hu, X. and Eberhart, R.,** Multiobjective optimization using dynamic neighborhood particle swarm optimization, in *Congress on Evolutionary Computation,* 2002b, 1677-1681.
56. **Hu, X. and Eberhart, R.,** Solving constrained nonlinear optimization problems with particle swarm optimization, in *World Multiconference on Systemics, Cybernetics and Informatics*, 2002c, 203–206.
57. **Hu, X. and Eberhart, R.,** Tracking dynamic systems with PSO: Where's the cheese, in *the Workshop on Particle Swarm Optimization*, Indianapolis, 2001, 80–83.
58. **Hu, X., Y. Shi, and R. Eberhart,** Recent advances in particle swarm, in *IEEE Congress on Evolutionary Computation,* 2004, 90-97.
59. **Huang, H. Z., Qu, J., and Zuo, M. J.,** A new method of system reliability multi-objective optimization using genetic algorithms, in *Annual Reliability and Maintainability Symposium,* 2006, 278-283.
60. **Jiao, B., Lian, Z., and Gu, X.,** A dynamic inertia weight particle swarm optimization algorithm, *Chaos, Solitons & Fractals*, 37, 2008, 698-705.
61. **Kennedy, J. and Eberhart, R.,** A discrete binary version of the particle swarm algorithm, in *IEEE International Conference on Systems, Man, and Cybernetics, Computational Cybernetics and Simulation.*, 5, 1997, 4104-4108.
62. **Kim, J. H. and Yum, B. J.,** A heuristic method for solving redundancy optimization problems in complex systems, *IEEE Transactions on Reliability,* 42, 1993, 572-578.
63. **Kishor, A., Yadav, S. P., and Kumar, S.,** A Multi-objective Genetic Algorithm for Reliability Optimization Problem, *International Journal of Performability Engineering*, 5, 2009, 227–234.
64. **Kishor, A., Yadav, S. P., and Kumar, S.,** Application of a Multi-objective Genetic Algorithm to solve Reliability Optimization Problem, in International Conference on Computational Intelligence and Multimedia Applications, 2007, 458-462.
65. **Knowles, J. D. and Corne, D. W.,** Approximating the nondominated front using the Pareto archived evolution strategy, *Evolutionary computation*, 8, 2000, 149-172.
66. **Kulturel-Konak, S., Smith, A. E., and Coit, D. W.,** Efficiently solving the redundancy allocation problem using tabu search, *IIE transactions*, 35, 2003, 515-526.
67. **Kumar, A., Pant, S., and Singh, S.B.,** Reliability Optimization of Complex System by Using Cuckoos Search algorithm , *Mathematical Concepts and Applications in Mechanical Engineering and Mechatronics*, IGI Global, 2016, 95-112.
68. **Kumar, A. & Singh, S.B. (2008).** Reliability analysis of an n-unit parallel standby system under imperfect switching using copula, *Computer Modelling and New Technologies*, 12(1), 2008, 47-55.
69. **Kuo, W. and Wan, R.,** Recent advances in optimal reliability allocation, *IEEE Transactions on Systems, Man, and Cybernetics, Part A: Systems and Humans*, 37, 2007, 1-36.
70. **Kuo,W. and Prasad,V. R.,** An annotated overview of system-reliability optimization, *IEEE Transactions on Reliability*, 49, 2000, 176-187.
71. **Laskari, E. C., Parsopoulos, K. E., and Vrahatis, M. N.,** Particle swarm optimization for integer programming, in *IEEE Congress on Evolutionary Computation,* 2002, 1582-1587.

72. **Lei, K., Qiu, Y., and He, Y.,** A new adaptive well-chosen inertia weight strategy to automatically harmonize global and local search ability in particle swarm optimization, in *International Symposium on Systems and Control in Aerospace and Astronautics,* 2006, 977-980.
73. **Leong, W. F. and Yen, G. G.,** PSO-based multiobjective optimization with dynamic population size and adaptive local archives, *IEEE Transactions on Systems, Man, and Cybernetics, Part B: Cybernetics,* 38, 2008, 1270-1293.
74. **Levitin, G., Hu, X., and Dai, Y. S.,** Particle Swarm Optimization in Reliability Engineering, *Intelligence in Reliability Engineering*, 2007, 83-112.
75. **Li, D. and Haimes, Y. Y.,** A decomposition method for optimization of large-system reliability, *IEEE Transactions on Reliability*, 41, 1992, 183-188.
76. **Li, X. and Deb, K.,** Comparing lbest PSO niching algorithms using different position update rules, in *IEEE Congress on Evolutionary Computation* ,2010, 1-8.
77. **Li, X.,** A non-dominated sorting particle swarm optimizer for multiobjective optimization, in *Genetic and Evolutionary Computation,* 2003, 198-198.
78. **Liang, Y. C. and Chen, Y. C.,** Redundancy allocation of series-parallel systems using a variable neighborhood search algorithm, *Reliability Engineering & System Safety*, 92, 2007, 323-331.
79. **Liu, D., Tan, K. C., Goh, C. K., and Ho, W. K.,** A multiobjective memetic algorithm based on particle swarm optimization, *IEEE Transactions on Systems, Man, and Cybernetics, Part B: Cybernetics*, 37, 2007, 42-50.
80. **Liu, X., Liu, H., and Duan, H.,** Particle swarm optimization based on dynamic niche technology with applications to conceptual design, *Advances in Engineering Software*, 38, 2007, 668-676.
81. **Luus, R.,** Optimization of system reliability by a new nonlinear integer programming procedure, *IEEE Transactions on Reliability,* 24, 1975, 14-16.
82. **Mahapatra, G. S. and Roy, T. K.,** Fuzzy multi-objective mathematical programming on reliability optimization model, *Applied mathematics and computation*, 174, 2006, 643-659.
83. **Mahapatra, G.S.,** Reliability optimization of entropy based series-parallel system using global criterion method, *Intelligent Information Management,* 1, 2009, 145-149.
84. **Majety, S. R.V., Dawande, M., and Rajgopal, J.,** Optimal reliability allocation with discrete cost-reliability data for components, *Operations Research*, 47, 1999, 899-906.
85. **Marseguerra, M., E. Zio, E., Podofillini, L., and Coit, D. W,** Optimal design of reliable network systems in presence of uncertainty, *IEEE Transactions on Reliability,*, 54, 2005, 243-253.
86. **Marseguerra, M., Zio, E., and Bosi, F.,** Direct Monte Carlo availability assessment of a nuclear safety system with time-dependent failure characteristics, *International Conference on Mathematical Methods in Reliability*, 2002, 429-432.
87. **Misra, K. B. and Sharma, U.,** An efficient algorithm to solve integer-programming problems arising in system-reliability design, *IEEE Transactions on Reliability,* 40, 1991a, 81-91.
88. **Misra, K. B. and Sharma, U.,** An efficient approach for multiple criteria redundancy optimization problems, *Microelectronics Reliability*, 31, 1991b, 303-321.
89. **Misra, K. B. and Sharma, U.,** Multicriteria optimization for combined reliability and redundancy allocation in systems employing mixed redundancies, *Microelectronics Reliability*, 31, 1991c, 323-335.
90. **Mohan, C. and Shanker, K.,** Reliability optimization of complex systems using random search technique, *Microelectronics Reliability*, 28, 1987, 513-518.
91. **Moore, J. and Chapman, R.,** Application of Particle Swarm to Multi-Objective Optimization: Department of Comput. Sci. Software Eng., Auburn University, 1999.
92. **Mostaghim, S. and Teich, J.,** Strategies for finding good local guides in multi-objective particle swarm optimization (MOPSO), in IEEE *Swarm Intelligence Symposium*, 2003, 26-33.

93. **Munoz, H. and Pierre, E.,** Interval arithmetic optimization technique for system reliability with redundancy, in *International Conference on Probabilistic Methods Applied to Power Systems,* 2004, 227-231.
94. **Nickabadi, A., Ebadzadeh, M. M., and Safabakhsh, R.,** A novel particle swarm optimization algorithm with adaptive inertia weight, Applied *Soft Computing*, 11, 2011, 3658-3670.
95. **Nickabadi, A., Ebadzadeh, M. M., and Safabakhsh, R.,** DNPSO: A dynamic niching particle swarm optimizer for multi-modal optimization, in. *IEEE Congress on Evolutionary Computation*, 2008, 26-32.
96. **Onishi, J., Kimura, S., James, R. J.W., and Nakagawa, Y.,** Solving the redundancy allocation problem with a mix of components using the improved surrogate constraint method, *IEEE Transactions on Reliability*, 56, 2007, 94-101.
97. **Padhye, N., Branke, J., and Mostaghim, S.,** Empirical comparison of MOPSO methods-guide selection and diversity preservation, in *IEEE Congress on Evolutionary Computation,* , 2009, 2516-2523.
98. **Pandey, M. K., Tiwari, M. K., and Zuo, M. J.,** Interactive enhanced particle swarm optimization: A multi-objective reliability application, in *Proceedings of the Institution of Mechanical Engineers, Part O: Journal of Risk and Reliability*, 221, 177-191, 2007.
99. **Panigrahi, B. K., Ravikumar Pandi, V., and Das, S.,** Adaptive particle swarm optimization approach for static and dynamic economic load dispatch, *Energy conversion and management*, 49, 2008, 1407-1415.
100. **Pant, S., Anand, D., Kishor, A., & Singh, S. B.**, A Particle Swarm Algorithm for Optimization of Complex System Reliability, *International Journal of Performability Engineering,* 11(1), 2015, 33-42.
101. **Pant, S., Singh, S. B.,** Particle Swarm Optimization to Reliability Optimization in Complex System, In the proceeding of *IEEE Int. Conf. on Quality and Reliability,* Bangkok, Thailand, 2011, 211-215.
102. **Pant, S., Kumar, A., Kishor, A., Anand, D., and Singh, S.B.**, Application of a Multi-Objective Particle Swarm optimization Technique to Solve Reliability Optimization Problem, In the proceeding of *IEEE Int. Conf. on Next generation Computing Technologies*, 2015, 1004-1007.
103. **Parsopoulos, K. E. and Vrahatis, M. N.,** Particle swarm optimization method for constrained optimization problems, *Intelligent technologies–theory and application: New trends in intelligent technologies*, 2002a, 214–220.
104. **Parsopoulos, K. E. and Vrahatis, M. N.,** Recent approaches to global optimization problems through particle swarm optimization, *Natural computing*, 1, 2002b, 235-306.
105. **Parsopoulos, K. E. and Vrahatis, M. N.,** Unified particle swarm optimization for tackling operations research problems, in *IEEE Swarm Intelligence Symposium,* 2005, 53-59.
106. **Parsopoulos, K. E., Tasoulis, D. K., and Vrahatis, M. N.,** Multiobjective optimization using parallel vector evaluated particle swarm optimization, in *International conference on artificial intelligence and applications,* 2004, 2, 823-828.
107. **Prasad, R. and Raghavachari, M.,** Optimal allocation of interchangeable components in a series-parallel system, *IEEE Transactions on Reliability*, 47,1998, 255-260.
108. **Prasad,V. R. and Kuo, W.,** Reliability optimization of coherent systems, *IEEE Transactions on Reliability*, 49, 2000, 323-330.
109. **Pulido, G. T. and Coello C.A.C.,** Using clustering techniques to improve the performance of a multi-objective particle swarm optimizer, in *Genetic and Evolutionary Computation Conference* , 2004, 225-237.
110. **Qin, Z., Yu, F., Shi, Z., and Wang, Y.,** Adaptive inertia weight particle swarm optimization, in *International conference on Artificial Intelligence and Soft Computing,* 2006, 450-459.
111. **Ramírez-Rosado, I. J. and Bernal-Agustín, J. L.,** Reliability and costs optimization for distribution networks expansion using an evolutionary algorithm, *IEEE Transactions on Power Systems*, 16, 2001, 111-118.

112. **Raquel, C. R. and Naval Jr, P. C.,** An effective use of crowding distance in multiobjective particle swarm optimization, in *Genetic and evolutionary computation conference*, 2005, 257-264.
113. **Ravi, V.,** Modified great deluge algorithm versus other metaheuristics in reliability optimization, *Computational Intelligence in Reliability Engineering*, 40, 2007, 21-36.
114. **Ravi, V., Murty, B. S. N., and J. Reddy,** Nonequilibrium simulated-annealing algorithm applied to reliability optimization of complex systems, *IEEE Transactions on Reliability*, 46, 1997, 233-239.
115. **Ravi, V.,** Optimization of complex system reliability by a modified great deluge algorithm, *Asia-Pacific Journal of Operational Research*, 21, 2004, 487–497.
116. **Ravi, V., Reddy, P. J., and Zimmermann, H. J.,** Fuzzy global optimization of complex system reliability, *IEEE Transactions on Fuzzy Systems*, 8, 2000, 241-248.
117. **Ray, T. and Liew, K. M.,** A swarm metaphor for multiobjective design optimization, *Engineering Optimization*, 34, 2002, 141-153.
118. **Reddy, M. J. and Kumar, D. N.,** An efficient multi-objective optimization algorithm based on swarm intelligence for engineering design, *Engineering Optimization*, 39, 2007, 49-68.
119. **Reibman, A. L. and Veeraraghavan, M.,** Reliability modeling: An overview for system designers, *Computer*, 24, 1991, 49-57.
120. **Reklaitis, G. V., Ravindran, A. and Ragsdell, K. M.,** Engineering optimization, methods and applications. John Wiley & Sons, 1983.
121. **Reyes-Sierra, M. and Coello, C. A.C.,** Multi-objective particle swarm optimizers: A survey of the state-of-the-art, *International Journal of Computational Intelligence Research*, 2, 2006, 287-308.
122. **Saber, A. Y., Senjyu, T., Yona, A., and Funabashi, T.,** Unit commitment computation by fuzzy adaptive particle swarm optimisation, *Generation, Transmission & Distribution, IET*, 1, 2007, 456-465.
123. **Sakawa, M.,** Multiobjective reliability and redundancy optimization of a series-parallel system by the Surrogate Worth Trade-off method, *Microelectronics and Reliability*, 17, 1978, 465-467.
124. **Sakawa, M.,** Optimal reliability-design of a series-parallel system by a large-scale multiobjective optimization method, *IEEE Transactions on Reliability,* 30, 1981, 173-174.
125. **Salazar, D, E., Rocco, S., and Claudio, M.,** Solving advanced multi-objective robust designs by means of multiple objective evolutionary algorithms (MOEA): A reliability application, *Reliability Engineering & System Safety*, 92, 2007, 697-706.
126. **Salazar, D., Rocco, C. M., and Galván, B. J.,** Optimization of constrained multiple-objective reliability problems using evolutionary algorithms, *Reliability Engineering & System Safety*, 91, 2006, 1057-1070.
127. **Shelokar, P. S., Jayaraman, V. K., and Kulkarni, B. D.,** Ant algorithm for single and multiobjective reliability optimization problems, *Quality and Reliability Engineering International*, 18, 2002, 497-514.
128. **Shi, Y. and Eberhart, R.,** A modified particle swarm optimizer, in *IEEE World Congress on Evolutionary Computational,* 1998, 69-73.
129. **Shi, Y. and Eberhart, R.,** Empirical study of particle swarm optimization, in *Congress on Evolutionary Computation,* 3, 1999a, 1945- 1950.
130. **Shi, Y. and Eberhart, R.,** Experimental study of particle swarm optimization, in *World Multiconf. Systematica, Cybernatics and Informatics*, 2000.
131. **Shi, Y. and Eberhart, R.,** Fuzzy adaptive particle swarm optimization, in *Congress on Evolutionary Computation,* 2001, 101-106.
132. **Shi, Y. and Eberhart, R.,** Parameter selection in particle swarm optimization, in *Annual Conference on Evolutionary Programming,* 1998b, 25-27.
133. **Sierra, M. R. and Coello, C. A.C.,** Improving PSO-based multi-objective optimization using crowding, mutation and e-dominance, in *International Conference on Evolutionary Multi-Criterion Optimization,* 2005, 505-519.

134. **Sivasubramani, S. and Swarup, K.,** Multiagent based particle swarm optimization approach to economic dispatch with security constraints, in *International Conference on Power Systems,* 2009, 1-6.
135. **Sun, C., Liang, H., Li, L., and Liu, D.,** Clustering with a Weighted Sum Validity Function Using a Niching PSO Algorithm, in *IEEE International Conference on, Networking, Sensing and Control,* 2007, 368-373.
136. **Sun, H., Han, J. J. and Levendel, H.,** A generic availability model for clustered computing systems, in *Pacific Rim International Symposium on Dependable Computing*, 2001, 241-248.
137. **Sun, L. and Gao, X.,** Improved chaos-particle swarm optimization algorithm for geometric constraint solving, in *International Conference on Computer Science and Software Engineering, 2008*, 992-995.
138. **Suresh, K., Ghosh, S., Kundu, D., Sen, A., Das, S., and Abraham, A.,** Inertia-adaptive particle swarm optimizer for improved global search, in *International Conference on Intelligent Systems Design and Applications*, 2008, 253-258.
139. **Taboada, H. and Coit, D. W.,** Data clustering of solutions for multiple objective system reliability optimization problems, *Quality Technology & Quantitative Management Journal*, 4, 2007, 35-54.
140. **Tillman, F. A., Hwang, C. L., and Kuo,W.,** Optimization of systems reliability, Marcel Dekker Inc., 1980.
141. **Tripathi, P. K., Bandyopadhyay, S., and Pal, S. K.,** Multi-objective particle swarm optimization with time variant inertia and acceleration coefficients, *Information Sciences*, 177, , 2007, 5033-5049.
142. **Twum, S. B.,** Multicriteria optimisation in design for reliability, Ph.D. Thesis, University of Birmingham, 2009.
143. **Vinod, G., Kushwaha, H. S., Verma, A. K., and Srividya, A.,** Optimisation of ISI interval using genetic algorithms for risk informed in-service inspection, *Reliability Engineering & System Safety*, 86, 2004, 307-316.
144. **Wang, J., Liu, D., and Shang, H.,** Hill valley function based niching particle swarm optimization for multimodal functions, in *International Conference on Artificial Intelligence and Computational Intelligence*, 2009, 139-144.
145. **Wattanapongsakorn, N. and Levitan, S. P.,** Reliability optimization models for embedded systems with multiple applications, *IEEE Transactions on Reliability,* 53, 2004, 406-416.
146. **Wattanapongsakorn, N. and Levitan, S.,** Reliability optimization models for fault-tolerant distributed systems, in *Reliability and Maintainability Symposium,* 2001,193-199.
147. **Wattanapongskorn, N. and Coit, D. W,** Fault-tolerant embedded system design and optimization considering reliability estimation uncertainty, *Reliability Engineering & System Safety*, 92, 2007, 395-407.
148. **Xu, Z., Kuo, W., and Lin, H. H.,** Optimization limits in improving system reliability, *IEEE Transactions on Reliability*, 39, 1990, 51-60.
149. **Yalaoui, A., Châtelet, E., and Chu, C.,** A new dynamic programming method for reliability & redundancy allocation in a parallel-series system, *IEEE Transactions on Reliability*, 54, 2005, 254-261.
150. **Yamachi, H., Tsujimura, Y., Kambayashi, Y., and Yamamoto, H.,** Multi-objective genetic algorithm for solving N-version program design problem, *Reliability Engineering & System Safety*, 91, 2006, 1083-1094.
151. **Yang, X., Yuan, J., Yuan, J., and Mao, H.,** A modified particle swarm optimizer with dynamic adaptation, *Applied Mathematics and Computation*, 189, 2007, 1205-1213.
152. **Yeh, W. C.,** A two-stage discrete particle swarm optimization for the problem of multiple multi-level redundancy allocation in series systems, *Expert Systems with Applications*, 36, 2009, 9192-9200.
153. **You, P. S. and Chen, T. C.,** An efficient heuristic for series-parallel redundant reliability problems, *Computers & Operations research*, 32, 2005, 2117-2127.

154. **Zafiropoulos, E. P. and Dialynas, E. N.,** Methodology for the optimal component selection of electronic devices under reliability and cost constraints, *Quality and Reliability Engineering International*, 23, 2007, 885-897.
155. **Zavala, A. E.M., Diharce, E. R.V., and Aguirre, A. H.,** Particle evolutionary swarm for design reliability optimization, in Evolutionary multi-criterion optimization. Third international conference, EMO 2005. Lecture notes in computer science, Coello Coello CA, Aguirre AH, Zitzler E (eds) , Springer, Guanajuato, Mexico, 3410, 2005, 856-869.
156. **Zhao, J. H., Liu, Z., and Dao, M. T.,** Reliability optimization using multiobjective ant colony system approaches, *Reliability Engineering & System Safety*, 92, 2007, 109-120.
157. **Zhao, S. Z., Liang, J. J., Suganthan, P. N., and Tasgetiren, M. F.,** Dynamic multi-swarm particle swarm optimizer with local search for large scale global optimization, in *IEEE Congress on Evolutionary Computation,* 2008, 3845-3852.
158. **Zheng, Y., Ma, L., Zhang, L. and Qian, J.,** On the convergence analysis and parameter selection in particle swarm optimization, in *International Conference on Machine Learning and Cybernetics,* 2003b, 1802-1807.
159. **Zheng, Y., Ma, L., Zhang, L., and Qian, J.,** Empirical study of particle swarm optimizer with an increasing inertia weight, in *IEEE Congress on* Evolutionary *Computation,* 2003a, 221-226.
160. **Zou, D., Wu, J., Gao, L., and Wang, X.,** A modified particle swarm optimization algorithm for reliability problems, in *IEEE Fifth International Conference on Bio-Inspired Computing: Theories and Applications (BIC-TA),* 2010, 1098-1105.

Reliability and Quality Control of Automated Diagnostic Analyzers

Ilias Stefanou, Alex Karagrigoriou and Ilia Vonta

Abstract The purpose of this work is to show how statistical quality control methods contribute to the quality control of diagnostic tests in clinical laboratories. This chapter presents the difficulties as well as the advantages of the implementation of quality control techniques in the daily routine of laboratories together with the disadvantages of non-application. It emphasizes the role of good and uninterrupted operation of laboratory equipment to ensure quality and how to reduce the effects of time and its use with a comprehensive maintenance program. It also documents the role of statistical quality control in early detection of a malfunction of the equipment and thereby in avoiding an incorrect result from the reference laboratory. Finally, it proposes ways to improve the reliability and maintainability of the equipment and highlights the contribution of statistical quality control to clinical laboratory equipment.

Keywords Automated diagnostic analyzers · Availability · Clinical laboratory · Maintainability · Reliability · Statistical quality control

1 Introduction

The World Health Organization (WHO) defines quality health as "the provision of diagnostic and therapeutic procedures to ensure the best possible result within the potential limits of modern medical science which should always focus on accurate

I. Stefanou
Hellenic Open University, Patras, Greece
e-mail: std090568@ac.eap.gr

A. Karagrigoriou
University of the Aegean, Samos, Greece
e-mail: alex.karagrigoriou@aegean.gr

I. Vonta (✉)
National Technical University of Athens, Athens, Greece
e-mail: vonta@math.ntua.gr

M. Ram and J.P. Davim (eds.), *Advances in Reliability and System Engineering*, Management and Industrial Engineering, DOI 10.1007/978-3-319-48875-2_8

results with the least possible risk for the patient's health and patient's satisfaction in terms of procedures, outcome and human contact" [23]. Quality in health care is a primary concern of any health organization or health system. In recent years, studies of health services turn increasingly towards the issue of ensuring the quality of service [18].

The clinical diagnostic laboratory is an integral part of medical science. Its role becomes daily more and more important. It contributes to the prognosis, diagnosis, and monitoring the progress of patients. The results of a clinical diagnostic test and the benefits it provide to clinicians depend directly on the quality of the analysis. Since any laboratory assay involves the unavoidable risk of a random or systematic error, the importance of statistical quality control becomes evident in every clinical laboratory to reduce, minimize, and hopefully eliminate the risk of nonrandom (human or machine) faults. It is the duty of a clinical laboratory to assure accurate and reliable test results and reports in addition to increase the laboratory's performance. An important role in this is played by the proper maintenance and reliability of all equipments used in the lab. To ensure the quality of analyses, the equipment is required to respond to as high as possible levels of reliability and availability.

The internal quality control in a clinical diagnostic laboratory is under a continuous, critical evaluation of laboratory analytical methods and routine analysis. The audit involves an analytical process/procedure starting with the sample entering the laboratory and ending with the analytical report. In this work, we make use of statistical tools and techniques that can be applied to a clinical laboratory to ensure the validity of the results produced. Section 2 presents the history of statistical quality control in clinical diagnostic laboratories and presents all the statistical tools that can be used daily by the professional team of a clinical diagnostic laboratory to increase its performance. We also briefly discuss the quality control issues associated with automated analyzers. Section 3 is devoted to the concepts of reliability and maintainability of the equipment of clinical laboratories which are recognized as being extremely important in ensuring the overall quality of the lab. The following section (Sect. 4) is devoted to the statistical analysis of data of actual failures and maintenance for an automated immunological analyzer and an automated hematology analyzer, and makes inferences on the reliability, maintainability, availability, and quality control. The analyzers are installed and operated in a large hospital in Thessaloniki, Greece. The last section is devoted to a discussion and some concluding remarks.

2 Literature Review

2.1 Quality Control in Clinical Laboratories

Statistical quality control is a daily routine performed by lab staff that handles automatic analyzers in a clinical laboratory. Quality control in a clinical laboratory is defined as the control of the analytical process to ensure that the results agree, as much as possible with the required specifications [6] and dates back to the early 1950s with the pioneer work of Levey and Jennings.

Levey and Jennings [12] were the first to introduce the Shewhart control chart (Levey–Jennings chart) in clinical laboratories in the 1950s. Archibald in [1] argued that the daily analysis of reference samples of unknown values in batches is a basic requirement in any quality control program of a clinical laboratory. The Shewhart control chart is considered to be ideal for the graphical presentation of the results of daily analyses of reference samples in a clinical laboratory. Henry and Segalove [8] proposed the use of three different control charts for routine quality control. One of which was the Shewhart chart. Freier and Rausch [6] used specific serum concentration in each batch of test samples, proposed the use of three standard deviations as a measure of accuracy and depicted in a daily basis, the results in a Shewhart control chart. Henry [7] is responsible for the implementation of Levey–Jennings chart in clinical chemistry and discussed in detail the use of such charts for the daily monitoring of quality control. Many control chart modifications have been proposed in recent years. Westgard et al. [27] for example, introduced in 1977 the use of Cusum (Cumulative Sum) chart in the quality control in laboratories and proposed the multiple controls chart as a quality control method in clinical chemistry [3, 19].

Westgard et al. [26] in an article on laboratory quality control laid the basis for assessing the quality of analyses in clinical laboratories. Westgard control rules are based on principles of statistical process control and are applied in the industry since the early 1950s. In total, there are six basic rules constituting the multi-standards process which could be applied either individually or cumulatively for the assessment of the accuracy of analyses [5].

Techniques used instead of control serum samples of patients have been proven to be very effective to the quality control laboratory. Among these techniques one could mention: the method of daily physiological medium ACN value (Average of Normals, [9]), the Bull's algorithm in hematology in 1973, the Delta check method [14], the use of the anion gap [29] and the use of retained samples of whole blood in hematology [4]. For an extended discussion please see Karkalousos and Evangelopoulos [11].

2.2 *Statistical Quality Control of Automated Analyzers*

The purpose of this section is the discussion of statistical quality control for automated analyzers in biomedical sciences, such as hematology, immunology, and biochemistry.

Statistical quality control can be applied to almost any automated process including automated analyzers and diagnostic tests. Unlike industry, where products have identical characteristics, in clinical laboratory tests, things are completely different, due to great biological diversity between individuals. As a result statistical quality control can be performed only on equipment and analytical methods and only rarely on laboratory results. Statistical quality control of automated analyzers does not rely on sampling from patient sample values, but instead on the use of special samples known as control samples for which the target value is known [10].

It should be noted that statistical quality control currently used in clinical diagnostic laboratories does not significantly differ from statistical techniques and methods first used in the 1950s by Levey and Jennings. They were the first to describe statistical quality control techniques based on the mean or average value and the range of duplicate measurements, which were in turn based on Shewhart's methods [20]. Later Henry and Segalove [8] used single instead of duplicate measurements and developed an alternative innovative procedure, where, instead of a patient's sample, a stable control sample is analyzed several times and the results are placed in a control chart. Thus, they simplified the Levey–Jennings method and introduced the single value control diagram known as "Shewhart chart" or "Levey and Jennings chart" [25]. This type of chart is currently widely used in clinical diagnostic laboratories worldwide.

Statistical quality control in a laboratory is applied either directly or retrospectively. In direct daily control, current analytical measurements are statistically tracked and the analysis includes the statistical analysis of standard samples, patients' background information, and electronic inspection of analyzers. Retrospective control provides information and background for performance of previous analyses and includes external quality assessment, proficiency testing, calibration, and the use of summary quality control data.

3 Quality Control and Reliability Indices

3.1 Basic Notation and Formulae

For a component, device or system, the failure time T is a continuous random nonnegative random variable $(T \geq 0)$. The distribution function expresses the probability that the observation value is contemplated to be less than a certain value of time t and it is defined as

$$F(t) = P(T \leq t).$$

The distribution function $F(t)$ expresses the probability of failure up to time t. On the other hand, the function $f(t)$ is known as the probability density function (pdf) and is defined as follows:

$$f(t) = \lim_{\Delta t \to 0} \frac{P(t \leq T \leq t + \Delta t)}{\Delta t}$$

or simply, as:

$$f(t) = \frac{\mathrm{d}F(t)}{\mathrm{d}t} = -\frac{\mathrm{d}R(t)}{\mathrm{d}t}$$

where $R(t)$ is the reliability function. The reliability is defined as the ability of a system to perform its task under specific conditions for a certain period of time. The concept of reliability is associated with the ability which is measured by the probability, the expected performance, the environmental or other pre-specified conditions, and the time of operation.

Mathematically speaking, the reliability is defined as the probability that the failure time T is greater or equal to time t (see, e.g., [15]):

$$R(t) = P(T \geq t).$$

The reliability function is associated with both the distribution as well as the probability density function

$$R(t) = P[T \geq t] = 1 - F(t)$$

and

$$R(t) = P[T \geq t] = \int_{t}^{\infty} f(u)\,\mathrm{d}u.$$

Figure 1 shows the relationship between reliability function, distribution function, and probability density function.

For a specified time interval (0, t) the reliability can be found to be equal to

$$R(t) = \exp\left[-\int_{t}^{\infty} h(u)\,\mathrm{d}u\right]$$

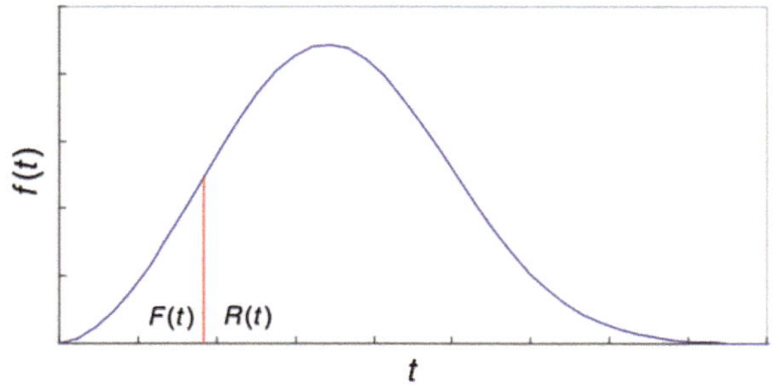

Fig. 1 Relationship between $R(t)$, $F(t)$ and $f(t)$

where $h(u)$ is the failure rate (or hazard function) of the system at a time $t = u$, for $u > 0$.

The failure rate which is usually denoted by $h(t)$ expresses the instantaneous probability that a product or item will fail in the time interval $(t,\ t + \Delta t)$ given that the item is functioning up to time t. It is defined as the probability threshold to fail a component in the next time interval (period) given that it has not failed until the beginning of the period when the width of the interval tends to zero [2] and it is denoted by

$$\begin{aligned} h(t) &= \lim_{\Delta t \to 0} \frac{P[t \le T \le t + \Delta t,\ T \ge t]}{P[T \ge t]\Delta t} = \lim_{\Delta t \to 0} \frac{P[t \le T \le t + \Delta t]}{R(t)\Delta t} \\ &= \frac{1}{R(t)} \lim_{\Delta t \to 0} \frac{F(t + \Delta t) - F(t)}{\Delta t} = \frac{F'(t)}{R(t)} = \frac{f(t)}{R(t)} \end{aligned}$$

The curve in Fig. 2 is known as "bathtub curve" after its characteristic shape and it represents the typical shape of the failure rate. From the shape of the bathtub curve, the lifetime of an item, component or a system is divided into three typical time periods (zones). Zone 1 is characterized by an initially high failure rate known as "infant mortality." This can be explained by the fact that there may be undiscovered defects in the system components; these show up as soon as the system's operation is initiated. In zone 2, the failure rate stabilizes at a level where it remains for a period of time known as the "useful life period" or "failure by chance period." Finally in zone 3, the failure rate starts to increase as the system components begin to wear out as a (natural) result of the aging process. The zone 3 indicates that the product should be replaced or scrapped as hazard rate starts increasing. Frequently, items are tested at the production level so that most if not all of the infant mortality is removed before the products are distributed for use. It should be noted that during the useful life period the failure rate of most mechanical items exhibit a slight upward tendency [16].

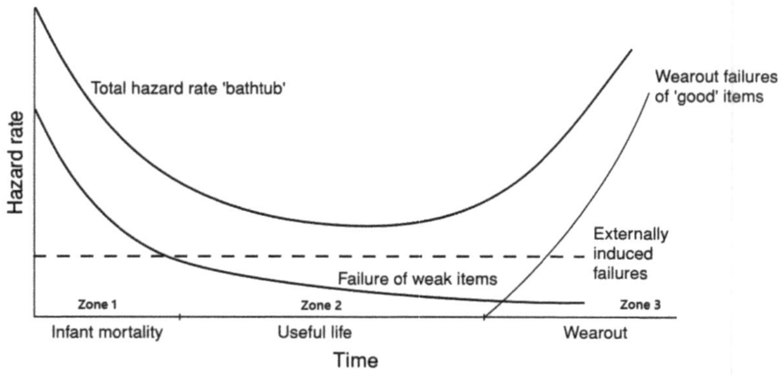

Fig. 2 Typical shape of the failure rate [15]

3.2 Reliability Indices

Reliability of Series Structures

The overall reliability R of a system consisting of subsystems $A, B, C, \ldots, Z$ connected in series is defined as the probability of success of all subsystems at the same time

$$R = P(A \cap B \cap C \cap \cdots \cap Z).$$

Assuming independence, the survival probability (reliability) of the entire system is equal to the product of individual reliabilities of all subsystems of the system

$$R = R_A \cdot R_B \cdot R_C \cdot \cdots \cdot R_z$$

where R_K is the reliability of the Kth subsystem.

Reliability of Parallel Structures

For a system of parallel structure to fail, all subsystems must fail at the same time. In case of a system with only two subsystems, A and B, the uncertainty of the system is the probability of failing both subsystems simultaneously. Hence the uncertainty F may be written as

$$F = P(\overline{A} \cap \overline{B})$$

where $\overline{A}$ and $\overline{B}$ represent the failure of the corresponding components. Assuming independence, then the overall reliability R of the system may be written as

$$\begin{aligned} R &= 1 - F = 1 - P(\overline{A}) \cdot P(\overline{B}) \\ &= 1 - (1 - R_A) \cdot (1 - R_B) = R_A + R_B - R_A \cdot R_B \end{aligned}$$

where R_A and R_B the reliabilities of A and B, respectively. Similarly if a system consists of n units (1, 2, …, n) connected in series then the overall reliability R of the system is defined as

$$R = 1 - (1 - R_1) \cdot (1 - R_2) \cdots (1 - R_n)$$

where R_i, $i = 1, 2, \ldots, n$ the reliability of the ith component.

Reliability of Mixed Structures

The previous relations can be combined to mixed systems, which consist of modules connected in series and in parallel as shown schematically in Figs. 3 and 4.

Thus, for the mixed system presented in Fig. 3 the overall reliability is defined as

$$R = \prod_{j=1}^{n} R_{1j} + \prod_{j=1}^{n} R_{2j} - \prod_{j=1}^{n} R_{1j} R_{2j}$$

while the overall reliability of the system in Fig. 4 is defined as

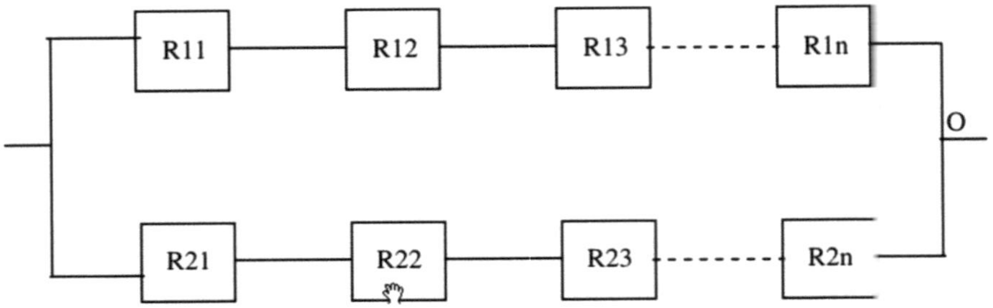

Fig. 3 Series–parallel structure [22]

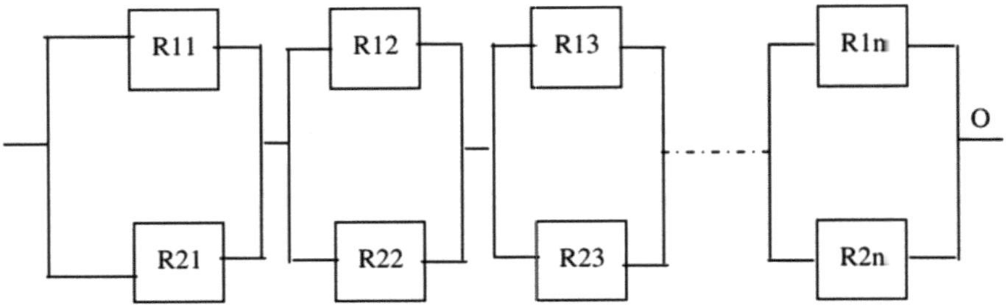

Fig. 4 Parallel–series structure [22]

$$R = \sum_{j=1}^{n} (R_{1j} + R_{2j} - R_{1j}R_{2j}).$$

Mean Time to Failure (MTTF)

The mean value μ_{T} of a random variable T is defined as the expected value of the random variable T, $E(T)$. So the mean time without failures (mean time to failure) is defined as

$$\begin{aligned}\mathrm{MTTF} = \mu_T = E(T) &= \int_0^\infty tf(t)\,\mathrm{d}t \\ &= -\int_0^\infty td[1 - F(t)] \\ &= -\int_0^\infty tdR(t).\end{aligned}$$

Under the assumption that the reliability function is positive through the entire positive part of the real line $[0, \infty)$ and also assume that $\lim_{t\to\infty} t^2 R(t) = 0$, then a simple application of integration by parts gives that

$$\mathrm{MTTF} = -tR(t)|_0^\infty + \int_0^\infty R(t)\,\mathrm{d}t.$$

For a reliability function $R(t)$ tending to zero faster than the time t tending to infinity the above expression simplifies to [2]:

$$\mathrm{MTTF} = \int_0^\infty R(t)\,\mathrm{d}t.$$

Availability
Availability is the probability of functioning for a product or a system over a stated instant of time. The term "functioning" means that the product or system is either in active operation or that it is able to operate if required

$$A(t) = P(\text{the item is functioning at time } t).$$

The availability coincides with reliability for non-repairable systems. A repairable system by the repair can be restored back to working state, after failure occurs, so the "result" of failure is minimized. By repairing the system, reliability remains the same but the availability not. The simplest representation of availability A is the availability over a period of time which is defined as the fraction of operating time over the total time period (which additionally includes the time to repair, TTR)

$$\mathrm{A} = \frac{\text{Uptime of the system}}{\text{Uptime of the system} + \text{Downtime of the system}}$$

It is clear that uptime depends on the reliability of the system while downtime depends on the maintainability of the system. Thus the availability is a function of both reliability and maintainability [22].

Maintainability
According to BS 4778 standard, the term of maintainability is "the ability of the item, under stated conditions of use, to be retained in, or restored to, a state in which it can perform its required functions, when maintenance is performed under stated conditions and using prescribed procedures and resources" (BS 4778). The maintainability is defined as the probability that the maintenance action can be carried out within a fixed time interval. Corrective maintenance is performed following the occurrence of failure while preventive or predictive maintenance is applied in order to reduce the chance of failures and associated inconvenience [22].

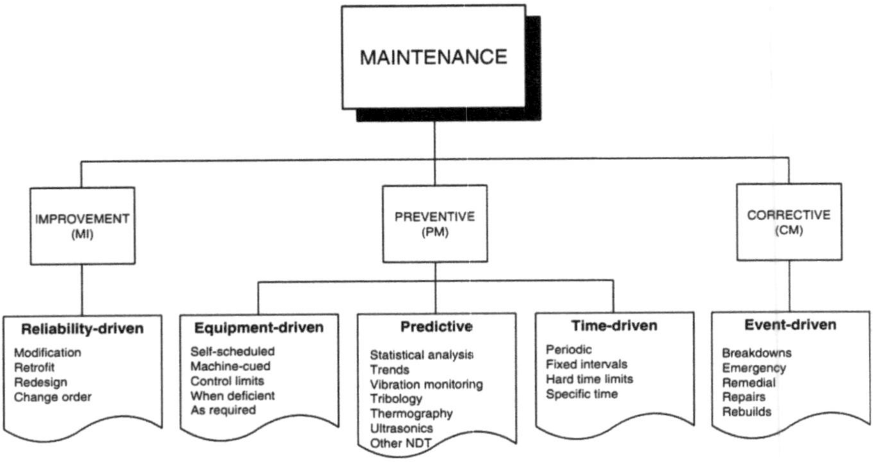

Fig. 5 Structure of maintenance [13]

Maintenance

Maintenance is defined as "the combinations of all technical and corresponding administrative actions, including supervision actions, intended to retain an entity in, or restore it to, a state in which it can perform its required function" [IEC50(191)]. The purpose of maintenance is to maximize the performance of production equipment efficiently and regularly, to prevent breakdown or failures, to minimize production loss from failures and finally to increase both the reliability and the availability of the system [2].

Maintenance may be classified into three main categories (see Fig. 5)

- Corrective or Breakdown maintenance
- Preventive maintenance
- Predictive (Condition-based) maintenance.

3.3 Maintenance—Medical Devices and Laboratory Performance

3.3.1 Maintenance and Medical Devices

Medical equipment directly affects human health and therefore the human life. It is thus necessary for the diagnosis of disease, treatment, and monitoring of patients. Medical devices are considerable investments and in many cases they have a high maintenance cost. As a result, it is important to have a well-designed and easily manageable maintenance program which will be able to retain the equipment at a high level of reliability, security, and availability. Furthermore, the proper maintenance program results in extending the useful life of the equipment and at the same time minimizing the cost of equipment ownership [24].

Medical equipment is highly specialized and if improperly maintained or repaired may have adverse consequences on human life. A maintenance program should include regular inspections and preventive or corrective maintenance procedures. Performance inspections ensure the proper functioning of the equipment, while safety inspections ensure the safety for both patients and equipment operators. Preventive maintenance is intended to extend the life of equipment and reduce failure rates as opposed to a scheduled inspection which detects certain malfunctions undetectable by the operator. It is though important to clarify that equipment inspections only ensure that the device is working properly at the time of inspection and cannot eliminate the possibility of any future failures. Finally, by corrective maintenance the functioning of a failed equipment is successfully restored and normal operation resumes.

The phases of an effective medical equipment maintenance program are: Designing, management, and implementation. The budget together with the physical and human resources required for the maintenance plan are the aspects thoroughly checked during the design of the program. During management and implementation, the staff constantly ensures that the program runs smoothly and actions for improvement are taken, when necessary. The exact application of the maintenance plan is the key to ensure the optimal operation of the equipment.

3.3.2 Clinical Lab Performance and Equipment

Recent technological advances should be considered in clinical laboratories for developing innovative process control techniques capable of providing decisions to release or repeat test results, evaluate the quality of testing processes and fix, replace or deactivate components of instruments with unsatisfactory or insufficient performance. At the same time, the proposed control procedures should ensure the optimization of the cost-effective operation of the system, the minimization of false results, and the maximization of the analytical run [28].

A malfunction of any laboratory equipment will lead to the release of erroneous results unless detected on time. Any erroneous clinical result may have adverse consequences on patient's life. Undoubtedly with the use of statistical quality control, all erroneous analysis results will be identified and rejected. In the preceding sections, we presented common statistical methods available to the manager of a clinical diagnostic laboratory, for timely detecting an incorrect value/result, verifying the failure, and determining its origin.

Standard equipment of a modern diagnostic clinical laboratory is automated diagnostic analyzers. A critical factor that contributes towards the maximization of any laboratory's performance is the operation of analyzers to stay within the performance specifications with the maintenance plan performed as defined by the manufacturer.

4 Applications

4.1 Introduction

Representative equipment of a modern diagnostic clinical laboratory includes automated diagnostic analyzers usually divided into three categories, immunological, biochemical, and hematological. These three types of analyzers perform most of the analyses in any clinical laboratory.

In this work, we will focus on a comparative study of reliability and maintainability of an automated immunological analyzer and an automated hematological analyzer. The system of analyzers studied in this work, are installed and operated in a large public hospital in Thessaloniki, Greece.

Immunological analysts are on constant operation disabled only in fault condition or at scheduled maintenance and are used for a wide range of tests. On the average, they operate for 18 h a day in basic routine laboratory operation while they are standby for the remaining six hours of the day. Their cleaning is done by the operator once daily (or more if necessary) and takes about 30 min.

Hematological analyzers are used for general blood tests 14 h a day, in the standard laboratory routine as well as in emergencies, while during the remaining 10 h are in state of purification and self-preservation and are disabled. They remain inoperative and disabled when experiencing a damage or during maintenance.

Immunological Analyzers
The automated immunological analyzers are medium-sized diagnostic devices used for the determination of concentrations of various residues present in human body fluids. They have the ability to analyze up to 100 samples per hour and consist of seven key subsystems:

1. Turntable Subsystem: Manages samples and reagents for examinations
2. Basic pipette Subsystem: chooses and splits the sample for analysis
3. Analysis Subsystem: Stir, dilutes, mixes, and incubates the sample
4. Fluid Subsystem
5. Electronic Subsystem
6. UPS Subsystem—uninterruptable power supply unit
7. Peripherals Subsystem.

If a subsystem fails, then the analyzer is considered inoperative so that the system is as if it is being connected in series.

Hematology Analyzers
Hematology analyzers are high-class top-bench analysts, enabled to analyze up to 600 samples per hour and provide a complete blood count of 26 parameters. Like the immunological analyzer, it is considered as a system connected in series and

consists of five key subsystems, the failure of any one of which results in failure of the analyzer/system. The five subsystems are

1. Reagent Subsystem: distribution and control circuit of reagents
2. Power Subsystem: Provides the power supply into the system
3. Dilution Subsystem: for the treatment of the test sample
4. Analysis Subsystem: for processing the data from the diluent
5. Workstation and peripheral devices Subsystem (computer, printer, etc.).

The analyzers under investigation are in a general hospital that keeps a record of failures and scheduled maintenance for each analyzer. Analysts store data and provide statistics of the daily quality control applied by each laboratory. Each analyst calculates quality control data and presents the operator with an error message when a control rule has been violated. The data for the immunological analyzer cover the period 2009–2016 while for the hematology analyzer the period 2005–2016. The available data include eight variables of interest

1. "Operating Time": "Repair" or "Maintenance"
2. "Quality Control": YES or NO if a warning message is displayed or not
3. "Subsystem" which was repaired
4. "Transition Time" for the technician to arrive
5. "Time to Repair—TTR" time to repair or maintenance
6. "Operating Hours" from installation to the moment of failure
7. "Time to Failure—TTF" operating time without failure or time to failure
8. "Inoperation Time" = TTR + Transition time.

4.2 *Reliability of Immunological Analyzer*

The basic statistical characteristics of TTF (in hours) of the analyzer are given in Table 1.

The mean TTF is reported to be equal to 888.8 h with a standard deviation of 92.7 h. Note also that as expected, the data do not follow the normal distribution and they exhibit a right skewness of 0.81. The high value of the coefficient of variation (82.17) is an indication of non-homogeneity. The Pareto Chart is displayed in Fig. 6 and clearly

Table 1 Basic statistics for TTF—immunological analyzer

N	Mean	SE mean	St dev	Coef-var	Min	*Q*1	Median	*Q*3	Max
62	888.8	92.7	730.3	82.17	24	282	804	1344	3024

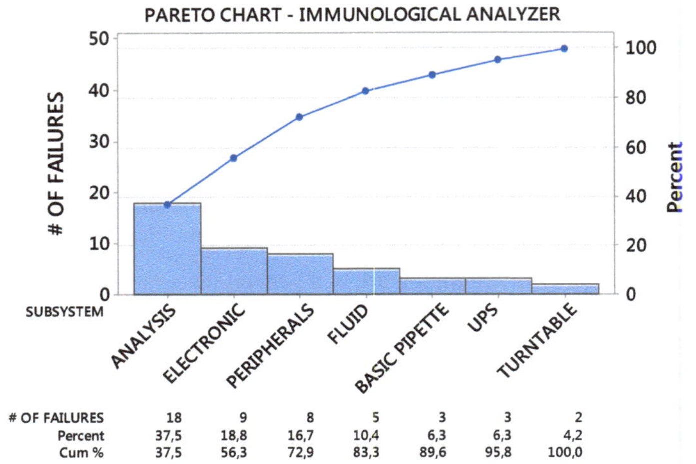

Fig. 6 Pareto chart of immunological analyzer

shows that the largest number of failures occurs at the "analysis" subsystem which accounts for 37.5 % of the analyzer failures followed by the "electronic" and "peripherals" subsystems with 18.8 and 16.7 % of the failures, respectively.

For the distributional analysis of TTF we implement the well-known pp plot (probability plot) and compare the most popular models of lifetime, namely the Weibull, the lognormal, the exponential and the loglogistic distribution. The analysis shows (see Fig. 7) satisfactory results indicating a good fit for all four models.

Indeed, according to the Anderson–Darling(AD) test the best fit to the data is provided by the Weibull distribution (AD statistic = 4.497) which is slightly better that the exponential distribution (4.607). The other two models stay close behind. Thus, we decide to apply to all candidate models the model selection (information) criteria AIC (Akaike Information Criterion) and BIC (Bayesian Information Criterion) that take into consideration not only the log-likelihood but also the complexity of the model

$$\mathrm{AIC} = -2\left[\log\left(\mathrm{Likelihood}\right)\right] + 2A$$
$$\mathrm{BIC} = -2\left[\log\left(\mathrm{Likelihood}\right)\right] + \ln\left(N\right)A$$

where A is the number of parameters of the distribution and N the number of observations. The model chosen is the one for which the value of the criterion is minimized [17]. The results for the best two models are presented in Table 2.

Although according to both criteria the exponential appears to be slightly preferable one should observe that the two distributions are very close and difficult to separate. Indeed the scale parameter of the Weibull distribution is almost equal to 1 (β = 1.13) indicating that the failure (or hazard) rate is almost constant which is a characteristic associated with the exponential distribution (with scale parameter

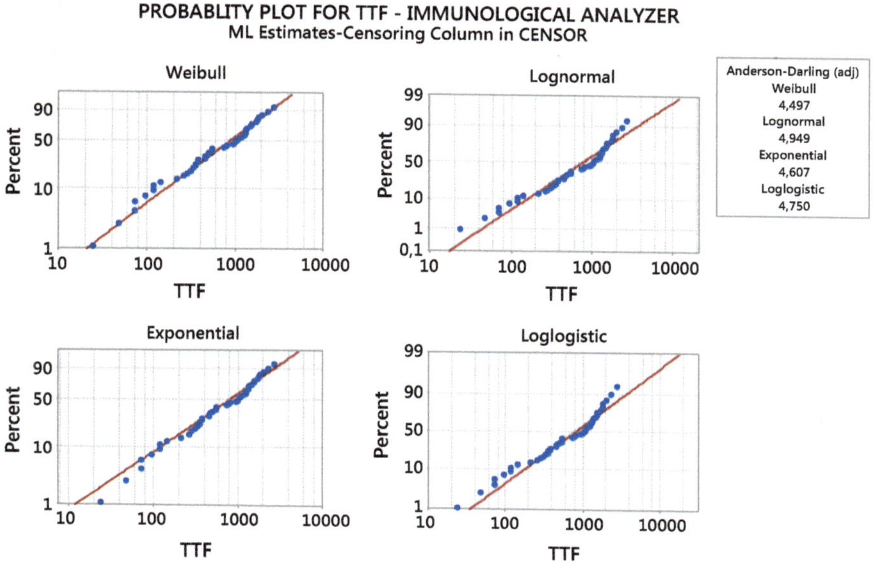

Fig. 7 Probability plot of TTF (immunological analyzer)

Table 2 Model selection for TTF—immunological analyzer

Distribution	Parameters	Log-likelihood	AIC	BIC
Weibull	$\beta = 1.13$ and $\lambda = 1164$	−385.6702	775.34	779.59
Exponential	$\beta = 1.00$ and $\lambda = 1148$	−386.1971	774.394	776.521

equal to 1). Considering the fact that the analyst was installed in 2005, it was in operation about four years till the beginning of the 11-year-period of this study. Based on the standard bathtub-shaped curve of hazard we expect the failure rate to be growing. The main issue that needs special attention is the constant rate of failures associated with the exponential distribution. In order for it to be accepted, one must assume that there is no aging process for immunological analyzers. Although regular maintenance is used to prevent failures and consequently result in delaying the aging process, it is not feasible to entirely eliminate it. Therefore, it is logical to assume that the failure rate is even slightly, growing with time which is in accordance with the logic of the process of natural aging. Based on the above, we come to accept that the scale parameter is not equal to 1 (exponential) but equal to 1.13344 thereby finally selecting the Weibull distribution as more appropriate for describing the TTF of the immunological analyzer.

The above model selection decision is further enforced by the fact that the reliability function $R(t)$ of the Weibull distribution is clearly higher than that of the exponential distribution for any time t (Table 3).

Table 3 Reliability function of the immunological analyzer

Weibull parameters	Exponential parameters	t	$R(t)$ Weibull	$R(t)$ Exponential
$\beta = 1.13344$	$\beta = 1$	100	0.939982	0.916578
		200	0.873031	0.840115
$\lambda = 1164.38$	$\lambda = 1148$	300	0.806542	0.770031
		500	0.681399	0.646915

Table 4 Availability of immunological analyzer

Year	Operating time	# of failures/maintenances	Inoperation time	$A(t)$
2010	6524.84	9	45.16	0.993126332
2011	6545.34	8	24.66	0.996246575
2012	6532.67	10	37.33	0.994318113
2013	6532.67	8	37.33	0.994318113
2014	6536.67	7	33.33	0.994926941
2015	6505.83	13	64.17	0.990232877

Table 5 Basic statistics for TTF—hematology analyzer

N	Mean	SE mean	St dev	Coef-var	Min	$Q1$	Median	$Q3$	Max
193	501.9	36.9	512.2	102.06	24	144	336	696	2712

Closing the analysis of the immunological analyzer, we note that an important aspect of reliability is the availability $A(t)$. Table 4 shows the operating times, inoperating times both in hours and the availability $A(t)$ per year.

We observe a slow downward trend in availability of the immunological analyzer with time which is considered natural and may be attributed to the aging process.

4.3 *Reliability of Hematology Analyzer*

The main statistical details of TTF (in hours) of the hematology analyzer are shown in Table 5.

Observe that the mean TTF is 501.9 h with a standard deviation of 36.9 h. Note also that as expected, the data do not follow the normal distribution and exhibit a right skewness of 1.88.

Figure 8 provides the Pareto Chart for the number of failures per subsystem of the hematology analyzer. It becomes clear that the "Operating dilution" subsystem is responsible for the largest number of failures accounting for 66 % of the analyzer's total number of failures, followed by the "Electronic Dilution" and "Power" subsystems with only 11.3 and 8.8 % of failures respectively.

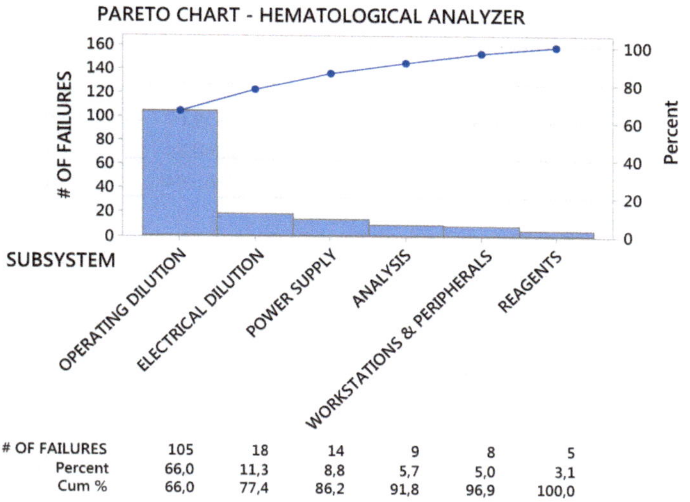

Fig. 8 Pareto Chart of TTF—hematology analyzer

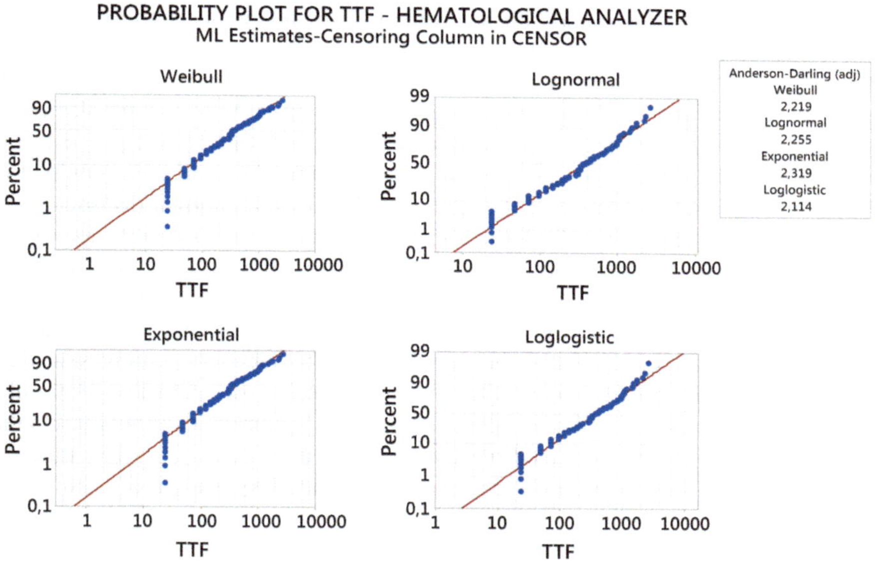

Fig. 9 Probability plot of TTF (hematology analyzer)

According to the Anderson–Darling test, all four models considered for describing TTF (see Fig. 9) provide good fits. The best fit to the data is provided by the loglogistic distribution (AD statistic = 2.114) followed by Weibull (2.219), Lognormal (2.255), and Exponential (2.319). The analysis will be concluded by

Table 6 Model selection for TTF—hematology analyzer

Distribution	Log-likelihood	*A*	AIC	BIC
Lognormal	−1176.954	2	2357.91	2364.43338
Exponential	−1178.533	1	2359.07	2362.32869
Loglogistic	−1178.420	2	2360.84	2367.36538
Weibull	−1178.468	2	2360.94	2367.46138

Table 7 Reliability function of the hematology analyzer

Lognormal parameters	Exponential parameters	t	$R(t)$ exponential	$R(t)$ lognormal
$\alpha = 5.90694$	$\lambda = 609.208$	100	0.848617	0.833818
$\beta = 0.713859$		200	0.720150	0.691139
		300	0.611132	0.588614
		500	0.440107	0.451211

implementing for all candidate models the model selection criteria AIC and BIC that take into consideration both the log-likelihood and the model's complexity, represented by the number A of parameters involved in the model (Table 6).

We observe that the criteria do not come to the same conclusion. So, the lognormal is selected according to AIC and the exponential according to BIC. It is noted here that the AIC has theoretically the tendency to choose more complex models than the actual (as for example is the lognormal as opposed to the simpler exponential) due to the property of overestimation [21]. For comparative purposes the reliability of the system for 100, 200, 300, and 500 h of operation is provided in Table 7 under the assumption of the exponential and the lognormal distributions.

Although for more than 500 h the lognormal distribution is slightly superior, as the hours of operation decrease the exponential distribution gives higher values of reliability. Taking into consideration the fact that the difference between the two distributions is not statistically significant, we conclude that either distribution is acceptable. Note though that if the decision will be relied on purely statistical criteria (like the log-likelihood, the selection criteria, or statistical tests like Anderson–Darling) one will most likely, choose the lognormal distribution. On the other hand, if a decision is based mostly on technical characteristics like the reliability function, then the best choice appears to be the Exponential distribution.

The section ends with the availability of the analyzer (Table 8).

4.4 Maintainability

Having available the times TTR for maintenance and repair, we proceed in this section to investigate the maintainability of the analyzers examined. For the immunological analyzer in the period of 77 months of operation we observed 62

Table 8 Availability of hematology analyzer

Year	Operating time	# of failures/maintenances	Inoperation time	$A(t)$
2005	4374.984	13	63.016	0.985800811
2006	4387.84	22	50.16	0.988697612
2007	4361	22	77	0.982649842
2008	4291.02	25	146.98	0.966881478
2009	4409.484	25	68.516	0.993574583
2010	4398.65	15	39.35	0.991133393
2011	4403.17	11	34.83	0.99215187
2012	4395.75	14	42.25	0.990479946
2013	4410,55	10	27.45	0.993814781
2014	4383.67	20	54.33	0.987757999
2015	4411.74	12	26.26	0.99408292

Table 9 Basic statistics for TTR—both analyzers

N	Mean	SE Mean	St dev	Coef-var	Min	$Q1$	Median	$Q3$	Max
62	2.989	0.270	2.125	71.09	0.5	1.5	2.5	4	12
193	2.024	0.105	1.457	71.99	0.333	1.0	1.5	2.5	12

events, 14 of which refer to scheduled maintenance. The corresponding values of the hematology analyzer are 193 events with 34 scheduled maintenances. The standard statistical characteristics are given in Table 9.

For both analyzers according to the Anderson–Darling (AD) test the two distributions that clearly differ from the others are the lognormal and the loglogistic with the lognormal being slightly superior (see Figs. 10 and 11). It is evident that both the Weibull and the exponential do not provide good fits. Thus, we choose for both analyzers the lognormal distribution for maintainability.

Table 10 provides the percentiles of the lognormal distribution for the immunological analyser. According to the table with 95 % confidence, 20 % of failures of the analyser will be resolved in at least 1.10457 h and at most 1.63876 h since the technician has started working on the problem. Note also that 40 % of failures will be repaired on the average, in 2 h (2.00828) (see bold values in Table 10).

Taking into consideration the reliability function $R(t)$ we observe (see bold values in Table 11) that approximately 10 % (100–89.9259 %) of the repairs of the immunological analyzer will be completed within one (1.0) hour while almost 40 % (100–60.2344 %) will be completed within two (2.0) hours. Note that the corresponding proportions for the hematology analyzer (results not shown) are significantly larger (approx. 20 % in one hour and approx. 60 % in 2 h). This is expected if one recalls the low TTR mean value (Table 9) but it is also confirmed by the estimated parameters of the lognormal distribution. Indeed, for the case of the hematology analyzer the location parameter is significantly smaller (0.50) than the corresponding

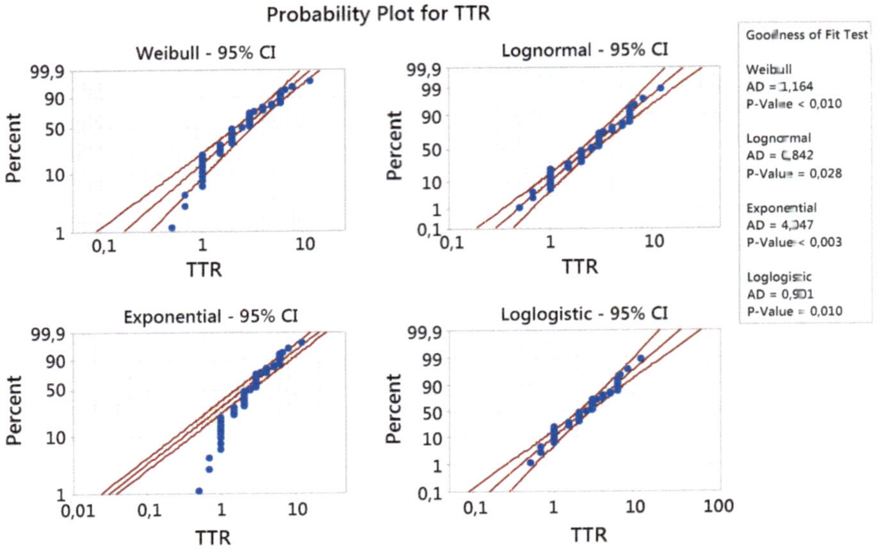

Fig. 10 Probability plot of TTR—immunological analyzer

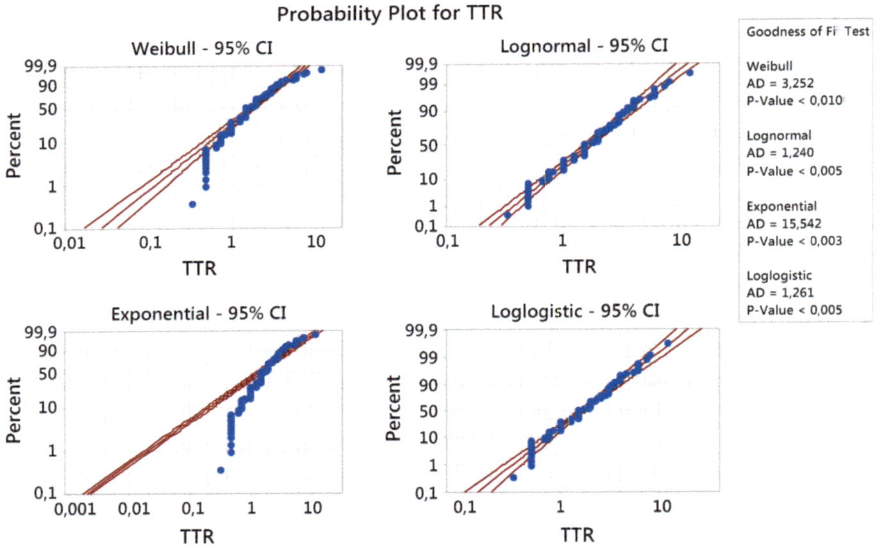

Fig. 11 Probability plot of TTR—hematology analyzer

parameter for the immunological analyser (0.87) resulting in faster repair times. Note that we have not observed significant differences between the scale parameters of the two lognormal distributions (0.63-hematological vs. 0.68—immunological).

Table 10 Distribution percentiles of TTR—immunological analyzer

%	Percentile	Error	Lower	Upper
10	0.99713	0.116372	0.79325	1.25341
20	1.34541	0.135396	**1.10457**	**1.63876**
30	1.66981	0.154013	1.39366	2.00067
40	**2.00828**	0.176441	1.69060	2.38566
50	2.38642	0.206378	2.01435	2.82722
60	2.83576	0.249141	2.38719	3.36863
70	3.41059	0.314572	2.84655	4.08638
80	4.23293	0.425983	3.47520	5.15587
90	5.71139	0.666553	4.54361	7.17930
95	7.31444	0.970257	5.63987	9.48620
97	8.58926	1.23597	6.47843	11.3878

Table 11 Reliability in terms of TTR of immunological analyzer with 95 % CI

Time	Probability	Lower	Upper
0.5	0.989141	0.965677	0.997195
1.0	**0.899259**	0.826883	0.946601
2.0	**0.602344**	0.502533	0.695845
5.0	0.138695	0.080747	0.219957

5 Discussion and Concluding Remarks

In this chapter, we thoroughly discussed statistical quality control and reliability, maintainability and availability of automated analyzers of clinical laboratories. Some brief concluded remarks are stated below.

Immunological analyzers should be better maintained and more detailed tests should be directed towards the "Analysis" and "Electronic" subsystems because our analysis showed that most failures occur in these subsystems. Regarding hematology analyzers, the technical staff should place extra emphasis on two subsystems of most concern, namely the "Operating dilution" and the "Electronic dilution."

Quite often, failures occur at specific components of a subsystem and almost every time they require similar handling. Such cases, as expected, are treated more than once by experienced technicians. This experience and expertise should be transmitted to the younger members of the technical team so that the handling of the most common and frequent failures could be treated both quickly and efficiently. Keeping detailed logs of all failures (especially the most common ones) and the specific actions taken will contribute significantly to the quick and efficient maintenance.

The stock of spare parts has a significant effect on improving the maintainability of a system. As it is well known, the cost of spare parts for medical devices is usually very high. As a result to a great extent, the liquidity of a medical equipment support company is directed towards the storage of at least some of the most commonly needed spare parts. Knowing the subsystems with the most frequent faults, one can easily arrange the management of stocks of spare parts. It is obvious that if the necessary parts are readily available, then not only the overall cost but also the time to repair (TTR) will be reduced.

We should finally be aware of the fact that in at least some instances the operator himself is responsible for some of the failures. The specified maintenance tasks should be fully performed according to the specifications of the manufacturer. For this purpose, the personnel should undergo thorough service training so that both operation and maintenance will be properly and timely performed.

All the above actions will contribute towards the increase of the laboratory's efficiency, since in order to ensure the quality of the analyzers, the equipment is required to respond to as high as possible levels of reliability and availability.

References

1. Archibald RM (1950) Criteria of analytical methods for clinical chemistry, Anal Chem, 22(5), 639-642
2. Bakouros I (2002) Reliability and Maintenance(in Greek), Patras, Greece: Hellenic Open University
3. Carroll AT, Pinnick AH, Carroll EW (2003) Brief Communication: Probability and the Westgard Rules, Ann Clin and Lab Sc, 33(1), 113-114
4. Cembrowski G, Luvetzsky E, Patrick C, Wilson M (1988) An optimized quality control procedure for haematology analyzers with the use of retained patient specimens, Am J Clin Pathol, 89(2), 203-210
5. Cooper G (2008) Basic Lessons in Laboratory Quality Control, Bio-Rad Laboratories, Inc
6. Freier EF, Rausch VL (1958) Quality control in clinical chemistry, Am J Med Tech, 24, 195-200
7. Henry RJ (1959) Use of the control chart in clinical chemistry, Clin Chem, 5(4), 309-319
8. Henry RJ, Segalove M (1952) The running of standards in clinical chemistry and the use of control chart, J. Clin Pathol, 5(4), 305-311
9. Hoffmann R, Waid M (1965) The "average of normals" method of quality control, Am J Clin Pathol, 43, 134-141
10. Karkalousos P, Evangelopoulos A (2011) Quality Control in Clinical Laboratories, Applications and Experiences of Quality Control, O. Ivanov (Ed.), ISBN: 978-953-307-236-4, Intech, Zagreb
11. Karkalousos P, Evangelopoulos A (2015) The History of Statistical Quality Control in Clinical Chemistry and Haematology (1950 – 2010), Intern J. Biomed Lab Sc, 4, 1-11
12. Levey S, Jennings ER (1950) The use of control charts in the clinical laboratory, Am J Clin Pathol, 20(11), 1059-1066
13. Mobley KR. (2004) Maintenance Fundamentals, 2nd ed., Elsevier
14. Nosanchuk JS, Gottmann AW (1974) CUMS and delta checks, Am J Clin Path 62, 707-712.
15. O'Connor PDT, Kleyner, A (2012) Practical Reliability Engineering, 5th ed.,Wiley
16. Rausand M, Hoyland, A (2004) System Reliability Theory: Models, Statistical Methods, and Applications, 2nd ed. Wiley
17. Renyan J (2015) Introduction to Quality and Reliability Engineering, Beijing: Springer, Science Press
18. Rikos N (2015) The concept of quality in health services (in Greek), To Vima tou Asklipiou, 14(4), 247-252

19. Rocco RM (2005) Landmark Papers in Clinical Chemistry, Elsevier
20. Shewhart WA (1931) Economic Control of Quality of Manufactured Product, D. Van Nostrand, Princeton, Eight Printing
21. Shibata R (1976) Selection of the order of an autoregressive model by Akaike's information criterion, Biometrika, 63, 117–126
22. Verma AK, Ajit S, Karanki DR (2016) Reliability and Safety Engineering, 2th ed. Hoang Pham, Piscataway, USA: Springer
23. W.H.O. (2011a) Laboratory quality management system: handbook, World Health Organization
24. W.H.O. (2011b) Medical equipment maintenance programme overview, World Health Organization
25. Westgard JO (2011) QC Past, Present and Future, Basic QC Practices, http://www.westgard.com/history-and-future-of-qc.htm
26. Westgard JO, Barry PL, Hunt MR, Groth T. (1981) A multi-rule Shewhart chart for quality control in clinical chemistry, Clin Chem 27, 493-501
27. Westgard JO, Groth T, Aronsson, T, Falk H, de Verdier CH (1977) Performance characteristics of rules for internal quality control: probabilities for false rejection and error detection, Clin Chem, 23(10), 1857-1867
28. Westgard JO, Smith FA, Mountain PJ, Boss S (1996) Design and assessment of average of normals (AON) patient data algorithms to maximize run lengths for automatic process control, Clin Chem, 42, 1683-1688
29. Witte D, Rodgers J, Barrett D (1976) The anion gap: its use in quality control, Clin Chem, 22, 643-646

Carburettor Performance Under Copula Repair Strategy

Nupur Goyal, Ajay Kaushik and Mangey Ram

Abstract This paper proposes an efficient study about the performance of the carburettor system for the analysis of the reliability characteristics subjected to the Gumbel–Hougaard family of copula with the required repairs facility. The problem has been solved by constructing the state transition model of the carburettor, considering together the supplementary variable technique and Markov process. In this research work, the authors tried to predict the various reliability indices to investigate the performance of the carburettor system with the deliberation of different types of failures occur in it. The performance of carburettor directly depends upon the working of some of its components. Furthermore, the sensitivity of the carburettor system has also been investigated for each reliability characteristic in context of their input parameters, which is very helpful and necessary for attaining a highly reliable carburettor system. At the end of the study, some numerical examples and their graphical representation have also been taken to highlight the practical utility of the model.

Keywords Automobile · Carburettor · Performance · Maintainability and cost analysis

N. Goyal · M. Ram (✉)
Department of Mathematics, Graphic Era University, Dehradun 248002, Uttarakhand, India
e-mail: drmrswami@yahoo.com

N. Goyal
e-mail: nupurgoyalgeu@gmail.com

A. Kaushik
Department of Mechanical Engineering, Graphic Era University, Dehradun 248002, Uttarakhand, India
e-mail: ajaykaushik0394@yahoo.in

M. Ram and J.P. Davim (eds.), *Advances in Reliability and System Engineering*, Management and Industrial Engineering, DOI 10.1007/978-3-319-48875-2_9

1 Introduction

In order to advance technology and growing complexity of engineering systems, and insistence on the system quality and performance, the reliability and maintainability have progressive importance in automobile engineering. Carburettor is an important part or subsystem within the numerous automobile applications. In daily life, there is a great requirement for a highly reliable carburettor. Nowadays, Carburettor is mostly used in motorcycles and some other related systems. In fact during mid to late 1990s, the use of carburettors ends in the new cars, but the carburettors are still used in motorcycles. Carburettor has a small place in engine and is used in some special vehicles, for example, vehicles that are used to build stock car racing, and also found in small equipment engines such as lawnmowers [1–3]. All new cars are equipped with fuel injection system instead of carburettor. Fuel injection systems are used widely, but most are operated electronically. While a carburettor system has no electronic part, so does not need an electricity supply. Therefore, it is easier to service and economically beneficial as compared to fuel injector [4].

Carburettor is also used in an internal combustion engine such as the type of an automobile. The aim or goal of the carburettor is to mix the fuel with air in an accurate proportion to fulfil the demands of engine during all phases of operation so that engine runs properly. If greater fuel is mixed with air, then the engine "runs rich" and either will not run (it floods), runs very smoky, runs poor fuel economy and if less fuel is mixed with the air, the engine "runs lean" and either will not run or potentially damages the engine. The working of carburettor depends upon the functioning of its components such as filter, choke, float chamber, throttle and accelerator pump [5, 6].

Today, in the competitive automobile world, the reliability and maintenance of the carburettor are the key factors of the carburettor's performance which faces many significant challenges. Reliability characteristics play an important role in measuring the performance of the carburettor. The reliability of the carburettor can be defined as the probability of the carburettor to work without failure for a specified time interval in the particular environment [7, 8].

In this novel research work, the designed Markov model of the carburettor estimates the reliability indices (such as availability, reliability MTTF) based on its failures and repairs [9]. Failures are the unavoidable phenomena of repairable carburettor for the practical use. The data for mathematical analysis of reliability measures are considered on the bases of previous research on the carburettor. Maintenance or repairs of the carburettor system increases the cost of engines and decrease the profit. Cost is also an important aspect of the use of carburettor in the automobile sector. In order to reduce the cost of carburettor system, it is necessary to increase the overall efficiency and reliability of the subsystems of carburettor system [10].

The present paper is organized in various sections as follows: Sect. 2 describes the literature review of this research, Sect. 3 gives the details of mathematical model, including the notations associated with the model, assumptions and descriptions of the model, and also describes the formulation and solution to the proposed model. Section 4 covers the numerical computation of reliability measures and their particular cases, Sect. 5 analysed the discussion results of the designed model and Sect. 6 presents the conclusions of the proposed analysis. Supported works are enclosed at the end of the work as reference section.

2 Literature Review

In automobile sector, the performance evaluation of a carburettor system in terms of reliability measure has not been extensively considered in the literature, while the reliability theory has been studied comprehensively. A brief review of appropriate literature would give the comprehensive information to the researchers about various applications of reliability and carburettor system. Ajav et al. [11] studied the performance of the carburettor, but only in the context of fuel. They discussed the effect of diesel and vaporized ethanol on the performance of carburettor and make some modification using dual fuelling in carburettor. They concluded that the vaporized ethanol degraded diesel fuel consumption, lowered exhausted temperature and lubricating oil temperatures, but increased total fuel delivery, power output, thermal efficiency and exhaust emissions. Air-to-fuel ratio and homogeneity much affect the carburettor's performance. Klimstra [12] discussed the performance of four carburettors, each of these based on different principles. Author found that an integrated system and an even initial distribution of fuel over the air stream are required for proper ratio of air-to-fuel ratio and homogeneity. Ma et al. [13] studied the performance of the petrol engine and developed the mathematical simulation and proposed model illustrated the effects of variation in moment of inertia and ignition advance.

As the above researchers have analysed the performance of carburettor or any other automobile system, authors does not get any deeper idea about the complete functioning of the system. A lot of failures can exist in the carburettor system and affect the overall performance of the system and consequently the behaviour of its parts or components. Many researchers have found the complete performance of other systems through reliability measures. El-Sebakhy [14] studied the complete performance of functional network to analyse the software reliability and presented some application and algorithms to demonstrate the functional network. After the comparison with other techniques such as neural network, the author found that it is more reliable, stable and accurate. Ahmad Niknam and Sawhney [15] and Narahari and Viswanadham [16] analysed the reliability of multistate manufacturing system at various performance stages to develop a new approach for examining the production systems under the consideration of partial failure as well as complete failure. Some other researchers, including Levitin et al. [17], Mo et al. [18] and Chao et al. [19] analysed the reliability of k-out-of-n systems by introducing various algorithms and redundancy techniques.

3 Mathematical Model Details

3.1 Notations

The notations associated with the model are shown in Table 1.

3.2 Assumptions

In the proposed model, the following assumptions have been used:

- Initially, the system is working failure free.
- With different transition, the system covers three states, namely, good, degraded and failed.
- Only one change is allowed at a time in the transition states.
- The system has three types of failures, namely equipment failure, waiting time to repair and common cause failure [20, 21].
- The sufficient repair facility is available to repair the failed components and maintain the carburettor.
- When the failed component is repaired, network is as good as a new network.

Table 1 Notations

Symbols	Description
t/s	Time scale/Laplace transform variable
$\bar{P}(s)$	Laplace transformation of $P(t)$
$\lambda_F/\lambda_{FC}/\lambda_{AP}/\lambda_T/$ λ_C/λ_{CC}	Failure rates for filter/float chamber/accelerator pump/throttle/choke/common cause
w_j	Waiting time to repair
$P_0(t)$	The probability of the state S_0
$P_i(t)$	The probability of the state S_i at time t when i = 1, 2, 3, 4
$P_j(y,t)$	The probability density function of the state S_j when j = 5, 6, 7, 8
$\mu(y)$	Repair rates for all the states except S_2 and S_3.
$u_1 = e^x$, $u_2 = \phi(x)$	The joint probability (failed states S_2, S_3 to normal state S_0) according to Gumbel–Hougaard family copula is given as $\exp[x^\theta + \{\log \phi(x)\}^\theta]^{\frac{1}{\theta}}$
$P_{up}(t)$	Up-state system probability at time t or availability of the system
$Rl(t)$	The reliability of the system at time t
K_1/K_2	Revenue/service cost per unit time

3.3 System Description

Carburettor, a most important part of the engine contains subparts such as float chamber, throttle, choke, filter and accelerator pump. In this model, three types of failure namely equipment failure, common cause failure, and waiting time to repair have been considered. On the bases of these three types of failures, the system covers three types of states, such as good, partially failed or degraded and completely failed, and contains total nine possible states. Partially or degraded state means system can work with less efficiency. Due to the failure of choke, filter, float chamber and accelerator pump, carburettor works partially that means it goes to the degraded state. If immediate repair is not provided to carburettor after the failure of filter and float chamber, the carburettor system will fail. The carburettor system goes to complete failed state after the failure of throttle and due to common cause failure also. The two different distributions that is to say general and exponential for the repairing are deliberated for the partially failed carburettor system due to filter and float chamber [22, 23]. For incorporation of this facility, the Gumbel–Hougaard family of copula technique has been used [24–26]. The configuration diagram and modelled state transition diagram of carburettor system are shown in Figs. 1 and 2, respectively, and state description of the designed carburettor system is described in Table 2.

Fig. 1 Configuration diagram of carburettor

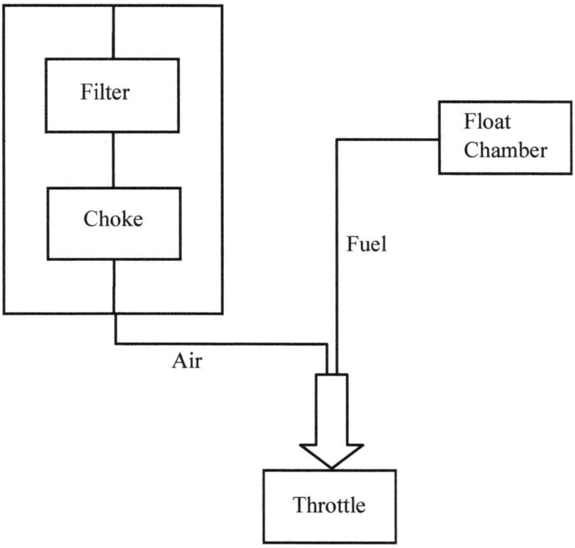

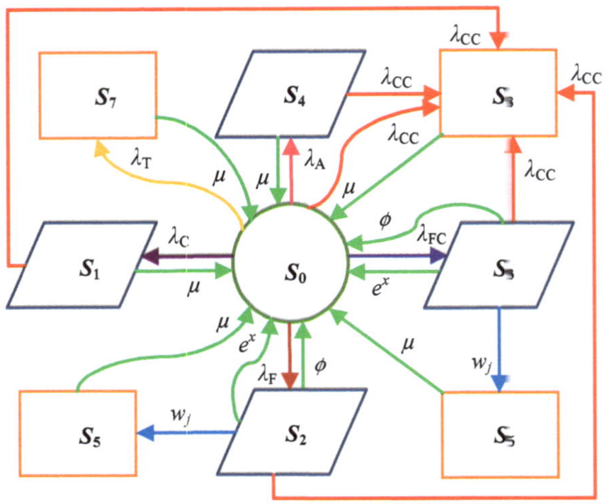

Fig. 2 State transition diagram

Table 2 State description

State	Description
S_0	Good working condition of the carburettor
S_1	The carburettor system is in the degraded state due to failure of the choke
S_2	The carburettor system goes to the degraded state due to filter failure
S_3	The carburettor system is in the degraded state due to failure of float chamber
S_4	The carburettor system is in the degraded state due to the failure of accelerator pump
S_5	The carburettor system is in the complete failed state due to the waiting in repair of filter
S_6	The carburettor system is in the complete failed state due to the waiting in repair of float chamber
S_7	State of the carburettor system when it is completely failed due to throttle
S_8	State of the carburettor system when it is completely failed due to any common cause failure

3.4 *Formulation of the Model*

By the probability consideration and continuity arguments, the following set of intro-differential equations possesses to the present model:

$$\left[\frac{\partial}{\partial t}+\lambda_{\mathrm{F}}+\lambda_{\mathrm{FC}}+\lambda_{\mathrm{C}}+\lambda_{\mathrm{AP}}+\lambda_{\mathrm{T}}+\lambda_{\mathrm{CC}}\right]P_0(t)$$
$$=\sum_{i=2}^{3}\exp\left[x^{\theta}+\{\log\phi(x)\}^{\theta}\right]^{\frac{1}{\theta}}P_i(t)$$
$$+\sum_{j=1,4}\mu(y)P_j(t)+\int_0^{\infty}\mu(y)\sum_{k=5}^{8}P_k(y,t)\mathrm{d}y \quad (1)$$

$$\left[\frac{\partial}{\partial t}+\lambda_{\mathrm{CC}}+\mu(y)\right]P_j(t)=\alpha P_0(t);\quad j=1,4;\quad \alpha=\lambda_{\mathrm{C}},\lambda_{\mathrm{AP}} \quad (2)$$

$$\left[\frac{\partial}{\partial t}+w_j+\lambda_{\mathrm{CC}}+\exp\left[x^{\theta}+\{\log\phi(x)\}^{\theta}\right]^{\frac{1}{\theta}}\right]P_i(t)=\beta P_0(t);$$
$$i=2,3;\quad \beta=\lambda_{\mathrm{F}},\lambda_{\mathrm{FC}} \quad (3)$$

$$\left[\frac{\partial}{\partial t}+\frac{\partial}{\partial y}+\mu(y)\right]P_k(y,t)=0;\quad k=5,6,7,8 \quad (4)$$

Boundary conditions

$$P_k(0,t)=\alpha P_i(t);\quad k=5,6,7;\quad i=2,3,0;\quad \alpha=w_j,w_j,\lambda_{\mathrm{T}} \quad (5)$$

$$P_8(0,t)=\lambda_{\mathrm{CC}}\sum_{i=0}^{4}P_i(t) \quad (6)$$

Initial condition

$$P_i(0)=\begin{cases}1 & i=0\\ 0 & i\geq 1\end{cases} \quad (7)$$

3.5 Solution of the Model

Taking the Laplace transformation of Eqs. (1–6), using the given initial condition in (7)

$$[s+\lambda_{\mathrm{F}}+\lambda_{\mathrm{FC}}+\lambda_{\mathrm{C}}+\lambda_{\mathrm{AP}}+\lambda_{\mathrm{T}}+\lambda_{\mathrm{CC}}]\overline{P}_0(s)=1+\sum_{i=2}^{3}\exp\left[x^{\theta}+\{\log\phi(x)\}^{\theta}\right]^{\frac{1}{\theta}}\overline{P}_i(s)+\sum_{j=1,4}\mu(y)\overline{P}_j(s)+\int_0^{\infty}\mu(y)\sum_{k=5}^{8}\overline{P}_k(y,s)\mathrm{d}y \tag{8}$$

$$[s+\lambda_{\mathrm{CC}}+\mu(y)]\overline{P}_j(s)=\alpha\overline{P}_0(s);\quad j=1,4;\ \ \alpha=\lambda_{\mathrm{C}},\lambda_{\mathrm{AP}} \tag{9}$$

$$\left[s+w_j+\lambda_{\mathrm{CC}}+\exp\left[x^{\theta}+\{\log\phi(x)\}^{\theta}\right]^{\frac{1}{\theta}}\right]\overline{P}_i(s)=\beta\overline{P}_0(s);\quad i=2,3;\ \ \beta=\lambda_{\mathrm{F}},\lambda_{\mathrm{FC}} \tag{10}$$

$$\left[s+\frac{\partial}{\partial y}+\mu(y)\right]\overline{P}_k(y,s)=0;\quad k=5,6,7,8 \tag{11}$$

Boundary conditions

$$\overline{P}_k(0,s)=\alpha\overline{P}_i(s);\quad k=5,6,7;\ \ i=2,3,0;\ \ \alpha=w_j,w_j,\lambda_{\mathrm{T}} \tag{12}$$

$$\overline{P}_8(0,s)=\lambda_{\mathrm{CC}}\sum_{i=0}^{4}\overline{P}_i(s) \tag{13}$$

After solving Eqs. (8–13), we get the state transition probabilities as

$$\overline{P}_0(s)=\frac{1}{(s+A)-\dfrac{\exp\left[x^{\theta}+\{\log\phi(x)\}^{\theta}\right]^{\frac{1}{\theta}}(\lambda_{\mathrm{F}}+\lambda_{\mathrm{FC}})}{s+w_j+\lambda_{\mathrm{CC}}+\exp\left[x^{\theta}+\{\log\phi(x)\}^{\theta}\right]^{\frac{1}{\theta}}}-\dfrac{\mu(y)(\lambda_{\mathrm{C}}+\lambda_{\mathrm{AP}})}{s+\lambda_{\mathrm{CC}}+\mu(y)}-K\cdot\overline{S}_{\mu}(s)} \tag{14}$$

$$\overline{P}_i(s)=\frac{\alpha}{s+\lambda_{\mathrm{CC}}+\mu(y)}\overline{P}_0(s);\quad i=1,4;\ \ \alpha=\lambda_{\mathrm{C}},\lambda_{\mathrm{AP}} \tag{15}$$

$$\overline{P}_j(s)=\varsigma\frac{\beta}{s+w_j+\lambda_{\mathrm{CC}}+\exp\left[x^{\theta}+\{\log\phi(x)\}^{\theta}\right]^{\frac{1}{\theta}}}\overline{P}_0(s);\quad j=2,3,5,6;\ \ \beta=\lambda_{\mathrm{F}},\lambda_{\mathrm{FC}},w_j\lambda_{\mathrm{F}},w_j\lambda_{\mathrm{FC}};\ \ \varsigma=1,1,\left(\tfrac{1-\overline{S}_{\mu}(s)}{s}\right),\left(\tfrac{1-\overline{S}_{\mu}(s)}{s}\right) \tag{16}$$

$$\overline{P}_7(s)=\left(\frac{1-\overline{S}_{\mu}(s)}{s}\right)\lambda_{\mathrm{T}}\overline{P}_0(s) \tag{17}$$

$$\overline{P}_8(s) = \left(\frac{1-\overline{S}_\mu(s)}{s}\right)\lambda_{CC}$$

$$\left[1+\frac{\lambda_F+\lambda_{FC}}{s+w_j+\lambda_{CC}+\exp\left[x^\theta+\{\log\phi(x)\}^\theta\right]^{\frac{1}{\theta}}}+\frac{\lambda_C+\lambda_{AP}}{s+\lambda_{CC}+\mu(y)}\right]\overline{P}_0(s) \quad (18)$$

where

$$A = \lambda_F+\lambda_{FC}+\lambda_C+\lambda_{AP}+\lambda_T+\lambda_{CC}$$

$$K = \lambda_T+\lambda_{CC}+\frac{\lambda_F+\lambda_{FC}}{s+w_j+\lambda_{CC}+\exp\left[x^\theta+\{\log\phi(x)\}^\theta\right]^{\frac{1}{\theta}}}(w_j+\lambda_{CC}) + \frac{\lambda_C+\lambda_{AP}}{s+\lambda_{CC}+\mu(y)}(\lambda_{CC})$$

The Laplace transformation of probability of upstate and downstate systems is given as

$$\overline{P}_{up}(s) = \left[1+\frac{\lambda_F+\lambda_{FC}}{s+w_j+\lambda_{CC}+\exp\left[x^\theta+\{\log\phi(x)\}^\theta\right]^{\frac{1}{\theta}}}+\frac{\lambda_C+\lambda_{AP}}{s+\lambda_{CC}+\mu(y)}\right]\overline{P}_0(s) \quad (19)$$

$$\overline{P}_{down}(s) = \left(\frac{1-\overline{S}_\mu(s)}{s}\right)\left\{\lambda_T+\lambda_{CC}+\frac{\lambda_F+\lambda_{FC}}{s+w_j+\lambda_{CC}+\exp\left[x^\theta+\{\log\phi(x)\}^\theta\right]^{\frac{1}{\theta}}}(w_j+\lambda_{CC})+\frac{\lambda_C+\lambda_{AP}}{s+\lambda_{CC}+\mu(y)}(\lambda_{CC})\right\}\overline{P}_0(s) \quad (20)$$

4 Particular Cases and Numerical Computations of Reliability Measures

4.1 Availability Analysis of the Carburettor

Availability of the designed carburettor system is the probability that performing its required function in a specified time period when operated and maintained in a prescribed discipline. Availability very much depends upon the both, its reliability and maintainability. So, to discuss the availability of the carburettor system, it is necessary to take attention on its failure and repair rates [27]. Availability of the carburettor system can be analysed by taking the inverse Laplace transformation of Eq. (19).

4.1.1 Comprehensive Case of Carburettor

When the carburettor system works properly, setting the value of input parameters as $\lambda_F = 0.25$, $\lambda_C = 0.09$, $\lambda_{AP} = 0.15$, $\lambda_{FC} = 0.065$, $\lambda_T = 0.045$, $\lambda_{CC} = 0.3$, $w_j = 0.1$ and repair rates as $\mu(y) = 1$, $x = 1$, $\theta = 1$ in availability, then the availability of the carburettor in terms of time t is

$$P_{up}(t) = 0.07486506302e^{(-1.553990933\,t)} - 0.0027407269e^{(-3.464289066\,t)} + 0.1831961854e^{(-1.3\,t)} + 0.7446794784 \tag{20a}$$

4.1.2 Availability of the Carburettor Without Waiting Time to Repair

When the carburettor system is not failed due to waiting time in repair, the availability of the carburettor system can be obtained, by taking the value of failure rates as $\lambda_F = 0.25$, $\lambda_C = 0.09$, $\lambda_{AP} = 0.15$, $\lambda_{FC} = 0.065$, $\lambda_T = 0.045$, $\lambda_{CC} = 0.3$, $w_j = 0$ and repair rates as $\mu(y) = 1$, $x = 1$, $\theta = 1$ as in terms of time

$$P_{up}(t) = 0.05355001005e^{(-1.535075208\,t)} + 0.00300458555e^{(-3.383204792\,t)} + 0.1943319838e^{(-1.3\,t)} + 0.7491134206 \tag{20b}$$

4.1.3 Availability of the Carburettor Without Common Cause Failure

Consider that the carburettor system is working perfectly without common cause failure. To determine the availability of carburettor, setting the failure rates as $\lambda_F = 0.25$, $\lambda_C = 0.09$, $\lambda_{AP} = 0.15$, $\lambda_{FC} = 0.065$, $\lambda_T = 0.045$, $\lambda_{CC} = 0$, $w_j = 0.1$ and repair rates as $\mu(y) = 1$, $x = 1$, $\theta = 1$ then availability of the carburettor system is

$$P_{up}(t) = 0.9601000486 + (0.0398999513\cosh(0.9551490666\,t) + 0.04517051844\sinh(0.9551490666\,t))e^{(-2.20914\,t)} \tag{20c}$$

Varying time unit t in Eq. (20a–c), the availability of the carburettor system is obtained as in Table 3 and graphically shown in Fig. 3.

Table 3 Availability of the carburettor system

Time (t)	Availability of the carburettor system $P_{\mathrm{up}}(t)$		
	Comprehensive stage	No waiting time to repair	No common cause failure
0	1.00000	1.00000	1.00000
1	0.81035	0.81371	0.97213
2	0.76163	0.76604	0.96356
3	0.74909	0.75358	0.96109
4	0.74584	0.75030	0.96038
5	0.74499	0.74943	0.96018
6	0.74476	0.74920	0.96012
7	0.74470	0.74914	0.96011
8	0.74468	0.74912	0.96010
9	0.74468	0.74911	0.96010
10	0.74468	0.74911	0.96010
11	0.74468	0.74911	0.96010
12	0.74468	0.74911	0.96010
13	0.74468	0.74911	0.96010
14	0.74468	0.74911	0.96010
15	0.74468	0.74911	0.96010

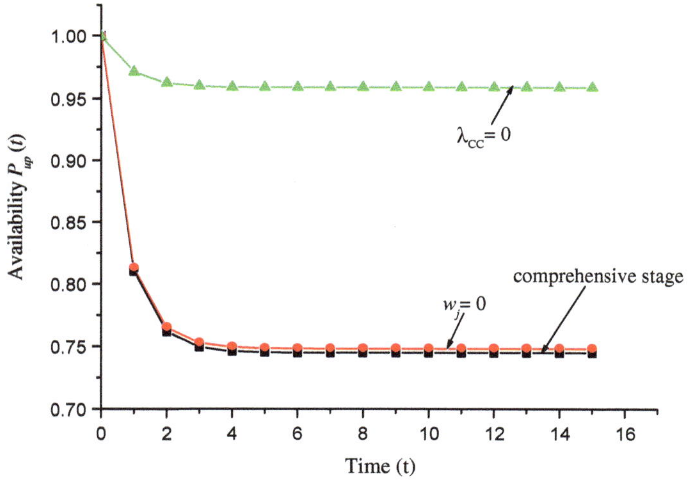

Fig. 3 Availability with respect to time

4.2 Reliability of the Carburettor

System reliability is one of the essential quality characteristics which contract with the behaviour of each apparatus of the system. Reliability represents the probability of non-failure components, subsystems and system to perform their required functions for a precise time period in specified environmental condition. Reliability does not account for any repair actions that may take place [28]. So, to determine the reliability of the designed carburettor system consider the repair rate zero in Eq. (19) and take the inverse Laplace transformation.

4.2.1 Reliability of Carburettor System in Comprehensive Stage

When the carburettor system works properly, setting the value of input parameters as $\lambda_F = 0.25$, $\lambda_C = 0.09$, $\lambda_{AP} = 0.15$, $\lambda_{FC} = 0.065$, $\lambda_T = 0.045$, $\lambda_{CC} = 0.3$, $w_j = 0.1$ in reliability function, then the reliability of the carburettor in terms of time t is

$$\mathrm{Rl}(t) = 0.63\mathrm{e}^{(-0.4\,t)} + 0.4\mathrm{e}^{(-0.3\,t)} - 0.03\mathrm{e}^{(-0.9\,t)} \tag{21a}$$

4.2.2 Reliability of the Carburettor System with Immediately Repair Facility

When the carburettor system is not failed due to waiting time in repair, the reliability of the carburettor system can be obtained, by taking the value of failure rates as $\lambda_F = 0.25$, $\lambda_C = 0.09$, $\lambda_{AP} = 0.15$, $\lambda_{FC} = 0.065$, $\lambda_T = 0.045$, $\lambda_{CC} = 0.3$, $w_j = 0$ as in terms of time

$$\mathrm{Rl}(t) = \mathrm{e}^{(-0.6\,t)}(\cosh(0.3\,t) + 0.85\sinh(0.3\,t)) \tag{21b}$$

4.2.3 Reliability of the Carburettor Without Common Cause Failure

Contemplate that there is no common cause failure in the carburettor system when it is working. To determine the reliability of the carburettor, setting the failure rates as $\lambda_F = 0.25$, $\lambda_C = 0.09$, $\lambda_{AP} = 0.15$, $\lambda_{FC} = 0.065$, $\lambda_T = 0.045$, $\lambda_{CC} = 0$, $w_j = 0.1$ in reliability function, then one can calculate the reliability of the carburettor system as

$$\mathrm{Rl}(t) = 0.63\mathrm{e}^{(-0.1\,t)} + 0.4 - 0.03\mathrm{e}^{(-0.6\,t)} \tag{21c}$$

Varying time unit t in Eq. (21a–c), the reliability of the carburettor system is obtained as revealed in Table 4 and graphically represented in Fig. 4.

Table 4 Reliability of the carburettor system

Time (t)	Reliability of the carburettor system Rl(t)		
	Comprehensive stage	No waiting time to repair	No common cause failure
0	1.00000	1.00000	1.00000
1	0.70643	0.71575	0.95358
2	0.49764	0.52005	0.90676
3	0.35036	0.38112	0.86176
4	0.24685	0.28065	0.81958
5	0.17418	0.20723	0.78062
6	0.12314	0.15324	0.74493
7	0.08724	0.11341	0.71240
8	0.06194	0.08397	0.68283
9	0.04409	0.06219	0.65600
10	0.03145	0.04606	0.63169
11	0.02249	0.03412	0.60967
12	0.01611	0.02527	0.58973
13	0.01157	0.01872	0.57168
14	0.00833	0.01387	0.55535
15	0.00600	0.01027	0.54057

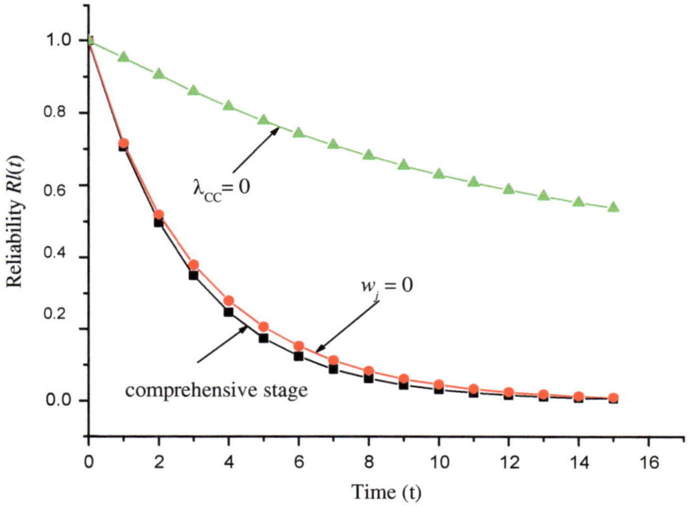

Fig. 4 Reliability with respect to time

4.3 Mean Time to Failure (MTTF)

The MTTF is the predicted elapsed time between inherent failures of a system during the operation. MTTF can be calculated as the average time between failures of the system [29]. By considering the repair rate as zero and the limits of s tends to zero in Eq. (19), one can evaluate the MTTF of the system as

$$\begin{aligned} \text{MTTF} &= \lim_{s \to 0} \overline{P}_{\text{up}}(s) \\ &= \frac{1 + \frac{\lambda_F + \lambda_{FC}}{w_j + \lambda_{CC}} + \frac{\lambda_C + \lambda_{AP}}{\lambda_{CC}}}{\lambda_F + \lambda_C + \lambda_{AP} + \lambda_{FC} + \lambda_T + \lambda_{CC}} \end{aligned} \tag{22}$$

4.3.1 MTTF of the Carburettor System in Comprehensive Stage

Varying the input parameters one by one at 0.1–0.9, respectively, and setting the other failure rate as $\lambda_F = 0.25$, $\lambda_C = 0.09$, $\lambda_{AP} = 0.15$, $\lambda_{FC} = 0.065$, $\lambda_T = 0.045$, $\lambda_{CC} = 0.3$, $w_j = 0.1$ in Eq. (22), one gets the variation in MTTF of the carburettor system with respect to failure rates that signify in Table 5 and graphical representation shown in Fig. 5.

4.3.2 MTTF of the Carburettor System with Immediate Repair Facility

When the system is not failed due to waiting time in the repair, one can get the variation in MTTF of the carburettor system with respect to failure rates by varying the input parameters one by one at 0.1–0.9, respectively, and setting the other failure rate as $\lambda_F = 0.25$, $\lambda_C = 0.09$, $\lambda_{AP} = 0.15$, $\lambda_{FC} = 0.065$, $\lambda_T = 0.045$,

Table 5 MTTF as function of failure rates in comprehensive stage

Variation in λ_F, λ_C, λ_{AP}, λ_{FC}, λ_T, λ_{CC}, w_j	MTTF with respect to failure rates						
	λ_F	λ_C	λ_{AP}	λ_{FC}	λ_T	λ_{CC}	w_j
0.1	2.95000	2.88004	2.84804	2.86096	2.70942	7.10714	2.87500
0.2	2.89706	2.92492	2.89912	2.82609	2.45261	4.06250	2.70000
0.3	2.85526	2.96171	2.94048	2.79736	2.24026	2.87500	2.58333
0.4	2.82143	2.99242	2.97464	2.77328	2.06175	2.23000	2.50000
0.5	2.79348	3.01845	3.00333	2.75281	1.90959	1.82273	2.43750
0.6	2.77000	3.04078	3.02778	2.73519	1.77835	1.54166	2.38889
0.7	2.75000	3.06015	3.04885	2.71987	1.66399	1.33585	2.35000
0.8	2.73276	3.07712	3.06720	2.70642	1.56344	1.17857	2.31818
0.9	2.71774	3.09210	3.08333	2.69452	1.47436	1.05444	2.29167

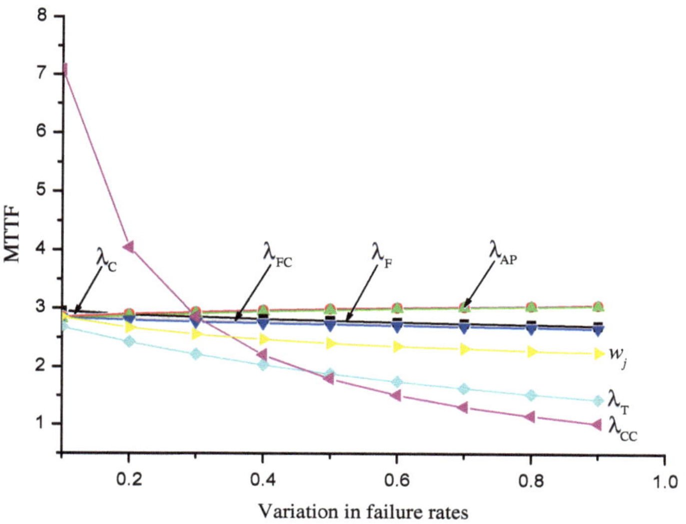

Fig. 5 MTTF with respect to variation in failure rates in comprehensive stage

$\lambda_{CC} = 0.3$, $w_j = 0$ in Eq. (22), that signify in Table 6 and graphical representation shown in Fig. 6.

4.4 *Expected Profit*

Expected profit has a great importance to maintain system reliability. Let the service facility be always available, the expected profit during the interval [0, *t*) is given as

Table 6 MTTF as function of failure rates with immediate repair

Variation in λ_F, λ_C, λ_{AP}, λ_{FC}, λ_T, λ_{CC}	MTTF with respect to failure rates						
	λ_F	λ_C	λ_{AP}	λ_{FC}	λ_T	λ_{CC}	w_j
0.1	3.13333	3.16850	3.15686	3.17290	2.98429	9.35714	3.16667
0.2	3.15686	3.18482	3.17544	3.18840	2.70142	4.71875	3.16667
0.3	3.17544	3.19820	3.19048	3.20117	2.46753	3.16667	3.16667
0.4	3.19048	3.20937	3.20290	3.21187	2.27092	2.38750	3.16667
0.5	3.20290	3.21883	3.21333	3.22097	2.10332	1.91818	3.16667
0.6	3.21333	3.22695	3.22222	3.22880	1.95876	1.60417	3.16667
0.7	3.22222	3.23399	3.22988	3.23561	1.83280	1.37912	3.16667
0.8	3.22988	3.24016	3.23656	3.24159	1.72205	1.20982	3.16667
0.9	3.23656	3.24561	3.24242	3.24688	1.62393	1.07778	3.16667

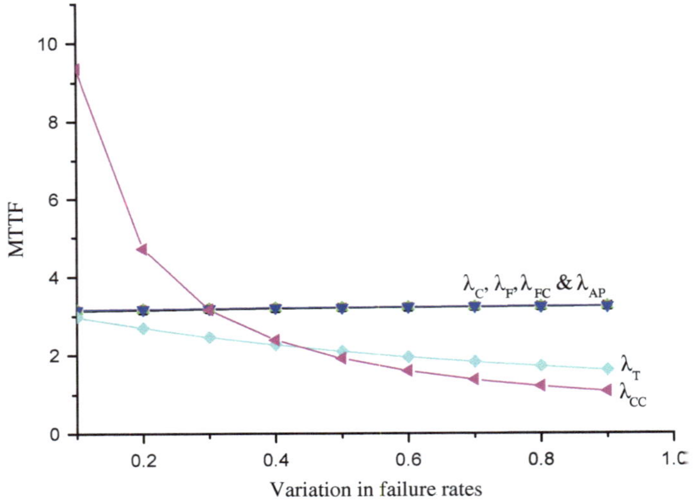

Fig. 6 MTTF with respect to variation in failure rates in immediate repair facility

$$E_{\mathrm{p}}(t) = K_1 \int_0^t P_{\mathrm{up}}(t)\mathrm{d}t - tK_2 \tag{23}$$

Taking the inverse Laplace transform of Eq. (19) and using in (23), we get the expected profit as mentioned in Sects. 4.4.1 and 4.4.2.

4.4.1 Expected Profit of the Carburettor System in Comprehensive Stage

By setting the value of input parameters as $\lambda_{\mathrm{F}} = 0.25$, $\lambda_{\mathrm{C}} = 0.09$, $\lambda_{\mathrm{AP}} = 0.15$, $\lambda_{\mathrm{FC}} = 0.065$, $\lambda_{\mathrm{T}} = 0.045$, $\lambda_{\mathrm{CC}} = 0.3$, $w_j = 0.1$, one can obtain the expected profit of the carburettor system in comprehensive stage as

$$\begin{aligned} E_{\mathrm{p}}(t) = K_1 \Big\{ & 0.188305003 - 0.4817599732\mathrm{e}^{(-1.553990933\,t)} + 0.0007911368964\mathrm{e}^{(-3.464289060\,t)} \\ & -0.1409201426\mathrm{e}^{(-1.3\,t)} + 0.7446794784\,t \Big\} - tK_2 \end{aligned} \tag{23a}$$

Setting $K_1 = 1$ and $K_2 = 0.1, 0.3, 0.5, 0.7, 0.9$, respectively, one can get expected profit for designed system as shown in Table 7 and graphically demonstrated in Fig. 7.

Table 7 Expected profit as function of time in comprehensive case

Time (t)	Expected profit $E_p(t)$				
	K_2 = 0.1	K_2 = 0.3	K_2 = 0.5	K_2 = 0.7	K_2 = 0.9
0	0.00000	0.00000	0.00000	0.00000	0.00000
1	0.78442	0.58442	0.38442	0.18442	−0.01558
2	1.46504	1.06504	0.66504	0.26504	−0.13495
3	2.11903	1.51903	0.91903	0.31903	−0.28096
4	2.76615	1.96615	1.16615	0.36615	−0.43385
5	3.41147	2.41147	1.41147	0.41147	−0.58853
6	4.05632	2.85632	1.65632	0.45632	−0.74368
7	4.70104	3.30104	1.90104	0.50104	−0.89895
8	5.34574	3.74574	2.14574	0.54574	−1.05426
9	5.99042	4.19042	2.39042	0.59042	−1.20958
10	6.63510	4.63510	2.63510	0.63510	−1.36490
11	7.27978	5.07978	2.87978	0.67978	−1.52022
12	7.92446	5.52446	3.12446	0.72446	−1.67554
13	8.56914	5.96914	3.36914	0.76914	−1.83086
14	9.21382	6.41382	3.61382	0.81382	−1.98618
15	9.85850	6.85850	3.85850	0.85850	−2.14150

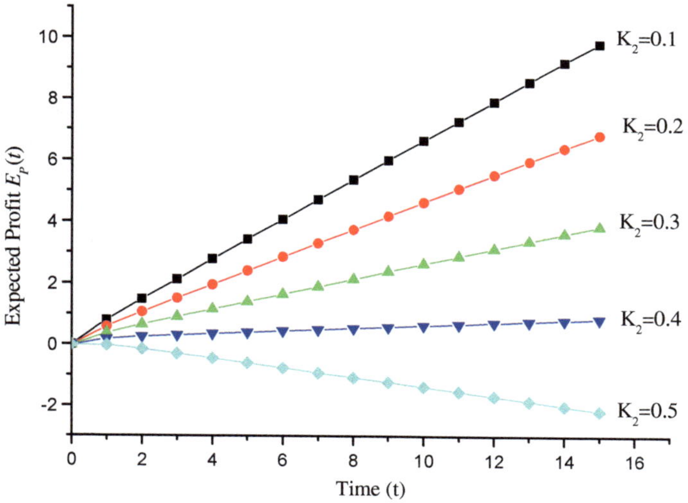

Fig. 7 Expected profit with respect to time in comprehensive case

Table 8 Expected profit as function of time in immediate repair facility

Time (t)	Expected profit $E_p(t)$				
	$K_2 = 0.1$	$K_2 = 0.3$	$K_2 = 0.5$	$K_2 = 0.7$	$K_2 = 0.9$
0	0.00000	0.00000	0.00000	0.00000	0.00000
1	0.78609	0.58609	0.38608	0.18608	−0.01391
2	1.47076	1.07076	0.67076	0.27076	−0.12923
3	2.12922	1.52922	0.92922	0.32922	−0.27077
4	2.78081	1.98081	1.18081	0.38081	−0.41919
5	3.43058	2.43058	1.43058	0.43058	−0.56941
6	4.07987	2.87987	1.67987	0.47987	−0.72012
7	4.72903	3.32903	1.92903	0.52903	−0.87096
8	5.37816	3.77816	2.17816	0.57816	−1.02184
9	6.02728	4.22728	2.42728	0.62728	−1.17272
10	6.67639	4.67639	2.67639	0.67639	−1.32361
11	7.32551	5.12551	2.92551	0.72551	−1.47449
12	7.97462	5.57462	3.17462	0.77462	−1.62538
13	8.62373	6.02373	3.42373	0.82373	−1.77626
14	9.27284	6.47284	3.67285	0.87285	−1.92715
15	9.92196	6.92196	3.92196	0.92196	−2.07804

4.4.2 Expected Profit of the Carburettor System with Immediate Repair Facility

Through taking the value of failure rates as $\lambda_F = 0.25$, $\lambda_C = 0.09$, $\lambda_{AP} = 0.15$, $\lambda_{FC} = 0.065$, $\lambda_T = 0.045$, $\lambda_{CC} = 0.3$, $w_j = 0$, expected profit of the designed carburettor system can be calculated as in terms of time

$$E_p(t) = K_1\Big\{0.1852585208 - 0.03488429086e^{(-1.535075208\,t)} + 0.0007888088583e^{(-3.383204792\,t)} - 0.1494861414e^{(-1.3\,t)} + 0.7491134206\,t\Big\} - tK_2 \tag{23b}$$

Setting $K_1 = 1$ and $K_2 = 0.1, 0.3, 0.5, 0.7, 0.9$, respectively, one can get expected profit for designed system as shown in Table 8 and Fig. 8.

4.5 *Sensitivity Analysis*

Sensitivity analysis finds out how the uncertainty in the output or sensitivity in the output of a system, can be assigned to different sources of uncertainty in its inputs or to change an input while keeping other inputs constant. Sensitivity of a measure

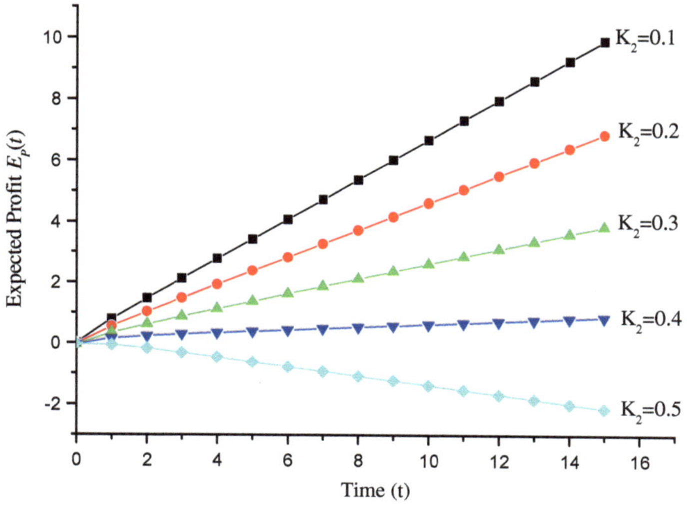

Fig. 8 Expected profit with respect to time in immediate repair facility

is defined as the partial derivative of the function with respect to their input factors [30, 31]. Here, these input factors are failure rates of the carburettor system.

4.5.1 Availability Sensitivity

The sensitivity of availability of the carburettor can be investigated by finding the partial differentiation of availability with respect to the input parameters of the carburettor system, by applying the value of parameters as $\lambda_F = 0.25$, $\lambda_C = 0.09$, $\lambda_{AP} = 0.15$, $\lambda_{FC} = 0.065$, $\lambda_T = 0.045$, $\lambda_{CC} = 0.3$, $w_j = 0.1$, $\mu(y) = 1$, $x = 1$, $w_j = 0.1$ as shown in Table 9 and their graphical representations are revealed in Fig. 9.

4.5.2 Reliability Sensitivity

The sensitivity in the reliability of carburettor resulting from the changes in the reliability of the carburettor subjected to their input parameters. It also can be analysed by partial differentiation of reliability function with respect to failure rates and setting the parameters as $\lambda_F = 0.25$, $\lambda_C = 0.09$, $\lambda_{AP} = 0.15$, $\lambda_{FC} = 0.065$, $\lambda_T = 0.045$, $\lambda_{CC} = 0.3$, $w_j = 0.1$, one may find the sensitivity in reliability as demonstrated in Table 10 and Fig. 10.

Table 9 Availability sensitivity as function of time

Time (*t*)	Availability sensitivity						
	$\frac{\partial P_{up}(t)}{\partial \lambda_F}$	$\frac{\partial P_{up}(t)}{\partial \lambda_C}$	$\frac{\partial P_{up}(t)}{\partial \lambda_{AP}}$	$\frac{\partial P_{up}(t)}{\partial \lambda_{FC}}$	$\frac{\partial P_{up}(t)}{\partial \lambda_T}$	$\frac{\partial P_{up}(t)}{\partial \lambda_{CC}}$	$\frac{\partial P_{up}(t)}{\partial w_j}$
0	0.00000	0.00000	0.00000	0.00000	0.00000	0.00000	0.00000
1	−0.00579	0.00777	0.00777	−0.00579	−0.40098	−0.47565	−0.03288
2	−0.00772	0.01273	0.01273	−0.00772	−0.43810	−0.55150	−0.04265
3	−0.00796	0.01404	0.01404	−0.00796	−0.43611	−0.55929	−0.04332
4	−0.00794	0.01425	0.01425	−0.00794	−0.43322	−0.55825	−0.04304
5	−0.00792	0.01425	0.01425	−0.00792	−0.43197	−0.55717	−0.04287
6	−0.00791	0.01424	0.01424	−0.00791	−0.43153	−0.55666	−0.04280
7	−0.00791	0.01423	0.01423	−0.00791	−0.43140	−0.55647	−0.04278
8	−0.00790	0.01422	0.01422	−0.00790	−0.43136	−0.55641	−0.04278
9	−0.00790	0.01422	0.01422	−0.00790	−0.43135	−0.55638	−0.04277
10	−0.00790	0.01422	0.01422	−0.00790	−0.43134	−0.55638	−0.04277
11	−0.00790	0.01422	0.01422	−0.00790	−0.43134	−0.55637	−0.04277
12	−0.00790	0.01422	0.01422	−0.00790	−0.43134	−0.55637	−0.04277
13	−0.00790	0.01422	0.01422	−0.00790	−0.43134	−0.55637	−0.04277
14	−0.00790	0.01422	0.01422	−0.00790	−0.43134	−0.55637	−0.04277
15	−0.00790	0.01422	0.01422	−0.00790	−0.43134	−0.55637	−0.04277

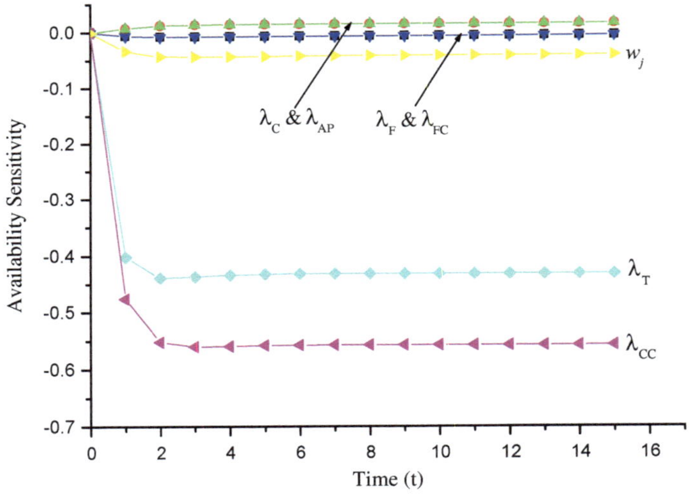

Fig. 9 Availability sensitivity with respect to time

4.5.3 MTTF Sensitivity

The sensitivity in MTTF of the carburettor can be studied through the partial differentiation of Eq. (22) with respect to the failure rates of the carburettor. By

Table 10 Reliability sensitivity as function of time

Time (t)	Reliability sensitivity						
	$\frac{\partial \mathrm{Rl}(t)}{\partial \lambda_F}$	$\frac{\partial \mathrm{Rl}(t)}{\partial \lambda_C}$	$\frac{\partial \mathrm{Rl}(t)}{\partial \lambda_{AP}}$	$\frac{\partial \mathrm{Rl}(t)}{\partial \lambda_{FC}}$	$\frac{\partial \mathrm{Rl}(t)}{\partial \lambda_T}$	$\frac{\partial \mathrm{Rl}(t)}{\partial \lambda_{CC}}$	$\frac{\partial \mathrm{Rl}(t)}{\partial w_j}$
0	0.00000	0.00000	0.00000	0.00000	0.00000	0.00000	0.00000
1	−0.01546	0.01412	0.01412	−0.01546	−0.54296	−0.70643	−0.08998
2	−0.03557	0.03555	0.03555	−0.03557	−0.60363	−0.99528	−0.20828
3	−0.04704	0.05059	0.05059	−0.04704	−0.51502	−1.05109	−0.27443
4	−0.05012	0.05719	0.05719	−0.05012	−0.39926	−0.98741	−0.28882
5	−0.04775	0.05716	0.05716	−0.04775	−0.29621	−0.87090	−0.26978
6	−0.04259	0.05298	0.05298	−0.04259	−0.21499	−0.73882	−0.23430
7	−0.03639	0.04670	0.04670	−0.03639	−0.15433	−0.61066	−0.19386
8	−0.03019	0.03973	0.03973	−0.03019	−0.11022	−0.49556	−0.15502
9	−0.02452	0.03294	0.03294	−0.02452	−0.07856	−0.39678	−0.12088
10	−0.01961	0.02678	0.02678	−0.01961	−0.05599	−0.31450	−0.09247
11	−0.01549	0.02144	0.02144	−0.01549	−0.03994	−0.24735	−0.06967
12	−0.01212	0.01697	0.01697	−0.01212	−0.02854	−0.19336	−0.05187
13	−0.00941	0.01330	0.01330	−0.00941	−0.02043	−0.15043	−0.03824
14	−0.00726	0.01034	0.01034	−0.00726	−0.01465	−0.11659	−0.02796
15	−0.00557	0.00799	0.00799	−0.00557	−0.01052	−0.09008	−0.02030

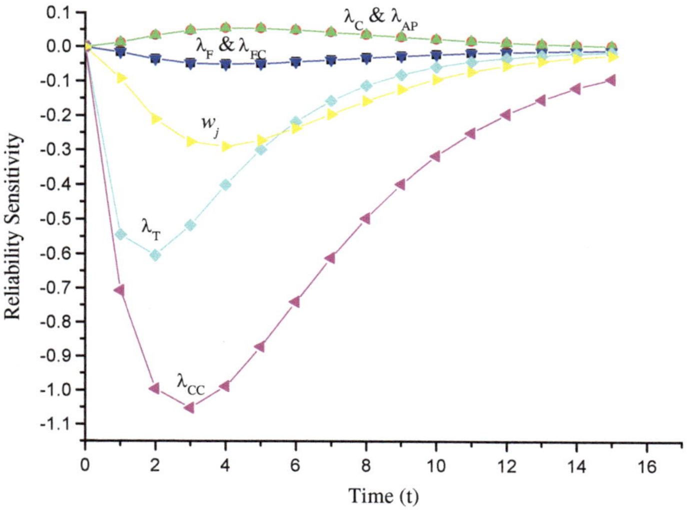

Fig. 10 Reliability sensitivity with respect to time

applying the set of parameters as $\lambda_F = 0.25$, $\lambda_C = 0.09$, $\lambda_{AP} = 0.15$, $\lambda_{FC} = 0.065$, $\lambda_T = 0.045$, $\lambda_{CC} = 0.3$, $w_j = 0.1$, in partial differentiation of MTTF, one can calculate the MTTF sensitivity as shown in Table 11 and corresponding graphs shown in Fig. 11.

Table 11 MTTF sensitivity as function of time

Variation in λ_F, λ_C, λ_{AP}, λ_{FC}, λ_T, λ_{CC}, w_j	MTTF sensitivity						
	λ_F	λ_C	λ_{AP}	λ_{FC}	λ_T	λ_{CC}	w_j
0.1	−0.60000	0.49813	0.57093	−0.38606	−2.83709	−55.68877	−2.18750
0.2	−0.46713	0.40437	0.45706	−0.31506	−2.32474	−16.95312	−1.40000
0.3	−0.37396	0.33479	0.37415	−0.26199	−1.93962	−8.34491	−0.97222
0.4	−0.30612	0.28174	0.31191	−0.22128	−1.64283	−4.99000	−0.71428
0.5	−0.25520	0.24037	0.26400	−0.18937	−1.40929	−3.32521	−0.54687
0.6	−0.21600	0.20748	0.22634	−0.16390	−1.22223	−2.37599	−0.43210
0.7	−0.18518	0.18091	0.19619	−0.14324	−1.07009	−1.78295	−0.35000
0.8	−0.16052	0.15914	0.17170	−0.12625	−0.94468	−1.38747	−0.28926
0.9	−0.14048	0.14107	0.15151	−0.11212	−0.84009	−1.11049	−0.24305

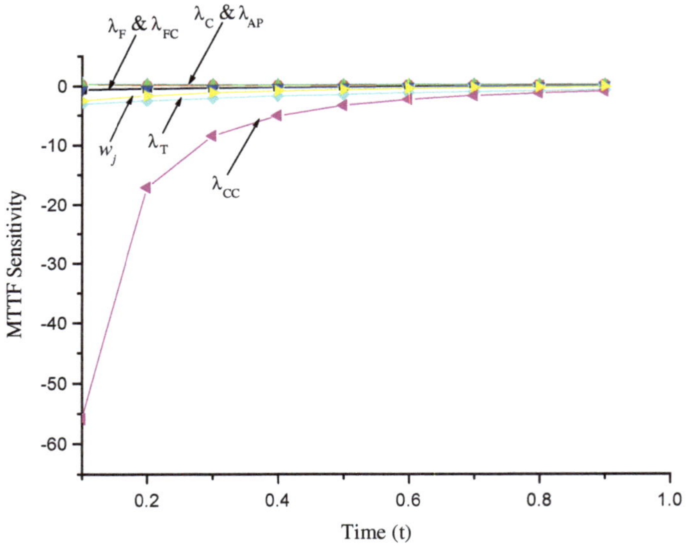

Fig. 11 MTTF sensitivity with respect to time

5 Results Discussion

For investigating the performance of carburettor, the reliability measures of carburettor system have been discussed. In order to analyse the most precarious part of the carburettor, sensitivity analysis has been conducted for various reliability parameters by varying the failure rates and time. Hence, overall the following results have been found.

Figure 3 gives the knowledge about the availability of the carburettor as shown in the graph of carburettor's availability with respect to time. The availability of the carburettor decreases with the slight increase in time. After some time, the availability of carburettor becomes constant as time increases. From the study of this graph, it can be concluded that with the immediate repairing facility, one can improve the availability of the carburettor. While removing the common cause failure in the carburettor, one can also attain a highly available carburettor system.

The critical examination of Fig. 4 gives the idea about how much the carburettor is reliable, with respect to three particular cases. One can see that the reliability of the carburettor system in comprehensive stage decreases quickly with the increment of time. The graph for the reliability of the carburettor system without the waiting time also decreases quickly with the time increment. But the reliability of the carburettor system without the common cause failure decreases less with the time increment as compared to other two cases. Finally, it can be concluded that without waiting in repair, one can predict the improvement in reliability only in its middle age, while taking care of common cause failure, one can acquire a more or most reliable carburettor system.

The trend of mean time to failure in the particular situations of carburettor is revealed graphically in Figs. 5 and 6. The graph of MTTF of the carburettor system in comprehensive stage shows that the MTTF of the carburettor decreases with the variation in failure rates of common cause, throttle, float chamber and fuel, while it slightly increases with the concern of variation in accelerator pump and choke and it is approximately equal for both. But from the study of MTTF graph with immediate repaired carburettor system, it is concluded that the MTTF decreases with the variation in throttle and common cause failure rates only, while it increases slightly with the variation in other failure rates and it is equal for all failure rates except throttle and common cause failure. Average failure time of carburettor cannot be determined if the carburettor is free from common cause failure. One can see that the MTTF of carburettor rapidly and highly decreases with the increment of common cause failure.

The maintenance cost of the carburettor system has much effect on the expected profit. Figures 7 and 8 epitomize the graph of expected profit verses time with the concern of service cost (maintenance cost, repairing and replacing cost). The grave examination of the graph expresses that the expected profit of the carburettor system increases as time increases and it decreases as service cost increases. It is surprising that if the service cost is greater than 40 % then expected profit decreases with the Passas age of time. So to attain the maximum profit controlling the service cost is necessary. From the comparative study of Figs. 7 and 8, it concluded that with the immediate repairing of failed or partially failed carburettor system, more profit can be attained.

The behaviour of sensitivity analysis of availability, reliability and MTTF of carburettor system is revealed by the graph in Figs. 9, 10 and 11, respectively. The vital inspection of Fig. 9 epitomizes that the availability sensitivity first decreases fastly with respect to common cause and throttle failure while the sensitivity of the availability is slightly decreasing with respect to waiting time to repair, float chamber and filter, and increases slightly with respect to accelerator pump and choke. After a short period, it would be established at some fixed value in context of each failure rate. Similarly, from the graph of reliability sensitivity, it can be seen that the waiting time to repair, common cause failure, failure of the throttle, float chamber and filter, can decrease the reliability of the carburettor with the increment in time, while the sensitivity of reliability also increases slightly with respect to the accelerator pump and choke. Sensitivity of reliability increases smoothly with respect to waiting time to repair, common cause failure and failure of the throttle, and slightly increases with respect to the failure rate of the accelerator pump and choke. With the precarious study of Fig. 11, one can see that the sensitivity of MTTF decreases slightly with respect to each failure of carburettor except common cause failure. In the situation of variation in common cause failure, sensitivity of MTTF increases rapidly for a little time period initially. From the inspection of sensitivity analysis, it is concluded that the carburettor system is most sensitive with respect to the common cause failure.

6 Conclusion

This research work proposed a Markov model of carburettor system and discussed how the functioning of each component affects the carburettor performance. For the practical use of carburettor in engines, the function of each component should be stochastic due to complete failure, partial failure or maintenance. With the consideration of all failures and repair, construct a state transition diagram of the carburettor to describe all the possible states of the carburettor. The performance of carburettor system is investigated through mathematical analysis of reliability measures. The analysis indicates that the carburettor system is the most sensitive in the case of common cause failure. By controlling the common cause failure in the carburettor, one can attain a highly reliable system which improves the availability of carburettor in automobiles. In order to improve the performance and life of carburettor, the Gumbel–Hougaard family of copula technique for maintenance or repair of the carburettor system is beneficial very much. Results obtained by this mathematical analysis of reliability measures for carburettor provide significant information to the automobile and mechanical engineers to improve the functioning of the designed carburettor.

References

1. Masaki, K., and Kato, S. (1972), U.S. Patent No. 3,693,947, Washington, DC: U.S. Patent and Trademark Office.
2. Masaki, K., and Kato, S. (1972), U.S. Patent No. 3,706,444, Washington, DC: U.S. Patent and Trademark Office.

3. Dashwood, F. J. (1970), U.S. Patent No. 3,493,217, Washington, DC: U.S. Patent and Trademark Office.
4. Mastanaiah, M. (2013), "Performance of Electronic Fuel Injection System Using Compressor And Controller", International Journal of Advanced Engineering Research and Studies, 2, 57-59.
5. Banapurmath, N.R., Tewari, P.G. (2009), "Comparative Performance Studies of a 4-Stroke CI Engine Operated on Dual Fuel Mode with Producer Gas and Honge Oil and Its Methyl Ester (HOME) with and Without Carburetor", Renewable Energy, 34 1009–1015.
6. Suryawanshi, S. J., and Yarasu, R. B. (2014), "Design and Simulation of a Producer Gas Carburetor–A Review", International Journal of Current Engineering and Technology, special issue 3, 10-13.
7. Elsayed, E. A. (2012), "Reliability Engineering", Vol. 88, John Wiley and Sons.
8. Yusuf, I., and Hussaini, N. (2012), "Evaluation of Reliability and Availability Characteristics of 2-out of-3 Standby System under a Perfect Repair Condition", American Journal of Mathematics and Statistics, 2(5), 114-119.
9. Denlinger, E. R., Irwin, J., and Michaleris, P. (2014), "Thermomechanical Modeling of Additive Manufacturing Large Parts", Journal of Manufacturing Science and Engineering, 136(6), 061007.
10. Singh, V. V., Ram, M., and Rawal, D. K. (2013), "Cost Analysis of an Engineering System Involving Subsystems in Series Configuration", IEEE Transaction on Automation and Science,
11. Ajav, E. A., Singh, B., and Bhattacharya, T. K. (1998), "Performance of a Stationary Diesel Engine Using Vapourized Ethanol as Supplementary Fuel", Biomass and Bioenergy, 15(6), 493-502.
12. Klimstra, J. (1989), "Carburettors for Gaseous Fuels-On Air-to-Fuel Ratio, Homogeneity and Flow Restriction", SAE Technical Paper, (No. 892141).
13. Ma, J., Allen, R., and Bowen, R. (1993), "Mathematical Simulation of the Transient Performance of a Petrol Engine", SAE Technical Paper, (No. 930855).
14. El-Sebakhy, E. A. (2009), "Software Reliability Identification Using Functional Networks: A Comparative Study", Expert systems with applications, 36(2), 4013-4020.
15. Ahmad Niknam, S., and Sawhney, R. (2014), "A Model for Reliability Analysis of Multi-State Manufacturing Systems", International Journal of Quality and Reliability Management, 31(8), 938-949.
16. Narahari, Y., and Viswanadham, N. (1994), "Transient Analysis of Manufacturing Systems Performance, "IEEE Transactions on Robotics and Automation, 10(2), 230-244.
17. Levitin, G., Xing, L., and Dai, Y. (2014), "Reliability and Mission Cost of 1-Out-of-: G Systems with State-Dependent Standby Mode Transfers", IEEE transaction on Reliability.
18. Mo, Y., Xing, L., Amari, S. V., and Dugan, J. B. (2015), "Efficient Analysis of Multi-State *k*-out-of-*n* systems", Reliability Engineering and System Safety, 133, 95-105.
19. Chao, M. T., Fu, J. C., and Koutras, M. V. (1995), "Survey of Reliability Studies of Consecutive-*k*-out-of-*n*: F and Related Systems", IEEE Transactions on Reliability, 44(1), 120-127.
20. Ram, M., and Singh, S. B. (2010), "Analysis of a Complex System with Common Cause Failure and Two Types of Repair Facilities with Different Distributions in Failure", International Journal of Reliability and Safety, 4(4), 381-392.
21. Dhillon, B. S., and Anude, O. C. (1994), "Common-Cause Failures in Engineering Systems: A Review", International Journal of Reliability, Quality and Safety Engineering, 1(01), 103-129.
22. Gupta, P. P., and S. C. Agarwal. (1984a), "A Parallel Redundant Complex System with Two Types of Failure under Preemptive-Repeat Repair Discipline", Microelectronics Reliability, 24 (3), 395–399.
23. Majeed, A. R., and N. M. Sadiq. (2006), "Availability and Reliability Evaluation of Dokan Hydro Power Station", In IEEE PES Transmission and Distribution Conference and Exposition Latin America, 1–6. Venezuela.

24. Nelsen, R.B. (2006), "An introduction to copulas (2nd edn.)", New York, Springer.
25. Ram, M., and Singh, S. B. (2010), "Availability, MTTF and Cost Analysis of Complex System under Preemptive-Repeat Repair Discipline using Gumbel-Hougaard Family Copula", International Journal of Quality and Reliability Management, 27(5), 576-595.
26. Patton, A. J. (2012), "A Review of Copula Models for Economic Time Series", Journal of Multivariate Analysis, 110, 4-18.
27. Avizienis, A., Laprie, J.C., Randell, B., and Landwehr, C. (2004), "Basic Concepts and Taxonomy of Dependable and Secure Computing", IEEE Transactions on Dependable and Secure Computing, 1(1), 11-33.
28. Ram, M. (2013), "On System Reliability Approaches: A Brief Survey", International Journal of System Assurance Engineering and Management, 4(2), 101-117.
29. Ram, M., and Manglik, M. (2014), "Stochastic Behaviour Analysis of A Markov Model under Multi-State Failures", International Journal of System Assurance Engineering and Management, 5(4), 686-699.
30. Saltelli, A., Chan, K., and Scott, E. M. (Eds.). (2000), "Sensitivity Analysis", Vol. 1, New York: Wiley.
31. Gandini, A. U. G. U. S. T. O. (1990), "Importance and Sensitivity Analysis in Assessing System Reliability", IEEE Transactions on Reliability, 39(1), 61-70.

Bayesian Inference and Optimal Censoring Scheme Under Progressive Censoring

Siddharth Vishwanath and Debasis Kundu

Abstract In recent times an extensive work has been done related to the different aspects of the progressive censoring schemes. Here we deal with the statistical inference of the unknown parameters of a three-parameter Weibull distribution based on the assumption that the data are progressively Type-II censored. The maximum likelihood estimators of the unknown parameters do not exist due to the presence of the location parameter. Therefore, Bayesian approach seems to be a reasonable alternative. We assume here that the location parameter follows a uniform prior and the shape parameter follows a log-concave prior density function. We further assume that the scale parameter has a conjugate gamma prior given the shape and the location parameters. Based on these priors the Bayes estimate of any function of the unknown parameters under the squared error loss function and the associated highest posterior density credible interval are obtained using Gibbs sampling technique. We have also used one precision criterion to compare two different censoring schemes, and it can be used to find the optimal censoring scheme. Since finding the optimal censoring scheme is a challenging problem from the computational view point, we propose suboptimal censoring scheme, which can be obtained quite conveniently. We have carried out some Monte Carlo simulations to observe the performances of the proposed method, and for illustrative purposes, we presented the analysis of one data set.

Keywords Credible intervals · Fisher information matrix · Gibbs sampling · Log-concave density function · Markov chain Monte Carlo · Optimum censoring scheme · Prior distribution

S. Vishwanath · D. Kundu (✉)
Department of Mathematics and Statistics, Indian Institute of Technology, Kanpur 208016, India
e-mail: kundu@iitk.ac.in

S. Vishwanath
e-mail: sidv@iitk.ac.in

M. Ram and J.P. Davim (eds.), *Advances in Reliability and System Engineering*, Management and Industrial Engineering, DOI 10.1007/978-3-319-48875-2_10

1 Introduction

In different areas of reliability and survival analysis researchers have used the Weibull distribution quite extensively. The main reason about the overwhelming popularity of an Weibull distribution is due to the fact that its probability density function (PDF) can take various shapes, and the cumulative distribution function (CDF) can be expressed in a closed and compact form. Since the CDF of a Weibull distribution can be written in a closed form, the Weibull distribution has been used quite conveniently when the data are censored. For detailed discussions on Weibull distribution and for different inferential issues see for example, Johnson et al. [9], Chap. 21 or Murthy et al. [14]. In any life-testing experiment often the data are censored. Type-I and Type-II are the two most common censoring schemes one encounters in practice. In the last two decades, since the appearance of the book by Balakrishnan and Aggarwala [2], progressive censoring scheme becomes quite popular. The review article by Balakrishnan [1] and the recent book by Balakrishnan and Cramer [3] provided the details about the development of this topic during this period.

In this article we consider the inference of the unknown parameters of a three-parameter Weibull distribution based on Type-II progressively censored data. The three-parameter Weibull distribution with the shape, scale, and location parameter as $\alpha > 0$, $\lambda > 0$, and $-\infty < \mu < \infty$, respectively, has the PDF

$$f(x; \alpha, \lambda, \mu) = \begin{cases} \alpha\lambda(x-\mu)^{\alpha-1} e^{-\lambda(x-\mu)^{\alpha}} & \text{if } x > \mu, \\ 0 & \text{if } x \leq \mu. \end{cases} \quad (1)$$

From now on a three-parameter Weibull distribution with the PDF (1) will be denoted by WE (α, λ, μ). It may be mentioned that although the two-parameter Weibull distribution (when $\mu = 0$) satisfies the regularity condition in the sense of Rao [16], the three-parameter Weibull distribution does not satisfy the regularity conditions. In fact it can be shown very easily that if all the three parameters are unknown, the maximum likelihood estimators (MLEs) do not exist even for complete sample. Due to this reason, several alternative estimators have been proposed in the literature, see for example, Nagatsuka et al. [15] and the references cited therein. They may not be as efficient as the MLEs. Due to this reason Bayesian inference seems to be a reasonable alternative.

In this paper we provide the Bayesian inference of a three-parameter Weibull when all the parameters are unknown, based on a Type-II progressively censored data. It may be mentioned that when $\mu = 0$, the three-parameter Weibull distribution reduces to a two-parameter Weibull distribution. For a two-parameter Weibull distribution, the classical and Bayesian inference of the unknown parameters, as well as the reliability sampling plans in presence of progressive censoring, have been considered by several authors. See for example, Viveros and Balakrishnan

[18], Balasooriya et al. [4] and Kundu [11]. Therefore, the present paper can be seen as an extension of these works to the three-parameter case.

One needs to assume some priors on the unknown parameters to perform the Bayesian inference. In this article, we have taken a set of fairly flexible priors on the shape (α), scale (λ), and location (μ) parameters. It can be easily verified that if the shape parameter α and the location parameter μ are known, then the scale parameter λ has a conjugate gamma prior. Therefore, it is quite natural to take a gamma prior on λ, for known α and μ. Even for known μ, there does not exist any conjugate joint priors on α and λ. Hence, based on the suggestion of Berger and Sun [5], no specific form of prior on α is assumed. It is simply assumed that α has a prior whose support is on $(0, \infty)$, and it has a log-concave PDF. It may be mentioned that many well-known common density functions such as, normal, log-normal, gamma (when the shape parameter is greater than one) and Weibull (when the shape parameter is greater than one) have log-concave PDFs. For the location parameter, it is assumed that μ has a uniform prior over a finite interval.

In this paper, based on the above priors on (α, λ, μ), we provide the posterior analysis of the unknown parameters. The Bayes estimator of any function of the unknown parameters cannot be obtained in explicit form under the squared error loss function. In this case, it is possible to generate samples directly from the joint posterior distribution of (α, λ, μ) given the *data*. Hence, the Gibbs sampling technique can be used quite conveniently to compute the simulation consistent Bayes estimate of any function of the parameters with respect to the squared error loss function, and also to construct the highest posterior density (HPD) credible interval. The second aim of this paper is to provide a methodology to compare two different progressive censoring schemes, hence, it can be used to compute the optimal sampling scheme, for a given set of prior distribution functions. During the last few years, Zhang and Meeker [19] and Kundu [11] discussed the Bayesian life testing plans for one-parameter (assuming the shape parameter to be known) Weibull and two-parameter Weibull distributions, respectively. In this article we extend the work to the three-parameter case. It is observed that finding the optimal censoring scheme is a discrete optimization problem, and it is computationally quite challenging. Due to this fact, we have suggested a suboptimal plan, which can be obtained very easily. Therefore, the implementation of our proposed procedure is quite simple in practice.

The rest of the paper is organized as follows. In Sect. 2, we provide the details of the model assumptions and prior distributions. Simulation consistent Bayes estimate and the associated HPD credible interval are provided in Sect. 3. In Sect. 4, we provide the simulation results and the analysis of a progressively censored data set. The construction of optimal progressive censoring scheme is provided in Sect. 5, and finally in Sect. 6, we conclude the paper.

2 Model Assumptions and Prior Distributions

2.1 Model Assumptions

It is assumed that n units which are identical in nature are put on a life testing experiment at the time point 0. We denote their life times as $T_1, \ldots, T_n$, where T_i's are independent and identically distributed (i.i.d.) random variables with PDF (1). Further, the integer $0 < m < n$, and also $R_1, \ldots, R_m$ are prefixed integers such that

$$R_1 + \cdots + R_m = n - m.$$

The experiment starts at the time point 0, and at the time of the first failure, say, t_1, R_1 out of the remaining $(n-1)$ units are chosen randomly and they are removed from the experiment. Similarly, at the time of the second failure, say, t_2, R_2 of the remaining units are chosen at random and they are removed, and so on. Finally, when the m-th failure takes place, the rest of the remaining R_m units are removed, and the experiment stops. Therefore, the usual Type-II censoring scheme can be obtained as a special case of the Type-II progressive censoring scheme. In a Type-II progressive censoring scheme the observed data will be of the form

$$\text{data} = \{t_1, \ldots, t_m\}$$

for a given $\{R_1, \ldots, R_m\}$.

2.2 Prior Distributions

In this section, we provide the prior distributions on the set of parameters. It is known that the scale parameter λ has a conjugate gamma prior when the shape and location parameters are known. Hence, it is quite reasonable to assume that the prior on λ, $\pi_1(\lambda|a,b) \sim$ Gamma (a, b), with PDF

$$\pi_1(\lambda|a,b) = \begin{cases} \frac{b^a}{\Gamma(a)} \lambda^{a-1} e^{-b\lambda} & \text{if } \lambda > 0 \\ 0 & \text{if } \lambda \leq 0 \end{cases} \tag{2}$$

Here the shape parameter $a > 0$ and the scale parameter $b > 0$ are the hyper-parameters, and

$$\Gamma(a) = \int_0^\infty x^{a-1} e^{-x} dx,$$

is the gamma function. Note that the above gamma prior is a very flexible prior as it can take variety of shapes depending on the shape parameter. In most practical applications with proper informative information on the Weibull scale parameter, the prior variance is usually assumed to be finite, see Congdon [6].

In practice, the shape parameter is usually unknown, and in this case the joint conjugate prior on α and λ do not exist even when μ is known, see for example, Kaminskiy and Krivtsov [10]. In this case, we do not take any specific prior on α. It is assumed that α has a prior $\pi_2(\alpha)$ which has a support on the $(0, \infty)$, and it has a PDF which is log-concave. This was originally suggested by Berger and Sun [5], see also Kundu [11]. For specific calculation, in this paper we have assumed that $\pi_2(\alpha|c, d) \sim$ Gamma (c, d). Finally following the approach of Smith and Naylor [17] it is assumed that μ has a prior $\pi_3(\mu|e,f) \sim \cup\ (e,f)$, i.e., the PDF of $\pi_3(\mu|e,f)$ is

$$\pi_3(\mu|e,f) = \begin{cases} \frac{1}{f-e} & \text{if } e \le \mu \le f \\ 0 & \text{if } \mu < e \text{ or } \mu > f \end{cases} \quad (3)$$

3 Bayes Estimates and Credible Intervals

In this section, we obtain the Bayes estimate of any function of the unknown parameters and the associated HPD credible interval, with respect to the set of prior distributions mentioned in Sect. 2. In calculating the Bayes estimates although it is assumed that the loss function is squared error, any other loss function also can be easily incorporated. Based on the observed data for a given $R_1, \ldots, R_m$, the likelihood function is

$$\mathcal{L}(\text{data}|\alpha, \lambda, \mu) \propto \alpha^m \lambda^m \prod_{i=1}^{m} (t_i - \mu)^{\alpha-1} \exp\left(-\lambda \sum_{i=1}^{m} (1 + R_i)(t_i - \mu)^{\alpha} \right); \quad (4)$$

for $\alpha > 0, \lambda > 0, \mu < t_1$, and 0, otherwise. From (4), it is easily observed that if we take $\lambda = 1$, $\alpha = 0.5$, and $\mu \uparrow t_1$, then $\mathcal{L}(\text{data}|\alpha, \lambda, \mu) \uparrow \infty$. It shows that when all the three parameters are unknown, then the MLEs do not exist. Therefore, Bayesian inference seems to be a reasonable choice in this case.

Based on the prior distribution mentioned above on α, λ, and μ, the joint distribution of the data, α, λ, and μ can be written as

$$\mathcal{L}(\text{data}, \alpha, \lambda, \mu) = \mathcal{L}(\text{data}|\alpha, \lambda, \mu) \cdot \pi_1(\lambda) \cdot \pi_2(\alpha) \cdot \pi_3(\mu). \quad (5)$$

Based on (5), the joint posterior distribution of the unknown parameters is given by

$$\mathcal{L}(\alpha, \lambda, \mu|\text{data}) = \frac{\mathcal{L}(\text{data}|\alpha, \lambda, \mu) \cdot \pi_1(\lambda) \cdot \pi_2(\alpha) \cdot \pi_3(\mu)}{\int_{-\infty}^{\infty} \int_{-\infty}^{\infty} \int_{-\infty}^{\infty} \mathcal{L}(\text{data}|\alpha, \lambda, \mu) \cdot \pi_1(\lambda|\alpha, \mu) \cdot \pi_2(\alpha|\mu) \cdot \pi_3(\mu)}. \quad (6)$$

Therefore, the Bayes estimate of any function of α, λ, and μ, say $g(\alpha,\lambda,\mu)$, under the squared error loss function is

$$\begin{aligned}\hat{g}_B(\alpha,\lambda,\mu) &= E_{\alpha,\lambda,\mu|\text{data}}(g(\alpha,\lambda,\mu))| \\ &= \int_0^\infty \int_0^\infty \int_e^f g(\alpha,\lambda,\mu)\mathscr{L}(\lambda|\alpha,\mu,\text{data})\cdot\mathscr{L}(\alpha|\mu,\text{data})\cdot\mathscr{L}(\mu|\text{data}) \\ &\qquad \mathrm{d}\mu\mathrm{d}\lambda\mathrm{d}\alpha.\end{aligned} \tag{7}$$

It is quite obvious that it will not be possible to compute (7) analytically for general $g(\alpha,\lambda,\mu)$. We propose to use Gibbs sampling technique to generate samples directly from the joint posterior distribution function (6), and based on the generated sample we provide a simulation consistent estimate of (7) and the associated highest posterior density credible interval of $\hat{g}_B(\alpha,\lambda,\mu)$.

We need the following results for further development.

Theorem 1 *The conditional density of* λ *given* α,μ *and* data is

$$\mathscr{L}(\lambda|\alpha,\mu,\text{data})\Gamma\left(a+m, b+\sum_{i=1}^{m}(1+R_i)(t_i-\mu)^\alpha\right).$$

Proof It is simple, hence the details are avoided.

Theorem 2 *The conditional density of* α *given* μ and data, $\mathscr{L}(\alpha|\mu,\text{data})$, *is*

$$\mathscr{L}(\alpha|\mu,\text{data}) = \kappa(\mu)\frac{\alpha^{c+m-1}\cdot \mathrm{e}^{-d\alpha}\prod_{i=1}^{m}(t_i-\mu)^{\alpha-1}}{\left[b+\sum_{i=1}^{m}(1+R_i)(t_i-\mu)^\alpha\right]^{a+m}}, \tag{8}$$

and (8) *is log-concave. Here* $\kappa(\mu)$ *is the normalizing constant, i.e.,*

$$\kappa(\mu) = \left[\int_0^\infty \frac{\alpha^{c+m-1}\cdot \mathrm{e}^{-d\alpha}\prod_{i=1}^{m}(t_i-\mu)^{\alpha-1}}{\left[b+\sum_{i=1}^{m}(1+R_i)(t_i-\mu)^\alpha\right]^{a+m}}\mathrm{d}\alpha\right]^{-1}.$$

Proof See in the Appendix.

Theorem 3 *The conditional density of* μ *given data is*

$$\mathscr{L}(\mu|\text{data}) \propto \kappa(\mu),$$

if $\mu \in [e,f]\cap(-\infty,t_1]$, *and* 0, *otherwise.*

Proof It is trivial.

Now using Theorems 1–3, and following the idea of Geman and Geman [8], we propose the following algorithm to generate samples (λ,α,μ) directly from the posterior distribution (6). The generated samples can be used to compute simulation consistent Bayes estimate of any function of the unknown parameters, and the associated HPD credible interval.

Algorithm 1

Step 1: Given the data $\{t_1, \ldots, t_m\}$, compute $\max\{\kappa(\mu)\}$ for $\mu \in [e,f] \cap (-\infty, t_1]$.
Step 2: Using acceptance rejection principle generate μ from $\mathscr{L}(\mu|\text{data})$.
Step 3: Generate $\alpha \sim \mathscr{L}(\alpha|\mu, \text{data})$ using the method proposed by Devroye [7] or Kundu [11].
Step 4: Generate $\lambda \sim \mathscr{L}(\lambda|\alpha, \mu, \text{data})$
Step 5: Repeat Step 2 to Step 4, N times to generate $\{(\lambda_1, \alpha_1, \mu_1), \ldots, (\lambda_N, \alpha_N, \mu_N)\}$ and obtain $g_i = g(\lambda_i, \alpha_i, \mu_i)$, for $i = 1, \ldots, N$.
Step 6: A simulation consistent estimate of $\hat{g}_B(\alpha, \lambda, \mu)$ is

$$\hat{g} = \frac{1}{N}\sum_{i=1}^{N} g_i.$$

Step 7: A $100(1 - \beta)$ % credible interval of $g(\alpha, \lambda, \mu)$ can be obtained as

$$\left(g_j, g_{j+[\beta N]}\right); \quad j = 1, \ldots, N - [\beta N].$$

Step 8: The HPD $100(1 - \beta)$ % credible interval of $g(\alpha, \lambda, \mu)$ can be obtained as

$$\left(g_{j^*}, g_{j^*+[\beta N]}\right); \quad \text{such that } g_{j^*+[\beta N]} - g_{j^*} \leq g_{j+[\beta N]} - g_j;\ j = 1, \ldots, N - [\beta N].$$

4 Simulation and Data Analysis

4.1 Simulation

In this section, we perform some simulation experiments to show the effectiveness of the proposed methods. All the simulations are performed using the statistical software **R**. We use different sample sizes n, different effective sample sizes m, different sets of parameter values, and different sampling schemes. We use the following set of parameter values; Set 1: $(\alpha = 1, \lambda = 1, \mu = 0)$ and Set 2: $(\alpha = 1.5, \lambda = 1, \mu = 1)$, and the following censoring schemes.

1. Scheme 1: $n = 20$, $m = 10$, $R_1 \ldots R_{10} = 1$,
2. Scheme 2: $n = 20$, $m = 10$, $R_1 = 5, R_2 \ldots R_9 = 0, R_{10} = 5$,
3. Scheme 3: $n = 30$, $m = 20$, $R_1 \ldots R_{19} = 0, R_{20} = 10$.

Further we have used three different sets of hyperparameters to see the effects of the priors on the Bayes estimates and the associated credible intervals.

1. Prior 0: $a = b = c = d = 0.0001$ and $e = -1.0, f = t_1$
2. Prior 1: $a = b = c = d = 5$ and e = -1.0, $f = t_1$
3. Prior 2: $a = b = 5$, $c = 2.25$, $d = 1.5$ and e = -1.0, $f = t_1$

Note that Prior 0 is a noninformative prior; Prior 1 is an informative prior when we use Set 1, and and Prior 2 is an informative prior when we use Set 2. Prior 1 is an informative prior as the means of the prior distributions of the shape and scale parameters are same with the original distribution when Set 1 is used. In addition, it has a low variance. Similarly, for Set 2, Prior 2 ensures that the means of the prior distributions for the scale and the shape parameters are same with the original distribution.

In each case we generated progressively censored samples based on the algorithm proposed by Balakrishnan and Aggarwala [2] and we computed simulation consistent Bayes estimate and the corresponding mean squared errors (MSEs). The results are presented in Tables 1, 2, 3, 4, 5 and 6. Some of the points are quite clear from these simulation results. It is observed that in all the cases considered as expected the biases and the MSEs of the Bayes estimators based on informative priors perform better than the noninformative priors. Moreover, as the effective sample size increases the biases and MSEs decrease in all cases considered.

Table 1 Average Bayes estimates and the MSEs for α, λ, μ: Set 1, Scheme 1

Par	Prior 0	Prior 1
α	1.1571 (0.6642)	1.0886 (0.1180)
λ	0.8299 (0.0988)	1.0809 (0.0459)
μ	−0.0514 (0.0017)	−0.0455 (0.0015)

Table 2 Average Bayes estimates and the MSEs for α, λ, μ: Set 1, Scheme 2

Par	Prior 0	Prior 1
α	1.3840 (0.7501)	1.2025 (0.0842)
λ	1.3714 (0.1499)	1.0213 (0.0935)
μ	−0.0498 (0.0016)	−0.0428 (0.0015)

Table 3 Average Bayes estimates and the MSEs for α, λ, μ: Set 1, Scheme 3

Par	Prior 0	Prior 1
α	1.2917 (0.6696)	1.2015 (0.2172)
λ	0.9364 (0.1092)	0.9947 (0.0611)
μ	−0.0505 (0.0016)	−0.0449 (0.0015)

Table 4 Average Bayes estimates and the MSEs for α, λ, μ: Set 2, Scheme 1

Par	Prior 0	Prior 1
α	1.5306 (0.1708)	1.4678 (0.1102)
λ	0.8083 (0.0727)	1.0305 (0.0526)
μ	0.9124 (0.0087)	0.9822 (0.0041)

Table 5 Average Bayes estimates and the MSEs for α, λ, μ: Set 2, Scheme 2

Par	Prior 0	Prior 1
α	1.7221 (0.3221)	1.5912 (0.1529)
λ	0.8617 (0.0866)	0.9694 (0.0633)
μ	0.9872 (0.0041)	0.9919 (0.0039)

Table 6 Average Bayes estimates and the MSEs for α, λ, μ: Set 2 Scheme 3

Par	Prior 0	Prior 1
α	1.5716 (0.2328)	1.6179 (0.1315)
λ	1.1093 (0.0852)	1.0376 (0.0691)
μ	0.9917 (0.0041)	1.0249 (0.0003)

We have also computed the 95 % highest posterior density (HPD) credible interval based on Prior 0, and we present the average lengths of the HPD credible intervals and the associated coverage percentages based on 10,000 MCMC samples. The results are presented in Table 7. From the simulation results it is clear that the proposed method is working quite satisfactorily in all the cases considered. It is

Table 7 Average HPD Bayes interval length and coverage percentage

Set	Scheme	μ	α	λ
$\mu = 0, \alpha = 1, \lambda = 1$	(20, 10, 10 * 1)	0.1191 0.91	0.9387 0.96	0.8349 0.92
	(20, 10, 5, 8 * 0, 5)	0.1200 0.92	1.1022 0.95	1.0421 0.96
	(20, 15, 14 * 0, 5)	0.1010 0.93	0.9074 0.96	0.8118 0.93
$\mu = 1, \alpha = 1.5, \lambda = 1$	(20, 10, 10 * 1)	0.1970 0.96	1.5352 0.98	0.6852 0.84
	(20, 10, 5, 8 * 0, 5)	0.1986 0.93	1.4811 0.95	1.0539 0.96
	(20, 15, 14 * 0, 5)	0.1955 0.94	1.0916 0.96	0.0512 0.96

observed that the coverage percentages of the HPD credible intervals are quite close to the nominal values in all the cases considered. Moreover as the effective sample size increases, the average lengths of the HPD credible intervals decrease.

4.2 Data Analysis

In this section, we perform the analysis of a data set for illustrative purposes. The original data is from Lawless [13], which represents the failure times for 36 appliances which are subject to an automatic life test. A progressively censored sample with the following censoring scheme: $n = 36$, $m = 10$, $R_1 = \ldots = R_9 = 2$, $R_{10} = 8$ was generated and and analyzed by Kundu and Joarder [12] using an exponential model. The data are presented below.

$$\{11, 35, 49, 170, 329, 958, 1925, 2223, 2400, 2568\}.$$

Kundu [11] used the same data set and analyzed it using a two-parameter Weibull model. Here, we use a three-parameter Weibull distribution to analyze this data set. For computational convenience, we divided all the values by 100 and it is not going to affect the analysis.

Since we do not have any prior knowledge of the parameters we have assumed noninformative priors, i.e., $a = b = c = d = 0.001$, as suggested by Congdon [6]. We have assumed $\mu \sim U\ (-0.5, 0.011)$. Based on the above priors, we compute the Bayes estimates and the associated 95 % HPD credible intervals. The Bayes estimates of α, λ, and μ are 0.6915, 0.2515, −0.0189, respectively. The associated 95 % HPD credible intervals are (0.5213, 0.7913), (0.1317,3112), (−0214, −0.0079), respectively.

5 Optimal Censoring Scheme

In practice, it is very important to choose the "optimum" censoring scheme among the class of all possible censoring schemes. Here possible censoring schemes mean, for fixed n and m, all possible choices of $\{R_1, \ldots, R_m\}$, such that

$$\sum_{i=1}^{m} R_i + m = n.$$

In this section, we use a similar precision criterion as used by Zhang and Meeker (2008) and Kundu [11] to choose the optimal censoring scheme from a class of possible schemes. We say that a censoring scheme $\Psi_1 = \{R_1^1 \ldots R_m^1\}$ is superior to $\Psi_2 = \{R_1^2 \ldots R_m^2\}$, if the information derived from Ψ_1 is more accurate than the information derived from Ψ_2. The method we employ here will be to examine the best censoring scheme for different combinations of n and m. The number of such possible schemes is $\binom{n-1}{m-1}$. Therefore, even when $n = 25$ and $m = 12$, the total number of possible schemes is 2,496,144, which is quite large. Hence, we search for a "suboptimum" censoring scheme by finding the best censoring scheme among those schemes where the entire weight *n-m* is on a single $R_j; j = 1, \ldots, m$. By doing so, we obtain a convex hull of all possible censoring schemes, and we search only along the corner points.

Note that the pth quantile point, T_p, of the three-parameter Weibull distribution is given by

$$T_p = \mu + \left[-\frac{1}{\lambda} \cdot \ln(1-p)\right]^{\frac{1}{\alpha}}.$$

The precision criterion we use here is the posterior variance of $\ln(T_p)$, see for example, Kundu [11] or Zhang and Meeker [19]. Since the posterior variance of $\ln(T_p)$ depends on the observed sample when the parameters are unknown, we eliminate any sample bias by taking the average posterior variance of $\ln(T_p)$ for multiple samples obtained from the same joint prior distribution. The precision criteria used is given as

$$C(\Psi) = \mathbb{E}_{\pi(\alpha,\lambda,\mu)}\left(\mathbb{V}_{\text{posterior}}\left(\ln(T_p)|\Psi\right)\right),$$

where $\Psi = \{R_1 \ldots R_m\}$ is the censoring scheme, and $\mathbb{V}_{\text{posterior}}\left(\ln(T_p)|\Psi\right)$ is the posterior variance of the plan Ψ. We expect to minimize the posterior variance of the pth quantile, and hence if $C(\Psi_1) < C(\Psi_1)$, we say that Ψ_1 is superior to Ψ_2. Therefore, we would like to choose that Ψ, a particular censoring scheme, for which $C(\psi)$ is minimum.

There are no explicit expressions known for the quantity $\mathbb{V}_{\text{posterior}}\left(\ln(T_p)|\Psi\right)$, hence we proceed to compute the value of the precision criterion using Monte Carlo method. The following algorithm can be used for that purpose.

ALGORITHM:

1. Fix *n*,*m* and the priors a, b, c, d, e, f
2. Generate $(\alpha_i, \lambda_i, \mu_i)$ from $\pi(\alpha, \lambda, \mu | a, b, c, d, e, f)$
3. Generate a progressively censored sample $\{t_{(1)} \ldots t_{(m)}\}$ from $(\alpha_i, \lambda_i, \mu_i)$
4. For the data $\{t_{(1)} \ldots t_{(m)}\}$, estimate $\mathbb{V}_{\text{posterior}}\left(ln(T_p)|\Psi\right)$ using Gibbs sampling method as suggested in Sect. 3.
5. Repeat Steps 1–4 for *N* iterations and their average will provide an estimate of $C(\Psi)$.

We have considered two different sets namely $(n = 20, m = 10)$ and $(n = 20, m = 15)$. We compute the precision criterion based on four different quantiles namely for $p = 0.50$, $p = 0.75, p = 0.95, p = 0.99$ for Prior 1, and the results are presented in Tables 8 and 9.

Table 8 Optimal censoring scheme for $n = 20, m = 10$

n = 20, m = 10	C(Ψ)	C(Ψ)	C(Ψ)	C(Ψ)
Scheme	p = 0.50	p = 0.75	p = 0.95	p = 0.99
10, 0, 0, 0, 0, 0, 0, 0, 0, 0	2.287547	3.033706	*3.456304*	*3.931685*
0, 10, 0, 0, 0, 0, 0, 0, 0, 0	1.409668	2.886745	3.584613	4.690873
0, 0, 10, 0, 0, 0, 0, 0, 0, 0	1.569803	*2.13762*	3.883272	4.941062
0, 0, 0, 10, 0, 0, 0, 0, 0, 0	1.816804	2.382721	4.044178	4.995784
0, 0, 0, 0, 10, 0, 0, 0, 0, 0	2.69913	2.621744	4.173664	5.163259
0, 0, 0, 0, 0, 10, 0, 0, 0, 0	2.268952	3.078297	4.330179	5.200208
0, 0, 0, 0, 0, 0, 10, 0, 0, 0	2.51713	3.623896	4.667893	5.300237
0, 0, 0, 0, 0, 0, 0, 10, 0, 0	2.15651	4.002551	4.954882	5.540816
0, 0, 0, 0, 0, 0, 0, 0, 10, 0	1.573069	3.407709	5.326496	6.04542
0, 0, 0, 0, 0, 0, 0, 0, 0, 10	*1.396492*	3.374444	4.979677	6.405011

Table 9 Optimal censoring scheme for $n = 20$, $m = 15$

n = 20, m = 15	C(Ψ)	C(Ψ)	C(Ψ)	C(Ψ)
Scheme	p = 0.50	p = 0.75	p = 0.95	p = 0.99
15, 0, 0, 0, 0, 0, 0, 0, 0, 0, 0, 0, 0, 0, 0	1.430799	2.0387188	*2.945988*	*3.527577*
0, 15, 0, 0, 0, 0, 0, 0, 0, 0, 0, 0, 0, 0, 0	1.588445	*1.472637*	3.909428	3.913904
0, 0, 15, 0, 0, 0, 0, 0, 0, 0, 0, 0, 0, 0, 0	1.096379	2.397453	6.1389017	4.038037
0, 0, 0, 15, 0, 0, 0, 0, 0, 0, 0, 0, 0, 0, 0	0.9984362	2.195926	3.237427	4.480111
0, 0, 0, 0, 15, 0, 0, 0, 0, 0, 0, 0, 0, 0, 0	*0.630354*	1.699538	4.917337	4.73495
0, 0, 0, 0, 0, 15, 0, 0, 0, 0, 0, 0, 0, 0, 0	0.8723458	2.405792	3.751554	4.7763508
0, 0, 0, 0, 0, 0, 15, 0, 0, 0, 0, 0, 0, 0, 0	0.8889859	2.637907	4.623516	5.238868
0, 0, 0, 0, 0, 0, 0, 15, 0, 0, 0, 0, 0, 0, 0	1.083057	3.242228	4.1193264	5.339358
0, 0, 0, 0, 0, 0, 0, 0, 15, 0, 0, 0, 0, 0, 0	1.410338	1.8871604	3.4573745	5.4188804
0, 0, 0, 0, 0, 0, 0, 0, 0, 15, 0, 0, 0, 0, 0	1.832797	2.494387	4.5991795	5.980342
0, 0, 0, 0, 0, 0, 0, 0, 0, 0, 15, 0, 0, 0, 0	1.083604	1.647242	3.170819	6.203538
0, 0, 0, 0, 0, 0, 0, 0, 0, 0, 0, 15, 0, 0, 0	1.289311	1.896631	3.816518	6.402096
0, 0, 0, 0, 0, 0, 0, 0, 0, 0, 0, 0, 15, 0, 0	1.011111	2.999492	3.665278	6.69655
0, 0, 0, 0, 0, 0, 0, 0, 0, 0, 0, 0, 0, 15, 0	0.7898133	2.0053203	4.8234677	6.92173
0, 0, 0, 0, 0, 0, 0, 0, 0, 0, 0, 0, 0, 0, 15	0.8662691	2.150556	5.021647	8.8057628

6 Conclusion

In this paper, we have provided the Bayesian inference of a three-parameter Weibull distribution based on the Type-II progressively censored data. It is well known that when all the three parameters are unknown the MLEs do not exist. Hence, Bayesian inference seems to be a reasonable choice. We have provided the Bayes estimates of the unknown parameters based on the squared error loss function when the shape and scale parameters have gamma priors and the location parameter has a uniform prior over a fixed interval. It is not possible to obtain the Bayes in closed form, hence, we have suggested to use Gibbs sampling procedure to compute the Bayes estimates and the associated HPD credible intervals. Some simulation experiments have been performed and it is observed that the Bayes estimates and the associated HPD credible intervals perform quite satisfactorily. It may be mentioned that although we have considered the squared error loss function our method can be easily generalized for any other loss function also.

We have further considered the problem of finding the optimum censoring scheme among the class of all possible censoring schemes based on a similar criterion proposed by Zhang and Meeker [19] and Kundu [11]. Since the total number of possible censoring schemes is quite high, we have provided an algorithm to find the obtained suboptimum censoring scheme. Some suboptimum sampling schemes have been presented. Finding an optimum censoring scheme from the all possible censoring schemes remains an open problem from the computational point of view. Efficient algorithm is needed to find the optimum censoring scheme particularly when n is large and $m \approx n/2$. More work is needed along that direction.

Appendix

Proof of Theorem 2: To prove Theorem 2, it is enough to prove that

$$\frac{d^2}{d\alpha^2} \ln\left[b + \sum_{i=1}^{m}(1+R_i)(t_i-\mu)^\alpha\right] > 0.$$

Now consider

$$g(\alpha) = b + \sum_{i=1}^{m}(1+R_i)(t_i-\mu)^\alpha;$$

then

$$g'(\alpha) = \frac{\mathrm{d}}{\mathrm{d}\alpha} g(\alpha) = \sum_{i=1}^{m} (1 + R_i)(t_i - \mu)^{\alpha} \ln(t_i - \mu)$$

and

$$g''(\alpha) = \frac{\mathrm{d}^2}{\mathrm{d}\alpha^2} g(\alpha) = \sum_{i=1}^{m} (1 + R_i)(t_i - \mu)^{\alpha} (\ln(t_i - \mu))^2.$$

Since

$$\begin{aligned} &\left(\sum_{i=1}^{m} (1 + R_i)(t_i - \mu)^{\alpha} (\ln(t_i - \mu))^2\right)\left(\sum_{i=1}^{m} (1 + R_i)(t_i - \mu)^{\alpha}\right) \\ &\quad - \left(\sum_{i=1}^{m} (1 + R_i)(t_i - \mu)^{\alpha} \ln(t_i - \mu)\right)^2 \\ &= \sum_{1 \le i < j \le m} (R_i + 1)(R_j + 1)\left(\ln(t_i - \mu) - \ln(t_j - \mu)\right)^2 \ge 0, \end{aligned}$$

the result follows.

References

1. Balakrishnan, N. (2007), ``Progressive censoring methodology: an appraisal'', *TEST*, vol. 16, 211-259.
2. Balakrishnan, N. and Aggarwala, R. (2000), *Progressive censoring: theory, methods, and applications*, Springer, New York.
3. Balakrishnan, N. and Cramer, E. (2014), *The art of progressive censoring: applications to reliability and quality*, Birkhauser, New York.
4. Balasooriya, U., Saw, S.L.C. and Gadag, V.G. (2000), ``Progressively censored reliability sampling plan for the Weibull distribution'', *Technometrics*, vol. 42, 160 - 168.
5. Berger, J.O. and Sun, D. (1993), ``Bayesian analysis for the Poly-Weibull distribution'', *Journal of the American Statistical Association*, vol. 88, 1412 - 1418.
6. Congdon, P. (2006), *Bayesian statistical modelling*, 2nd. edition, Wiley, New Jersey.
7. Devroye, L. (1984), ``A simple algorithm for generating random variates with a log-concave density'', *Computing*, vol. 33, 247-257.
8. Geman, S. and Geman, A. (1984), ``Stochastic relaxation, Gibbs distribution and the Bayesian restoration of images'', *IEEE Transactions of Pattern Analysis and Machine Intelligence*, vol. 6, 721 - 740.
9. Johnson, N.L., Kotz, S. and Balakrishnan, N. (1995), *Continuous univariate distribution*, (2nd edition), New York, Wiley.
10. aminskiy, M. P., and Krivtsov, V. V. (2005), ``A Simple Procedure for Bayesian Estimation of the Weibull Distribution'', *IEEE Transactions on Reliability Analysis*, vol. 54, 612¡V616.
11. Kundu, D. (2008), ``Bayesian inference and life testing plan for the Weibull distribution in presence of progressive censoring'', *Technometrics*, vol. 50, 144-154.

12. Kundu, D. and Joarder, A. (2006), ``Analysis of type-II progressively hybrid censored data", *Computational Statistics abd Data Analysis*, vol. 50, 2509 - 2528.
13. Lawless, J.F. (1982), *Statistical models and methods for lifetime data*, Wiley, New York.
14. Murthy, D.N.P., Xie, M. and Jiang, R. (2004), *Weibull Models*, John Wiley and Sons, New York.
15. Nagatsuka, H., Kamakura, T. and Balakrishanan, N. (2013), ``A consistent method of estimation for the three-parameter Weibull distribution", *Computational Statistics and Data Analysis*, vol. 58, 210 -226.
16. Rao, C.R. (1945), ``Information and the accuracy attainable in the estimation of statistical parameters", *Bulletin of Calcutta Mathematical Society*, vol. 37, 81 - 91.
17. Smith, R. L. and Naylor, J. C. (1987), ``A comparison of maximum likelihood and Bayesian estimators for the three-parameter Weibull distribution", *Applied Statistics*, 358-369.
18. Viveros, R. and Balakrishnan, N. (1994), ``Interval estimation of parameters of life from progressively censored data", *Technometrics*, vol. 36, 84 - 91.
19. Zhang, Y. and Meeker, W. Q. (2005), ``Bayesian life test planning for the Weibull distribution with given shape parameter", *Metrika*, vol. 61, 237-249.

Vulnerability Discovery Modelling for Software with Multi-versions

Adarsh Anand, Subhrata Das, Deepti Aggrawal and Yury Klochkov

Abstract Security vulnerabilities have been of huge concern as an un-patched vulnerability can potentially permit a security breach. Vulnerability Discovery Modelling (VDM) has been a methodical approach that has helped the developers to effectively plan for resource allocation required to develop patches for problematic software releases; and thus improving the security aspect of the software. Many researchers have proposed discovery modelling pertaining to a specific version of software and talked about time window between the discovery (of vulnerability) and release of the patch as its remedy. In today's cut throat and neck to neck competitive market scenario, when every firm comes up with the successive version of its previous release; fixing of associated susceptibilities in the software becomes a more cumbersome task. Based on the fundamental of shared code among multi-version software system, in this chapter of the book, we propose a systematic approach for quantification of number of vulnerabilities discovered. With the aim of predicting and scrutinising the loopholes the applicability of the approach has been examined using various versions of Windows and Windows Server Operating Systems.

Keywords Multi-version software · Patch · Vulnerability discovery model (VDM)

A. Anand (✉) · S. Das
Department of Operational Research, University of Delhi,
New Delhi 110007, India
e-mail: adarsh.anand86@gmail.com

S. Das
e-mail: shus.das@gmail.com

D. Aggrawal
Keshav Mahavidyalaya, University of Delhi, New Delhi 110034, India
e-mail: deepti.aggrawal@gmail.com

Y. Klochkov
St. Petersburg Polytechnic University, St. Petersburg, Russia
e-mail: y.kloch@gmail.com

M. Ram and J.P. Davim (eds.), *Advances in Reliability and System Engineering*,
Management and Industrial Engineering, DOI 10.1007/978-3-319-48875-2_11

Notation

$\Omega_i(t)$	Expected number of vulnerabilities discovered by time $t(i = 1, 2, 3...n)$
$F_i(t)$	Probability distributions function for vulnerability discovery process $(i = 1, 2, 3...n)$
a_i	Total number of vulnerabilities in the software $(i = 1, 2, 3...n)$
b_i	Vulnerability detection rate function of software $(i = 1, 2, 3...n)$
β_i	Learning parameter for vulnerability discovery process $(i = 1, 2, 3...n)$

1 Introduction

Today's virtually connected business network infrastructure is frequently changing and dependency on networked systems has fetched the concern about security of software systems to a very good extent. Addition of new services, servers, connections and ports under complex environment has brought the threshold of software system to a new level. With the continuous evolution of software system, consequences, in terms of the increasing count of vulnerabilities and loopholes are of great concern because an un-patched loophole can potentially harm the system. The growing number of adopters of service is keen to use the facilities provided by the software services and might be trapped due to unsecured code usage. Security is one of the major aspects during the transition of code to a well formed application: If a proper encryption is provided while coding; the likelihood of getting exploits are fewer. During the course of software development, due to various issues; like lack of resources and time; developer unintentionally generates some loop holes and most of them might result in exploits. There are many examples that can be quoted in respect to the effect of vulnerabilities in real life situations; some examples of which were discovered in the most widely known is the Code-Red as one of the highly exploited vulnerability ever contained in windows operating system which infected more than three million computers entering through internet connected computers. Due to which there was a massive financial loss and affected many business activities [1]. Similar sort of vulnerability named 'Slammer worm' was detected in the year 2003 in windows system; it spread so quickly that within 15 min it targeted around 75 thousand computers. Similarly, in October 2014, a critical vulnerability in Windows Operating system was exploited which allows attackers to download and install malware into the target computer [2].

Vulnerability is a security flaw in a computing system, which can be consequently attacked and exploited by an attacker [3]. Alternatively, a vulnerability may be defined as a defect "which enables an attacker to bypass security measures" [4]. Thus, lesser count of vulnerabilities will be considered as less risky compared to a software system which has more number of vulnerabilities. From Software's perspective, vulnerability may be defined as a "flaw or weakness in the security system which might be exploited by a harmful user causing loss or harm" [5]. Thus, it can

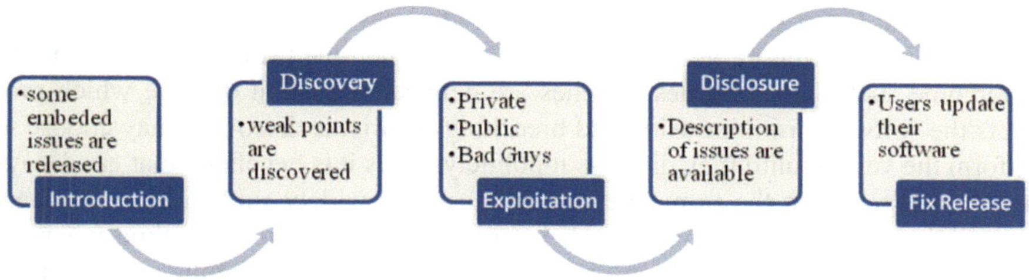

Fig. 1 Vulnerability life cycle [7]

be articulated that vulnerability is one of the major outcomes due to the improper handling of code in software development process. Vulnerability discovery process is nothing but a security checker for the software developers. It helps in enhancement of software quality. The idea of vulnerability discovering is very important for defenders. For this very reason, it is important to inspect the procedures behind a loop hole being discovered and it getting exploited by intruders. Life cycles of vulnerability comprises of introduction, discovery, exploitation, disclosure and fix release as shown in Fig. 1. In literature [6]; the process has been described comprising of following steps:

- Introduction: In this phase the vulnerability is first released as ingredient of the software.
- Discovery: The loopholes are discovered either by good guys or bad guys.
- Private Exploitation: The discovered vulnerability is exploited by the team of software developer.
- Disclosure: A detail report of the vulnerability is released.
- Public Exploitation: The discovered vulnerability is exploited by the bad guys for their own benefit.
- Fix Release: A patch or update is supplied to fix the cause of exploit, i.e. loopholes which were discovered are fixed.

For the average end user, software vulnerability fixation has always been a simplistic view. Most of the software keeps on updating themselves as an update is available in repository. But the other side of the coin is much more complicated. Here comes the characterization of vulnerabilities and its disclosure policies. In modelling of vulnerability, it has been an important task to know who discovered the vulnerability; is it benign or malicious? The lifecycle of vulnerabilities may vary depending on the type of disclosing policies, the basic three types of discovery are: public, private and bad actors [7]. Bad actors are those who may discover the security loopholes and can take advantage of it. Treating the software unfairly may cause heavily to the firm in terms of monetary resources, loss of goodwill, etc. A bad guys or a malicious user is always keen to know the loopholes in order to exploit the vulnerability. Public discovery is somewhat analogous to bad actor finding, except the detector is unconcerned about whether the vulnerability is exploited or not. A private discovery or benign detector always wants to prevent the exploitation. Good guys are pool of people who may find the weak points but do not exploit it and report back to the firm. At the same time firm issue the patches to overcome such vulnerabilities. Patches are defined as portion of code which protects the software from intrusion and breeching. A private discovery may choose to inform the vendor until they disclose it publicly. Thus it is beneficial that good guys find it and report to the firm. The concept has been widely studied by researchers across the globe [8] and much work is in the pipeline. In today's time when no software comes in single version; the need to look in the vulnerability aspect of the concept of multiple versions is an interesting task.

Survival in this competitive scenario calls for attracting customer with new features and enhanced quality of software. Adding new functionalities periodically is one way of penetrating in the market. Feature enhancement is also one of the major attribute of this change, upgrading a software system provides an opportunity to stay ahead of the competition. With successive releases, innovative features and functionalities are added to the existing software which possibly increases the breaching points in the system. In multi-versions of Vulnerability Discovery Modelling (VDM), all versions of software are used concurrently. As name implies, the succeeding versions are expected to have better reliability and performability than the earlier versions. As software design is changing at a high pace and also new codes are being added from time to time, multi-version VDM is required to understand the exact scenario. A newer version entrenches some parts of code of prior ones. Even when the significance of previous version decreases, the usage of code is still important in the succeeding versions. Due to code sharing, it might happen that the number of vulnerabilities might be discovered in current release. The number of vulnerabilities found in the existing version will be accountable to its earlier versions because of code sharing.

Many models in literature of VDM are based on single version concept but little or no attention is paid to the concept of successive versions of software product competing in the market. In this chapter, we have come up with an idea of modelling vulnerability discovery for multiple versions under the assumption that newer version embeds a portion of code from its predecessor. The plus point with coming of successive releases is code sharing among different versions, which facilitate the vulnerability discovery of the older versions even if they are no longer in usage. Rest of the chapter is divided into following sections: Sect. 2 follows brief backgrounds of vulnerability discover models. Discovery modelling framework is discussed in Sect. 3. Section 4 illustrates the analysis part of the model. Finally conclusion is given in Sect. 5.

2 Literature Review on Vulnerability Discovery Models

The allied risk in the vulnerable software needs great efforts to find out, publish and fix the same in software merchandise. The discovery of the vulnerabilities is done throughout its existence by both the producers and users. There is vast literature on software reliability, which is concerned about total number of faults removed and not on safety measures of software. Global digitalization in all sectors requires software system to perform every single operation. In today's technocrat world, each and every individual is highly dependent on *E-systems*. As an example, every sector's dependence on software system made it mandatory for firms to have highly secure software, which can be noticed from the fact that every sector's admission processes and recruitment procedures are also getting internet-based. Thus, computer system need to be secure and should function perfectly in dynamic environments. Software security is a method of structural secure software. Few researchers focused on quantitative aspect of security in software reliability modelling. Chou et al. [9] discussed the rate of faults finding and fixing loop holes but did not differentiate them in terms of vulnerabilities and some other bugs. Anderson [10] discussed a software failure model to fetch out the trend in vulnerability discovery. Rescorla [11] classified the trends in vulnerability data by applying different models. As pointed out by Alhazmi and Malaiya [12, 13], there is the relationship between the cumulative vulnerabilities with time. Further they proposed an effort-based exponential model.

Miller et al. [14] modelled a formula which calculates the count of defects in software, even when testing discloses no flaws. They have done it with random testing results, without using any the software's history. A dynamic model of security loopholes incorporating the concept of uncertainty is developed by Tofts et al. [16]. In their study, they have not highlighted the detailed predictions of vulnerable software components. Yin et al. [15] paid attention to the security issues in large software system, but did not describe about its implementation or evaluation in field. Rescorla [7] has tried to convince that finding and updating security weak points did not show the way of enhancement in quality of the software. He focused that exploring of loop holes by black hat finder rather than developing team of firms. Andy Ozment [17] proposed a theoretical explanation of susceptibilities and their following finding process. An optimal timing for software vulnerable disclosure under the role a social planer and vendor was formulated by Arora et al. [18]. Further, Kapur et al. [19] explained the utilisation of logistic detection rate in the discovery of vulnerabilities. Recently Anand and Bhatt [8] proposed a hump-shaped vulnerability discovery model, which is an improvisation of Kapur et al. [19]. Further, they used a raking-based approach to compare the proposed model with the existing VDM models in the literature. Not much attention has been given to discovery of vulnerabilities in multi-version of the software which coexists in the market. One of such work has been done by Kim et al. [20] in which the emphasis has been given to the modelling of vulnerability discovery process quantitatively, based on shared source code measurements among multi-version software systems. Contrasting above mentioned studies, in this chapter, we have proposed the modelling of vulnerability discovered in multi-versions of software and analysed how the shared code increases the count of vulnerabilities in earlier version.

3 Vulnerability Discovery Modelling Framework

The growing need of users and vendors in the market forces a software developing firm to come up with successive versions of their offerings (i.e. computer applications, software, etc.). The major idea behind launching the successive version is code sharing among different releases which helps the firms to test the vulnerability and associated risk both before and after the release of the software. Due to the notion of code sharing among different releases, it is assumed that number of vulnerabilities discovered in the shared portion of the code will be accountable to the older versions which contain the shared code. One of the major assumption is that shared portion of the code is tested under usage even if the older version is no longer in use. Emergent need of highly reliable software requires a model which can predict the number of loopholes that are probable to be discovered while usage. With this aim we have come with a proposal that caters the need of developing secure software and provide an estimate for the number of vulnerabilities that can result in various loopholes. The proposed model accounts for number of vulnerabilities that might arise in case when firm releases successive versions of the software to maintain its competitive grip in the market.

Considering the case that if only one version of the software exists in the market then the number of susceptibility that can occur can be modelled in the following fashion:

$$\Omega_1(t) = a_1 \cdot F_1(t) \tag{1}$$

where a_1 is the total number of vulnerability that can be revealed under the field and $F_1(t)$ is the cumulative distribution function which accounts for the rate of vulnerabilities that will be discovered and $F_1(t) = \left(1 - \frac{(1+\beta_1)\cdot e^{-b_1 t}}{1+\beta_1 \cdot e^{-b_1 t}}\right)$.

The rigid competition among firms results into a state in which it becomes necessary for a firm to launch software version to uphold their position in the market. Thus, there may be a possibility that two versions of the software coexist in the market. When some new features are added to the software for the first time then it might come across some new loopholes along with some loophole that can transpire in shared portion of the code. This type of scenarios can be mathematically modelled as:

$$\Omega_1(t) = a_1.F_1(t) + a_1 \cdot (1 - F_1(t)).F_2(t-\tau_1) \tag{2}$$

$$\Omega_2(t) = a_2 \cdot F_2(t-\tau_1) \tag{3}$$

where a_2 is the total number of vulnerability that can be discovered in second version and $F_2(t-\tau_1)$ follows logistic distribution, i.e. $F_2(t-\tau_1) = \left(1 - \frac{(1+\beta_2)\cdot e^{-b_2.(t-\tau_1)}}{1+\beta_2 \cdot e^{-b_2.(t-\tau_1)}}\right)$.

In Eq. (2), the first component illustrates the total number of vulnerabilities of first version and second component demonstrates the count of vulnerabilities discovered in second version but it is accounted in vulnerabilities of first version due to code sharing. Equation (3) represents the number of vulnerabilities in second version.

When the new functionalities are added in the software for the second time again new lines of code are developed. This new code is integrated with the existing code. There are chances that some vulnerability might get encountered in newly added portion of the code and at the same time it is likely that some ambiguity might get encountered in the shared portion code.

$$\Omega_1(t) = a_1 \cdot F_1(t) + a_1 \cdot (1 - F_1(t)) \cdot F_2(t - \tau_1) + a_1 \cdot (1 - F_1(t)) \cdot (1 - F_2(t - \tau_1))F_3(t - \tau_2) \tag{4}$$

$$\Omega_2(t) = a_2 \cdot F_2(t - \tau_1) + a_2 \cdot (1 - F_2(t - \tau_1)) \cdot F_3(t - \tau_2) \tag{5}$$

$$\Omega_3(t) = a_3 \cdot F_3(t - \tau_2) \tag{6}$$

where a_3 is the total number of vulnerability that can be discovered third version and $F_3(t - \tau_2)$ follows logistic distribution, i.e. $F_3(t - \tau_2) = \left(1 - \frac{(1+\beta_3) \cdot e^{-b_3 \cdot (t - \tau_2)}}{1 + \beta_3 \cdot e^{-b_3 \cdot (t - \tau_2)}}\right)$.

Equation (4) comprises of the total number of vulnerabilities of first version and the count of vulnerabilities discovered in second version and third version but it is accounted in vulnerabilities of first version due to code sharing. In Eq. (5), the first component represents the total number of vulnerabilities of second version and second component shows the number of vulnerabilities discovered in third version but it is added in vulnerabilities of previous versions due to code sharing. Equation (6) represents the number of vulnerabilities in third version. Here we have analysed the proposed model for three versions.

The above described model can also be generalised till n-versions of the software.

$$\begin{aligned} \Omega_1(t) = {} & a_1 \cdot F_1(t) + a_1 \cdot (1 - F_1(t))F_2(t - \tau_1) + a_1 \cdot (1 - F_1(t)) \\ & \cdot (1 - F_2(t - \tau_1)) \cdot F_3(t - \tau_2) \\ & + \cdots + a_1 \cdot (1 - F_1(t)) \cdots F_n(t - \tau_{n-1}) \end{aligned} \tag{7}$$

$$\begin{aligned} \Omega_2(t) = {} & a_2 \cdot F_2(t - \tau_1) + a_2 \cdot (1 - F_2(t - \tau_1)) \cdot F_3(t - \tau_2) \\ & + a_2 \cdot (1 - F_2(t - \tau_1)) \cdot (1 - F_3(t - \tau_2)) \cdot F_4(F_3(t - \tau_3)) \\ & + \cdots + a_2 \cdot (1 - F_2(t - \tau_1)) \cdots F_n(t - \tau_{n-1}) \end{aligned} \tag{8}$$

$$\Omega_n(t) = a_n \cdot F_n(t - \tau_{n-1}) \tag{9}$$

Table 1 Parameter estimates for DS-I and DS-II

Parameters	DS-I			DS-II		
	Version 1	Version 2	Version 3	Version 1	Version 2	Version 3
a	546.201	302.325	340.395	408.919	686.341	424.647
b	0.30466	0.999	0.999	0.825	0.459	0.999
β	2.8287	64.3281	21.2802	433.82	14.469	52.785

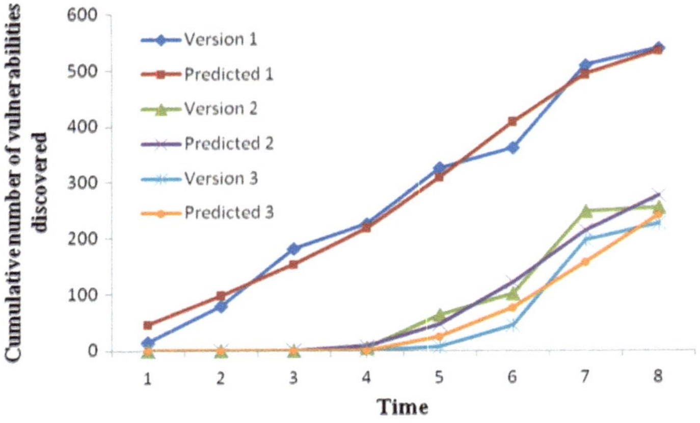

Fig. 2 Goodness fit curve for proposed methodology of DS-I

The proposed model quantifies the vulnerabilities discovered in current version and successive versions as given by the aforesaid equations.

4 Data Analysis

Inspecting the prediction capability and accuracy of these models has been analysed by two operating system data sets. Data set I comprise of 538 numbers of vulnerabilities for version 1, 254 for version 2 and 226 for version 3 while data set II consist of 414 number of vulnerabilities for version 1, 673 for version 2 and 290 for version 3. These data obtained from Windows and Window Server Operating System [21] and used to validate the model using the SAS software for analysing the parameters of proposed model on multi-versions of the software [22]. For analysing the model, we have used SAS software package [23]. The estimated values of parameters on both data sets are supplemented in Table 1. Table 2 comprises of comparison criterion for the proposed model.

From Figs. 2 and 3 shows the graphical representations of the multi-version vulnerability discover modelling. From Table 2 the higher value of R^2 for Version-1 for both the data sets implies that vulnerability discovery happens in an accurate

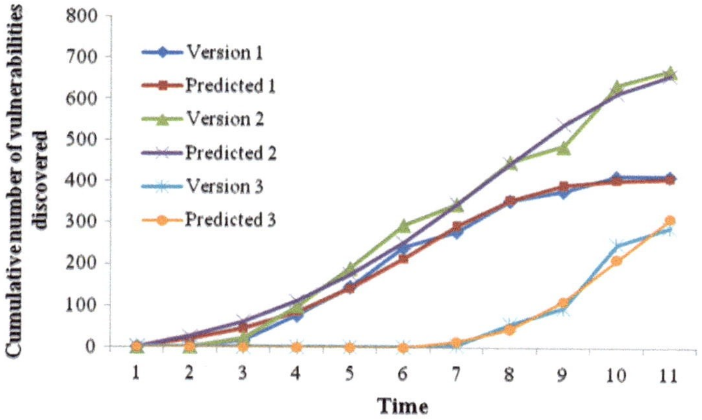

Fig. 3 Goodness fit curve for proposed methodology of DS-II

Table 2 Comparison criteria for DS-I and DS-II

Criterion	DS-I			DS-II		
	Version 1	Version 2	Version 3	Version 1	Version 2	Version 3
SSE	4906.1	2269.9	2982.5	2435.6	7789.1	2308.4
MSE	1177.5	368.1	447.4	365.3	898.7	238.8
Root MSE	34.3143	19.1859	21.1514	19.114	29.979	15.4531
R^2	0.9804	0.9729	0.953	0.9914	0.9875	0.9798

manner due to the terms which are added in the Eq. (4). Thus it implies that the concept of code sharing facilitate the discovery of large number of vulnerabilities in parent version which were transferred to later versions. Further, the values of R^2 for version 2 and 3 are in acceptable range, i.e. their values are 0.9729, 0.953 for DS-I and 0.9875, 0.9798 for DS-II and the values of other comparison attributes, i.e. SSE, MSE and Root MSE are also satisfactory.

5 Conclusion

Effective control and providing a secure interface while coding, has shown a clear-cut protection from the attackers over the time period of software development. VDM has proven to be a conclusive approach in order to capture the loopholes present in the software. Predicting the potential number of vulnerable points would mitigate the threats before they are exploited by the hackers. Due to the competitive environment and the attractiveness of software in the market, firms release new versions of software over the time period with upgraded features.

Improvisation of new functions results in the exploitation of security breach. The number of vulnerabilities present in the shared portion of code of the product which is in the field can be discovered in any version of the software. But eventually it is accountable for the original version. Keeping this frame in mind, we have modelled a multi-version vulnerability discovery approach that facilitates the inheritance of code amongst software versions. The predictability capabilities of the model have been verified on two real life operating system data sets and the outcomes are satisfactory.

References

1. Vulnerability Examples, https://securelist.com/threats/vulnerabilities-examples/, July 10, 2016.
2. Symantec Security Response, http://www.symantec.com/connect/blogs/sandworm-windows-zero-day-vulnerability-being-actively-exploited-targeted-attacks, July 10, 2016.
3. Schneier B., "Full Disclosure and the Window of Vulnerability", Crypto-Gram http: //www.counterpane.com/cryptogram-0009.html#1, September 15, 2000.
4. Schultz E. E., Brown Jr., D. S. and Longstaffs T. A., "Responding to Computer Security Incidents", Lawrence Livermore National Laboratory, 165 ftp://ftp.cert.dfn.de/pub/docs/csir/ihg.ps.gz , July 23, 1990.
5. Pfleeger C. P. and Pfleeger S. L., "Security in Computing", 3rd ed. Prentice Hall PTR, 2003.
6. Browne H. K., Arbaugh W. A., McHugh J., and Fithen W. L., "A Trend Analysis of Exploitations", University of Maryland and CMU Technical Reports, 2000.
7. Rescorla E., "Is finding security holes a good idea?" IEEE Security and Privacy, 3(1):14–19, 2005.
8. Anand A., and Bhatt N., "Vulnerability Discovery Modeling and Weighted Criteria Based Ranking", Journal of Indian Society for Probability and Statistic, Springer, 17, pp. 1–10, 2016.
9. Chou, A., Yang, J., Chelf, B., Hallem, S., and Engler, D., "An Empirical Study of Operating Systems Errors", Symposium on Operating Systems Principles, 2001.
10. Anderson R. J., "Security in Opens versus Closed Systems-The Dance of Boltzmann, Coase and Moore" Open Source Software: Economics, Law and Policy, Toulouse, France, June 20-21, 2002.
11. Rescorla E., "Security holes…Who Cares?", USENIX Security, 2003.
12. Alhazmi O. H. and Malaiya Y. K., "Modeling the vulnerability discovery process", In Proceedings of 16th IEEE International Symposium on Software Reliability Engineering (ISSRE'05), pp. 129–138, 2005.
13. Alhazmi O. H. and Malaiya Y. K., "Quantitative Vulnerability Assessment of Systems Software," in Proc. Annual Reliability and Maintainability Symposium, pp. 615–620, January 2005.
14. Miller K.W., Morell L.J., Noonan R.E., Park S.K., Nicol D.M., Murrill B.W., and M. Voas. "Estimating the Probability of Failure when Testing Reveals no Failures", IEEE Transactions on Software Engineering, 18(1):33–43, January 1992.
15. Yin J., Tang C., Zhang X., and McIntosh M., "On Estimating the Security Risks of Composite Software Services" In Proc. PASSWORD Workshop, June 2006.
16. Tofts C. and Monahan B., "Towards an Analytic Model of Security Flaws" Technical Report 2004-224, HP Trusted Systems Laboratory, Bristol, UK, December 2004.
17. Ozment A., "Improving Vulnerability Discovery Models: Problems with Definitions and Assumptions", ACM, Alexandria, Virginia, USA, 2007.

18. Arora A., Telang R., and Xu H., "Optimal Policy for Software Vulnerability Disclosure", Management Science, 54(4), pp. 642-656, 2008.
19. Kapur P. K., Sachdeva N. and Khatri S. K., "Vulnerability Discovery Modeling", International Conference on Quality, Reliability, Infocom Technology and Industrial Technology Management, pp. 34-54, ISBN 978-93-84588-57-1, 2012.
20. Kim J., Malaiya Y. K., and Ray I., "Vulnerability Discovery in Multi-version Software Systems", In 10th IEEE High Assurance Systems Engineering Symposium, pp. 141–148, 2007.
21. Windows, "Vulnerability Statistics", http://www.cvedetails.com/product/739/Microsoft-Windows.html?vendor_id=26. Accessed 20 Feb 2016.
22. Windows Server, "Vulnerability Statistics", http://www.cvedetails.com/product/739/Microsoft-Windows.html?vendor_id=26. Accessed 20 Feb 2016.
23. SAS Institute Inc., "SAS/ETS user's guide version 9.1", Cary, NC: SAS Institute Inc., 2004.

MIX
Papier aus verantwortungsvollen Quellen
Paper from responsible sources
FSC® C105338

If you have any concerns about our products,
you can contact us on
ProductSafety@springernature.com

In case Publisher is established outside the EU,
the EU authorized representative is:
Springer Nature Customer Service Center GmbH
Europaplatz 3, 69115 Heidelberg, Germany

Printed by Libri Plureos GmbH
in Hamburg, Germany